AF598131

METHODS IN MOLECULAR BIOLOGY™

For further volumes:
http://www.springer.com/series/7651

In Vivo Cellular Imaging Using Fluorescent Proteins

Methods and Protocols

Edited by

Robert M. Hoffman

AntiCancer Inc., Department of Surgery, University of California, San Diego, CA, USA

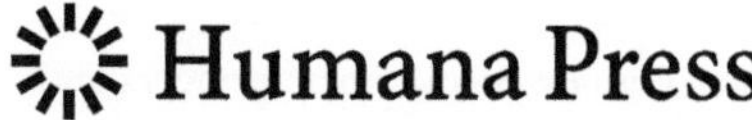

Editor
Robert M. Hoffman
AntiCancer Inc., Department of Surgery
University of California
San Diego, CA, USA

ISSN 1064-3745 ISSN 1940-6029 (electronic)
ISBN 978-1-61779-796-5 ISBN 978-1-61779-797-2 (eBook)
DOI 10.1007/978-1-61779-797-2
Springer New York Dordrecht Heidelberg London

Library of Congress Control Number: 2012936123

Printed on acid-free paper

Humana Press is part of Springer Science+Business Media (www.springer.com)

Dedication

This volume is dedicated to Charlene M. Cooper who has devoted 16 years of way-beyond the call-of-duty to AntiCancer Inc. Without Charlene's devotion, superb administration, and thoughtfulness, this volume could not have been written.

Preface

The discovery and genetic engineering of fluorescent proteins has revolutionized cell biology. What was previously invisible in the cell often can be made visible with the use of fluorescent proteins. This volume presents state-of-the-art research contributing to the revolution fluorescent proteins brought the visualization of biological processes in the live animal. This is the first volume in the new field of in vivo cell biology. The chapters in this volume are highlighted below.

Chapter 1 describes the use of the chick CAM model to visualize cancer cell migration and metastasis in a physiologically-relevant, but simple, in vivo setting using fluorescent proteins and other fluorescent probes.

Chapter 2 describes intravital fluorescent imaging of the real-time behavior of the individual cells of mammary tumors labeled with fluorescent proteins using multiphoton microscopy.

Chapter 3 describes the use of window chambers for cellular and subcellular imaging of cancer cells in mice.

Chapter 4 describes imaging of tumor–host interaction between pancreatic cancer cells and host-derived stroma and vasculature in which cancer cells and the host mice are color-coded with fluorescent proteins.

Chapter 5 describes stable transformation of cancer cells with fluorescent protein genes, using lentiviral vectors, which can be used for whole-body imaging on essentially any organ in mice.

Chapter 6 describes an in vivo imaging system consisting of mouse-implanted fluorescent protein-tagged metastatic cancer cell lines and a hand-held detection device for external, noninvasive and real-time monitoring of the therapeutic effects of drugs.

Chapter 7 describes three-dimensional imaging of tumors in mice expressing red fluorescent protein.

Chapter 8 describes real-time high-resolution imaging of angiogenesis and vascular response to anticancer and antiangiogenic therapy in live mice with orthotopic breast cancer labeled with fluorescent proteins.

Chapter 9 describes a tumor-specific, replication-competent, telomerase-dependent, GFP-expressing adenovirus to label tumors and metastasis with GFP in mice for detection and surgical navigation.

Chapter 10 describes a replication-competent, tumor-specific herpes simplex virus expressing GFP to label cancer cells in mice for visualization by endoscopy and in vivo microscopy.

Chapter 11 describes tumor-targeting GFP-expressing vaccinia viruses and bacteria to label tumors in mice for high-resolution imaging.

Chapter 12 describes genetic engineering of rats, rabbits, and pigs to express GFP which can be used for cell therapy and transplantation.

Chapter 13 describes the matching of exogenous fluorophores and endogenous fluorescent proteins in cancer cells to develop sensitive and specific cancer-targeting probes.

Chapter 14 describes embryo culture and fluorescent proteins to image developing vasculature and hemodynamics.

Chapter 15 describes new fluorescent proteins, with a wide range of spectral colors, including those that switch colors and kindle, isolated from coral reefs.

Chapter 16 describes how new improved far-red and infrared fluorescent proteins can be designed.

Chapter 17 describes imaging the effects of siRNA and microRNA in vivo.

Chapter 18 describes the use of different color fluorescent proteins to image the nuclear-cytoplasmic dynamics of cancer cells in vivo.

San Diego, CA, USA ***Robert M. Hoffman***

Contents

Contributors

LYAYSAN R. ARSLANBAEVA • *Laboratory of Physical Biochemistry, A.N. Bach Institute of Biochemistry of Russian Academy of Sciences, Moscow, Russia*

ANATOLY YU BARYSHNIKOV • *Institute of Experimental Diagnostics and Therapy of Tumors, N.N. Blokhin Russian Cancer Research Center of Russian Academy of Medical Sciences, Moscow, Russia*

MICHAEL BOUVET • *GI Cancer Unit, Moores Cancer Center, University of California San Diego, La Jolla, CA, USA*

OLGA S. BUROVA • *Institute of Experimental Diagnostics and Therapy of Tumors, N.N. Blokhin Russian Cancer Research Center of Russian Academy of Medical Sciences, Moscow, Russia*

SUSANNE CARPENTER • *Department of Surgery, Memorial Sloan-Kettering Cancer Center, New York, NY, USA*

ANN F. CHAMBERS • *The London Regional Cancer Center, London, ON, Canada*

PETER L. CHOYKE • *Molecular Imaging Program, Center for Cancer Research, National Cancer Institute, National Institutes of Health, Bethesda, MD, USA*

JOHN CONDEELIS • *Department of Anatomy and Structural Biology, Gruss-Lipper Biophotonics Center, Albert Einstein College of Medicine, Bronx, NY, USA*

MARK W. DEWHIRST • *Department of Radiation Oncology, Duke University, Durham, NC, USA; Department of Biomedical Engineering, Duke University, Durham, NC, USA*

MARY E. DICKINSON • *Molecular Physiology & Biophysics, Baylor College of Medicine, Houston, TX, USA*

DAVID ENTENBERG • *Department of Anatomy and Structural Biology, Gruss-Lipper Biophotonics Center, Albert Einstein College of Medicine, Bronx, NY, USA*

ILYA I. FIKS • *Biophotonics Laboratory, Institute of Applied Physics of Russian Academy of Sciences, Nizhny Novgorod, Russia*

YUMAN FONG • *Department of Surgery, Memorial Sloan-Kettering Cancer Center, New York, NY, USA*

ANDREW N. FONTANELLA • *Department of Biomedical Engineering, Duke University, Durham, NC, USA*

TOSHIYOSHI FUJIWARA • *Center for Gene and Cell Therapy, Division of Surgical Oncology, Department of Surgery, Okayama University Graduate School of Medicine & Dentistry, Okayama, Japan*

BOJANA GLIGORIJEVIC • *Department of Anatomy and Structural Biology, Gruss-Lipper Biophotonics Center, Albert Einstein College of Medicine, Bronx, NY, USA*

ROBERT M. HOFFMAN • *AntiCancer, Inc., Department of Surgery, University of California San Diego, CA, USA*

JAMES HULIT • *Department of Anatomy and Structural Biology, Albert Einstein College of Medicine, Bronx, NY, USA; Centre for Tumour Biology, Barts and the London Queen Mary's Medical and Dental School, London, UK*

SEIJI ITO • *Department of Gastroenterological Surgery, Aichi Cancer Center Hospital, Nagoya, Japan*

YUICHI ITO • *Department of Gastroenterological Surgery, Aichi Cancer Center Hospital, Nagoya, Japan*
DMITRIY KEDRIN • *Department of Anatomy and Structural Biology, Albert Einstein College of Medicine, Bronx, NY, USA*
MICHAEL S. KLESHNIN • *Biophotonics Laboratory, Institute of Applied Physics of Russian Academy of Sciences, Nizhny Novgorod, Russia*
EIJI KOBAYASHI • *Division of Development of Advanced Treatment, Center for Development of Advanced Medical Technology, Jichi Medical University, Tochigi, Japan*
HISATAKA KOBAYASHI • *Molecular Imaging Program, Center for Cancer Research, National Cancer Institute, National Institutes of Health, Bethesda, MD, USA*
YASUHIRO KODERA • *Department of Surgery II, Nagoya University School of Medicine, Nagoya, Japan*
NOBUYUKI KOSAKA • *Molecular Imaging Program, Center for Cancer Research, National Cancer Institute, National Institutes of Health, Bethesda, MD, USA*
IRINA V. LARINA • *Molecular Physiology & Biophysics, Baylor College of Medicine, Houston, TX, USA*
HON SING LEONG • *The London Regional Cancer Center, London, ON, Canada*
JOHN D. LEWIS • *University of Alberta, Edmonton, AB, Canada*
MAKOTO MATSUI • *Division of Oncological Pathology, Aichi Cancer Center Research Institute, Nagoya, Japan*
THOMAS E. MCCANN • *Molecular Imaging Program, Center for Cancer Research, National Cancer Institute, National Institute of Health, Bethesda, MD, USA*
IRINA G. MEEROVICH • *Laboratory of Physical Biochemistry, A.N. Bach Institute of Biochemistry of Russian Academy of Sciences, Moscow, Russia*
KAZUNARI MISAWA • *Department of Gastroenterological Surgery, Aichi Cancer Center Hospital, Nagoya, Japan*
TAKASHI MURAKAMI • *Laboratory of Tumor Biology, Takasaki University of Health and Welfare, Gunma, Japan*
HAYAO NAKANISHI • *Division of Oncological Pathology, Aichi Cancer Center Research Institute, Nagoya, Japan*
TAKAHIRO OCHIYA • *Division of Molecular and Cellular Medicine, National Cancer Center Research Institute, Tokyo, Japan*
JUN ONODERA • *Division of Molecular and Cellular Medicine, National Cancer Center Research Institute, Tokyo, Japan*
ANNA G. ORLOVA • *Biophotonics Laboratory, Institute of Applied Physics of Russian Academy of Sciences, Nizhny Novgorod, Russia*
GREGORY M. PALMER • *Department of Radiation Oncology, Duke University, Durham, NC, USA*
ANYA SALIH • *Confocal Bio-Imaging Facility (CBIF), School of Science and Health, University of Western Sydney, Sydney, NSW, Australia*
ALEXANDER P. SAVITSKY • *A.N. Bach Institute of Biochemistry of Russian Academy of Sciences, Moscow, Russia*
JEFFREY E. SEGALL • *Anatomy and Structural Biology, Albert Einstein College of Medicine, Bronx, NY, USA*
ALEXANDER M. SERGEEV • *Institute of Applied Physics of Russian Academy of Sciences, Nizhny Novgorod, Russia*
SIQING SHAN • *Department of Radiation Oncology, Duke University, Durham, NC, USA*

DARINA V. SOKOLOVA • *Institute of Experimental Diagnostics and Therapy of Tumors, N.N. Blokhin Russian Cancer Research Center of Russian Academy of Medical Sciences, Moscow, Russia*

ALADAR A. SZALAY • *Rudolf-Virchow-Center, DFG-Research Center for Experimental Biomedicine, University of Wuerzburg, Wuerzburg, Germany; Genelux Corporation, San Diego Science Center, San Diego, CA, USA*

RYOU-U TAKAHASHI • *Division of Molecular and Cellular Medicine, National Cancer Center Research Institute, Tokyo, Japan*

FUMITAKA TAKESHITA • *Division of Molecular and Cellular Medicine, National Cancer Center Research Institute, Tokyo, Japan*

ELENA M. TRESHCHALINA • *Institute of Experimental Diagnostics and Therapy of Tumors, N.N. Blokhin Russian Cancer Research Center of Russian Academy of Medical Sciences, Moscow, Russia*

ILYA V. TURCHIN • *Biophotonics Laboratory, Institute of Applied Physics of Russian Academy of Sciences, Nizhny Novgorod, Russia*

STEPHANIE WEIBEL • *Department of Biochemistry, Biocenter, University of Wuezburg, Wuezburg, Germany*

JEFFREY WYCKOFF • *Department of Anatomy and Structural Biology, Gruss-Lipper Biophotonics Center, Albert Einstein College of Medicine, Bronx, NY, USA*

H. ROSIE XING • *Department of Pathology & Radiation Oncology, University of Chicago, Chicago, IL, USA*

YONG A. YU • *Genelux Corporation, San Diego Science Center, San Diego, CA, USA*

ZEQIAN YU • *Department of General Surgery, Zhongda Hospital, Southeast University, Nan Jing City, Jiang Su Province, China*

QINGBEI ZHANG • *Department of Pathology, Cellular and Radiation Oncology, University of Chicago, Chicago, IL, USA*

VICTORIA V. ZHERDEVA • *Laboratory of Physical Biochemistry, A.N. Bach Institute of Biochemistry of Russian Academy of Sciences, Moscow, Russia*

JIAHUA ZHOU • *Department of General Surgery, Zhongda Hospital, Southeast University, Nan Jing City, Jiang Su Province, China*

MARC ZIMMER • *Hale Laboratory, Connecticut College, New London, CT, USA*

Chapter 1

Assessing Cancer Cell Migration and Metastatic Growth In Vivo in the Chick Embryo Using Fluorescence Intravital Imaging

Hon Sing Leong, Ann F. Chambers, and John D. Lewis

Abstract

Cell migration and metastasis are key features of aggressive tumors. These processes can be difficult to study, as they often occur deep within the body of a cancer patient or an experimental animal. In vitro assays are able to model some aspects of these processes, and a number of assays have been developed to assess cancer cell motility, migration, and invasion. However, in vitro assays have inherent limitations that may miss important aspects of these processes as they occur in vivo. The chick embryo provides a powerful model for studying these processes in vivo, facilitated by the external and accessible nature of the chorioallantoic membrane (CAM), a well-vascularized tissue that surrounds the embryo. When coupled with multiple fluorescent approaches to labeling both cancer cells and the embryonic vasculature, along with image analysis tools, the chick CAM model offers cost-effective, rapid assays for studying cancer cell migration and metastasis in a physiologically-relevant, in vivo setting. Here, we present recent developments of detailed procedures for using shell-less chick embryos, coupled with fluorescent labeling of cancer cells and/or chick vasculature, to study cancer cell migration and metastasis in vivo.

Key words: Chick embryo, Chorioallantoic membrane, Fluorescence, Embryonic vasculature, Cancer cell migration, Metastasis, In vivo, Shell-less, GFP, RFP, Lectin LCA-fluorescein/rhodamine

1. Introduction

The ability of cancer cells to migrate, invade into surrounding tissues, and spread to distant organs—metastasis—is an important contributor to cancer cell aggressiveness and patient mortality (1). The metastatic process is a complex one, requiring interactions of cancer cells with multiple host tissues and cell types (selected, e.g., (2–13)). Because metastasis generally occurs over time and within organs generally inaccessible to direct observation, it can be diffi-

Robert M. Hoffman (ed.), *In Vivo Cellular Imaging Using Fluorescent Proteins: Methods and Protocols*, Methods in Molecular Biology, vol. 872, DOI 10.1007/978-1-61779-797-2_1,

cult to study. One approach, which has provided a wealth of information, has been to use a variety of in vitro assays to model specific aspects of the metastatic process such as cell motility, migration, and invasion (14, 15). These approaches, however, lack the full spectrum of tissue interactions that occur physiologically in vivo.

In vivo metastasis assays in experimental animals provide this context, but often suffer from the inherent "hidden" nature of the metastatic process. One in vivo model, the chick embryo, overcomes this limitation, due to the accessibility of the chorioallantoic membrane (CAM), a well-vascularized tissue that forms around the embryo. The chick CAM is a thin tissue, accessible on the outside of the embryo, and ideally suited for microscopic observation and analyses. Chick CAM metastasis assays have been used with success for decades, have been combined with intravital imaging approaches, and have illuminated many aspects of the metastatic process (16–29). Recent developments in fluorescent labeling capabilities, for both cancer cells and the chick CAM vasculature, coupled with improved methods for "shell-less" embryo preparations that further enhance the imaging capabilities of the chick embryo model (30), provide an excellent approach to imaging and studying cancer cell migration, invasion, and metastasis in vivo.

Because the CAM is a translucent, accessible tissue, fluorescent cancer cells can be readily imaged over long periods using various intravital imaging techniques (28, 29). Unlike rodent models, intravital imaging in the CAM requires no surgery, anesthesia, or specialized lenses. Instead, an imaging system for the shell-less chicken embryo model can be integrated into a standard upright fluorescent microscope that maintains the proper temperature and humidity as well as immobilizing the tissue to be visualized. Using widefield or confocal microscopy, fluorescently-labeled cancer cells can be followed longitudinally and their migration paths analyzed using image analysis software. This approach allows for in vivo evaluation of cancer cell migration that is compelling due to its relative cost, ease of use, and biophysical context. Details of fluorescence imaging protocols for this approach are provided here.

2. Materials

2.1. Cancer Cell Line Preparation

Please see Note 1.

1. 1× PBS pH 7.4.
2. 2.5% Trypsin (10×).
3. 15-mL Falcon tubes.
4. 1.5-mL Eppendorf tubes.
5. Benchtop centrifuge.

6. Culture media appropriate for the cell lines used.
7. Hemocytometer for cell counting.

2.2. Preparation of Chick Embryo Chorioallantoic Membrane

1. Fine point forceps.
2. Circular coverslips 18 mm.
3. Fertilized White Leghorn eggs, incubated to appropriate age, as described (28–30). Please see Note 2.
4. Egg incubator, many commercially available models, such as model 1550E from G.Q.F. MGF Company Inc., Savannah, GA, or Marsh Farms Roll-X Flowing Air Incubators, Lyon Electric Company Inc., Chula Vista, CA.
5. Avian embryo imaging unit (Quorum Technologies, Guelph, ON) (Fig. 1e). This is a specialized microscope-mounted enclosure that maintains the avian embryo in a humidified (>90% humidity) environment, while stabilizing the area of CAM to be imaged using a standard coverslip fixed into the lid of the unit. This allows for long-term noninvasive intravital imaging of the CAM using an upright fluorescence microscope.

2.3. Intravenous Injection of Cancer Cells or Agents to Visualize Vessels and CAM Plexus

1. Lectin *Lens culinaris* Agglutinin (LCA) conjugated with fluorescein or rhodamine (Vector Labs Inc. RL-1042, FL-1041) (31).
2. 1-mL disposable syringes for injections.
3. 18-gauge disposable hypodermic needles for injections.
4. Tygon R-3603 laboratory tubing, 50 ft, for injections (1/32 in. inner diameter, 3/32 in. outer diameter, 1/32 in. wall thickness).
5. Vertical pipette puller (David Kopf Instruments, Tujunga, CA; Model 720).
6. Sodium borosilicate glass capillary tubes, outer diameter 1.0 mm, inner diameter 0.58 mm, 10 cm length (Sutter Instrument, Novato, CA; Cat. No. BF100-58-10).
7. Fine point forceps.
8. Dextran, fluorescein, 70 kDa (Invitrogen Inc., Carlsbad, CA; Cat. No. D1822). Dilute fluorescent dextran with 1× PBS pH 7.4 to 0.5 mg/mL.
9. Dextran, rhodamine, 70 kDa (Invitrogen Inc., Carlsbad, CA; Cat. No. D1819). Dilute fluorescent dextran with 1× PBS pH 7.4 to 0.5 mg/mL.
10. Kimwipes.
11. Fertilized chicken eggs and egg incubator, as in Subheading 2.2.
12. Appropriate microscope(s) and image analysis software (please see Note 3).

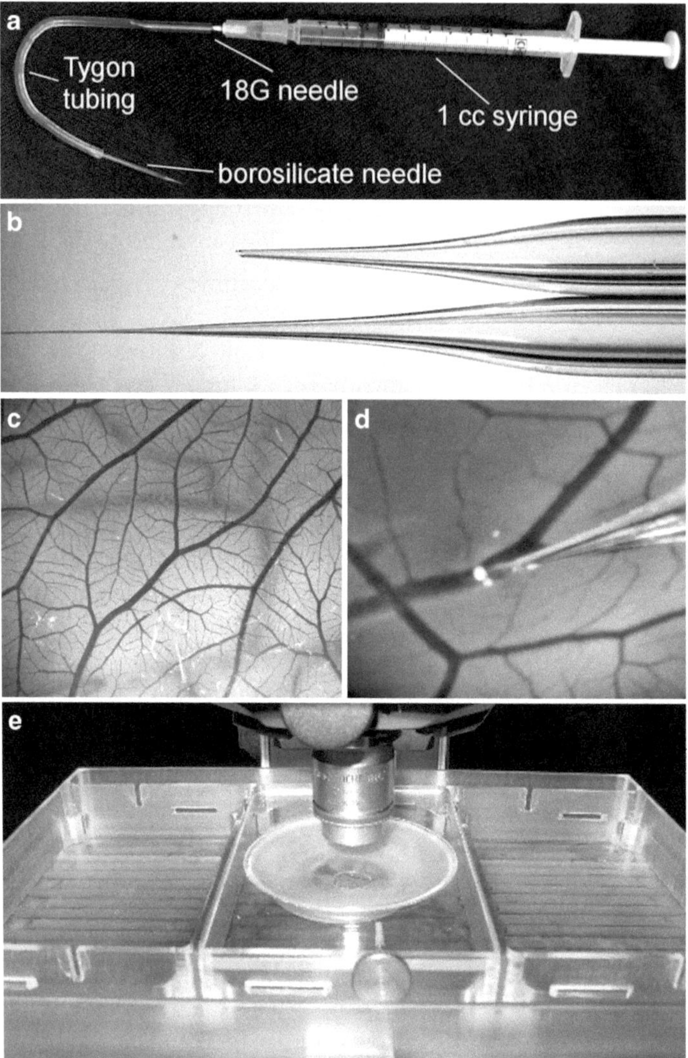

Fig. 1. Preparation of equipment for intravenous injection of cancer cells and fluorescent agents into the avian embryo chorioallantoic membrane (CAM) for intravital imaging. (**a**) Assembled injection apparatus for IV injection of labeling agents and/or cancer cells. (**b**) Example of pulled sodium borosilicate needles for intravenous injection of cancer cells or labeling agents. A pulled borosilicate needle is presented at the *bottom*. Pulled tips are modified to look like the upper borosilicate needle in (**b**) prior to being inserted into the Tygon tubing of the assembled injection apparatus in (**a**). (**c**) Representative image of CAM and vessels. (**d**) Depicts the borosilicate needle of the microinjector prior to IV injection into the CAM. (**e**) Example of custom-built incubation unit for in vivo fluorescence imaging of shell-less chick embryos. The unit consists of a removable lid and chamber tray that can house up to three shell-less chick embryos. Shown is the chamber on the stage of a microscope which is also housed by a temperature-regulated enclosure.

3. Methods

3.1. Cancer Cell Line Preparation

1. Culture cancer cell line of interest grown in the absence of antibiotics or selective medium to 80% confluency. Higher confluency negatively impacts tumor take and extravasation efficiency. Routinely check for absence of mycoplasma

contamination. (See Welch (32) for technical considerations on growth of cancer cell lines for in vivo assays).

2. To trypsinize cells, wash twice with 1× PBS pH 7.4. Aspirate remaining PBS then add 0.5% Trypsin–EDTA (e.g., 2 mL to T75 flask, 3 mL to T175 flask, 3 mL to 150 mm culture dish) and incubate at 37°C for 2–5 min until all cells detach.
3. Add 7–8 mL of PBS and transfer cell suspension to 15-mL Falcon tube. Optional: Use another 10 mL of PBS to wash and collect remaining cells and transfer to a 15-mL Falcon tube to use as a balance.
4. Centrifuge at room temperature at $200 \times g$ for 5–10 min.
5. Pour out supernatant and resuspend with 5 mL PBS. Combine contents of both tubes. Repeat step 4.
6. Pour out supernatant and resuspend cells with 1,000 μL of PBS with a P1000 micropipette and transfer to a 1.5-mL Eppendorf tube.
7. Take 10 μL of suspension and dilute into 490 μL PBS. Count the number of cells in this diluted suspension.
8. For intravenous (IV) injection of cells (see Subheading 3.5), concentrate cells to 0.5×10^6 to 1.0×10^6 cells/mL. Only use 1× PBS to dilute/resuspend cell concentrates (please see Note 4).

3.2. Preparation of Fluorescently Labeled Lectin to Visualize Endothelium of the Chorioallantoic Membrane

1. To label endothelium, inject lectin LCA-labeled with fluorescein or rhodamine. Select the lectin based on the fluorescence of your tumor cell line; inject lectin LCA-rhodamine if using cancer cells expressing GFP and LCA-fluorescein if using RFP or its equivalent. Listed below is a protocol for preparing lectin (50 μL of lectin/animal) for injection into four embryos.
2. Dilute 30 μL of lectin LCA-fluorescein/rhodamine in 270 μL of 1× PBS pH 7.4. Mix well; there is no need to sonicate. Shield from ambient light.
3. Assemble a 1-mL syringe with 18-gauge needle. Draw diluted lectin into syringe.
4. Cut a 2–3-in. long piece of Tygon tubing (1/32 in. inner diameter, 3/32 in. outer diameter, 1/32 in. wall thickness) and carefully insert bevel of needle into lumen of tubing. Slowly slide the tubing all the way into the needle. There should be 1–2 in. of free tubing (Fig. 1a).
5. Using the needle puller, pull 10–20 needles from sodium borosilicate pipettes. Make needles that are as long and tapered as possible (Fig. 1b, bottom needle).
6. Using a dissecting microscope and fine forceps, take a needle and use forceps to bend the needle tip until it breaks. Grasp the

apex of the tip, finding the earliest break point of the needle (Fig. 1b, top needle).

7. Insert the blunt end of the needle into the exposed end of the Tygon tubing attached to the syringe (Fig. 1a). Slowly eject all air in the needle-syringe to prevent injection of air bubbles into animal.
8. Microinject 50 μL of diluted lectin into embryos. Refer to Subheading 3.4 for microinjection protocol.

3.3. Preparation of Fluorescently-Labeled Dextrans to Visualize Lumen of the Chorioallantoic Membrane

1. To illuminate vessel lumen, inject embryo intravenously with fluorescent dextran. Listed below is a protocol for injecting dextran into four embryos.
2. Mix diluted dextrans well, there is no need to sonicate. Shield from ambient light.
3. Assemble a 1-mL syringe with 18-gauge needle. Draw ~500 μL of diluted dextran into syringe.
4. Cut a 2–3 in. long piece of Tygon tubing (1/32 in. inner diameter, 3/32 in. outer diameter, 1/32 in. wall thickness) and carefully insert bevel of needle into lumen of tubing. Slowly slide the tubing all the way into the needle. There should be 1–2 in. of free tubing.
5. Using the needle puller, pull 10–20 needles with sodium borosilicate pipettes. Make needles that are as long and tapered as possible (Fig. 1b, bottom needle).
6. Using a dissecting microscope and fine forceps, take a needle and use forceps to bend the end of the needle tip until it breaks. Bend your way from the very apex of the tip towards the base, trying to find the earliest break point of the needle (Fig. 1b, top needle).
7. Insert the blunt end of this needle into the tubing of the syringe containing diluted lectin (Fig. 1a). Slowly eject all air in the needle-syringe to prevent injection of air bubbles into animal.
8. Microinject 50 μL of diluted dextrans into days 9–12 embryos. Inject 100 μL into days 13–19 embryos.

3.4. Intravenous Injection of Lectin/Dextrans into Chorioallantoic Membrane

1. First, distinguish veins from arteries on surface of CAM using a dissecting scope. The arteries and veins interdigitate on the surface of the CAM (Fig. 1c), terminating in the capillary plexus. Since the CAM acts as a gas exchange organ, the arteries appear dark red because they deliver deoxygenated blood to the CAM while veins are bright red because they transport freshly oxygenated blood back to the embryo. Under a dissecting scope, this subtle color difference allows veins or arteries to be distinguished. One can also examine the direction of blood flow and follow the flow from the capillary bed

towards larger, wider vessels (veins). It is important to inject into a vein. If an artery is injected, the vessel will bleed profusely and compromise the viability of embryo for imaging experiments.

2. Identify the vein to be injected. With a sufficiently tapered microneedle, very narrow veins can be injected, which will minimize bleeding during and after injection. Avoid injecting into major vessels, as this will impact embryo viability. It is recommended to target only vessels that are tributaries or secondary tributaries of the major veins. Additionally, it is technically easier to pierce the vascular wall of smaller veins compared to larger veins.
3. Increase magnification of dissecting scope to view only the vein of interest. Using the assembled injection apparatus (Fig. 1a), press the tip of the borosilicate needle against the surface of a vein and gently press forward (Fig. 1d) in the same direction as blood flow. As you press forward, use your other hand to depress the plunger lightly. When the needle tip successfully enters the vessel lumen, the (clear) solution will stream through the vein away from the tip.
4. Minimize movement of the needle and continue to depress plunger until desired volume is injected as indicated by the syringe markings. This may take 1–5 min depending on the quality of vessel cannulation. If there is excessive clear fluid buildup at the site of injection, pick another site of injection.
5. After injection, slowly pull needle out of the vessel. Clean up blood or excess injection fluid by dabbing lightly with a Kimwipe.
6. Needles can be reused for multiple injections, but the sharpness will decrease with each injection. If injection becomes difficult, replace the needle.

3.5. Intravenous Injection of Cancer Cells for Migration Assay

1. An "experimental metastasis" assay approach (28, 32) is utilized whereby cancer cells are injected intravenously and allowed to extravasate into the stroma of the CAM. The injection of cancer cells can be performed using steps similar to those described in Subheading 3.2/3.3.
2. Use day 10 embryos for injection of cancer cells, prepared as described (29).
3. When preparing needles for injection of cancer cells (as prepared in Subheading 3.2, steps 3–7), the needle bore must be slightly wider in order to avoid shearing of the cancer cells.
4. Cells must be homogenously suspended. Between injections, look for cell aggregation. If clumping is observed, remove the borosilicate needle and use the syringe plunger to mix the

suspension until clumps are dispersed. Ensure that any air bubbles are removed prior to injection.

5. Depending on the cell type, cell aggregates may form and clog the needle head. If this occurs, discard the needle and tubing and replace with new tubing and needle. It is generally necessary to change the needle every 2–4 injections.
6. During injections, it is recommended to gently pulse the plunger. This will result in more uniform distribution of cells throughout the CAM. Expect an average injection time of 2–5 min per embryo.
7. After removing needle from CAM, dab the injection site with a Kimwipe to remove blood and cancer cells that have spilled onto the surface of the CAM. Cells left behind on the CAM surface may be mistaken for micrometastases during imaging.
8. Use a fluorescence dissection microscope to verify successful injection and to assess uniform distribution of cancer cells throughout the capillary plexus of the CAM. Return embryo to incubator until Days 15–19. For imaging of single extravasated cells, image between Days 12 and 15.
9. If imaging vessels is required, complete steps in Subheadings 3.1 and 3.2/3.3 prior to imaging.

3.6. Imaging of Cell Migration In Vivo

1. Set temperature-regulated microscope chamber to 37°C for 6 h prior to imaging. This will stabilize temperature and help minimize Z drift during imaging.
2. To image micrometastatic colonies or single cells (see Subheading 3.5), a 10× or 20× objective is recommended.
3. At least 10 h of continuous imaging is required for the in vivo cell migration assay.
4. Apply a thin layer of vacuum grease to underside of the lid to secure coverslip. Gently position a coverslip into the lid and wipe away any excess vacuum grease.
5. Position the embryo in the imaging unit such that the coverslip can be lowered down directly onto an open area of the CAM. Slowly lower the lid onto the embryo until the coverslip just makes contact with the CAM. Tighten the screws to secure the lid in place, ensure the lid is as level as possible and that the coverslip is not putting any pressure on the CAM.
6. Fill the outer jacket of the embryo imaging unit with water heated to 37°C and then place the unit onto the microscope stage (Fig. 1e). The embryo imaging unit can be fixed onto the stage with tape to minimize XY drift.
7. Acquire time-lapse images, preferably with multiple XY points every 5–15 min.

3.7. Image Analysis of In Vivo Cell Migration

1. Specialized software can be used to define cells and their motion paths without user bias. We suggest software such as Volocity (Perkin Elmer) and ImageJ (NIH) to do this. Outlined below are general parameters and steps that will assist in quantitation of cell migration.
2. First, generate a time-based image sequence of only the cell-signal channel (i.e., GFP for HEp3-GFP cells as demonstrated in Fig. 2). Export as a .tif or movie file.
3. Open this file in ImageJ. The StackReg plugin (http://big-www.epfl.ch/thevenaz/stackreg/) can be used to align all the images and correct global *XY*-movement artifacts typically encountered during acquisition.
4. Save this file as a .tif and then import into Volocity as an Image Sequence. Ensure that you organize the imported image sequence based on timepoints.
5. In the measurements tab, select the "Find Objects Using SD Intensity." Set the threshold parameters such that you include all objects without selecting background noise.

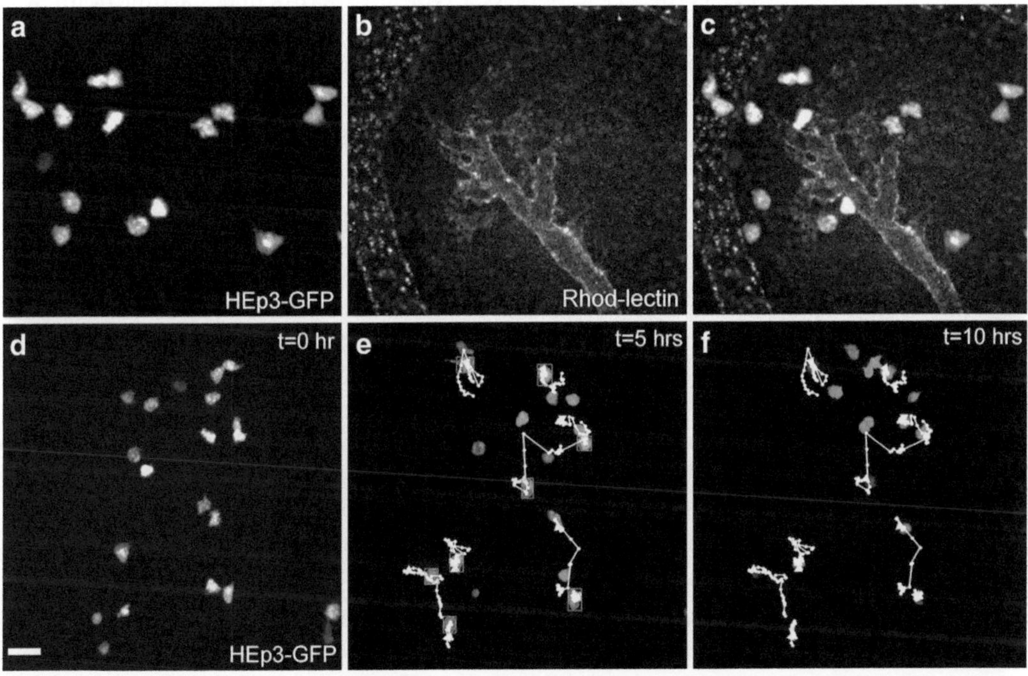

Fig. 2. In vivo migration assay using HEp3-GFP cells. (**a–c**) Representative HEp3-GFP micrometastatic colony with CAM vasculature labeled with rhodamine lectin. (**a**) Represents the GFP channel image of cells at $t=0$ h. (**b**) Represents the vessel architecture surrounding the cells as labeled by rhodamine lectin at $t=0$ h. (**c**) Represents merged image of both (**a**) and (**b**). (**d–f**) Another representative HEp3-GFP micrometastatic colony as imaged at $t=0$ (**d**), $t=5$ h (**e**), and $t=10$ h (**f**). Nine cell migration tracks are shown in (**e**) and (**f**) and quantitated in Table 1. The *arrowhead* denotes the end of the cell migration track. There is a bounding box for the cell during each migration track at $t=5$ h. Scale bars represent 25 μm.

6. Then select the “Exclude Objects by Size” and define the size range of the cells of interest. This will help eliminate noise or large cell clumps from being analyzed.
7. Often, cells will migrate near or off the edge of the field of view. To exclude these cells, select the “Exclude Objects Touching Edge of Image.”
8. There is always cell–cell interaction during migration; in order to prevent cell motion tracks from being eliminated, select the “Separate Touching Objects.”
9. To visualize cell motion tracks of all cells as defined by this protocol, select “Track Objects” and vary the “maximum distance between objects” as needed. Then click “Measure all Timepoints” in the main measurements tab. In the data field, you will see a filter selector; select “tracks” and you will see all the tracks listed as analyzed by Volocity. It lists all relevant parameters such as displacement and length of motion track for each track.
10. *Important*: Calibrate the scale using a stage micrometer before each experiment in order to generate accurate cell migration data. We present quantitation of such data in Fig. 2 and Table 1.
11. Figure 2a–c depicts a typical experimental starting point with cancer cells labeled with GFP and the CAM vasculature labeled with rhodamine lectin. Figure 2d–f depicts the migration tracks of cancer cells over a 10-h timelapse at $t=0$ (d), $t=5$ h (e), and $t=10$ h (f). Table 1 lists cell migration parameters as analyzed by Volocity.

4. Notes

1. Use of cancer cells either transiently or stably transfected with a construct that constitutively expresses a fluorescent protein such as GFP or tdTomato, or are otherwise fluorescently labeled (e.g., fluorescent nanobeads or fluorescent fusion protein that labels the nucleus such as H1-mCherry), is required. The reader is referred to numerous references on this subject (e.g., (33–41)).
2. Details on incubation of chick embryos and the shell-less chick embryo preparation and their use in imaging studies are beyond the scope of this review. Use of the specialized embryo imaging unit (Fig. 1e) is absolutely required, as the complete restriction of movement is necessary for microscopic imaging of cancer cells. The reader also is referred to refs. (29, 30, 35). Embryo imaging units are available from Quorum Technologies (Guelph, ON).

Table 1
Tabulated in vivo migration data of representative cell migration tracks

Track ID	Length (μm)	Track velocity (μm/h)	Displacement (μm)	Displacement rate (μm/h)	Meandering index	Angle (°)	Bearing (°)
Track A	152.28	15.23	23.09	2.31	0.14	90.98	269.02
Track B	235.45	23.55	41.00	4.10	0.16	9.63	350.37
Track C	148.06	14.81	27.85	2.78	0.17	89.93	270.07
Track D	201.35	20.13	82.88	8.29	0.37	6.27	6.27
Track E	81.58	8.16	12.72	1.27	0.14	159.40	159.40
Track F	196.47	19.65	76.33	7.63	0.35	33.96	326.04
Track G	102.90	10.29	20.48	2.05	0.18	1.83	1.83
Track H	267.29	26.73	109.49	10.95	0.37	130.19	229.81
Track I	215.78	21.58	29.23	2.92	0.12	137.59	222.41

Volocity parameters—Length: Calibrated length of the track. *Track velocity*: Average speed over migration track. *Displacement*: Length of straight line from the first position in the track to the last. *Displacement rate*: Displacement/(time of the last time point in the track – time of the first time point in the track). *Meandering index*: This is the displacement rate/velocity, which provides a measure of a track's deviation from a straight line. The meandering index is always a value less than or equal to 1. A meandering index of 1 indicates that the track is a perfect straight line; the smaller the value of the meandering index, the greater the meandering of the track. *Angle*: A measure of the spatial relationship between the vector of the displacement of the track and the vector 0, –1, 0 (up). The angle of the shortest distance between these two vectors. Ranges from 0 to 180. 0 means pointing "up the screen," along the Y axis. 180 means pointing "down" the screen. *Bearing*: A measure of the spatial relationship between the vector of the displacement of the track and the vector 0, –1, 0 (up). The angle disregarding the Z component. Ranges from 0 to 360. 90 means pointing "right," along the X axis. 270 means pointing "left," negatively along the X axis

3. For confocal microscopy, we use a Zeiss AxioExaminer upright microscope fitted with a Yokogawa spinning disk unit and a Hamamatsu EM2 EM-CCD 512X512 camera. It is equipped with both a mercury arc lamp and diode-based lasers (405, 491, 561, 647, 730 nm). For widefield microscopy, we use a Zeiss Examiner upright microscope fitted with a Hamamatsu 9100-02 EM-CCD camera. We use Volocity 5 (Perkin Elmer) and ImageJ (http://rsbweb.nih.gov/ij/) software for image acquisition and image processing.
4. Do not leave prepared cells on ice for long as this will negatively impact cell viability. (Different cell lines are differentially sensitive, so this should be checked for your cell lines). Proceed to injection or implantation immediately after preparation of cells.

Acknowledgments

This work is supported by the Canadian Cancer Society Research Institute (grant #18176) and the Canadian Institutes of Health Research (grants #845351 and #42511). H.S.L. is funded by a Post-Doctoral Fellowship from the Canadian Breast Cancer Foundation (Ontario Division). A.F.C. is Canada Research Chair in Oncology, supported by the Canada Research Chairs Program.

References

1. Hanahan, D., and Weinberg, R. A. (2000) The hallmarks of cancer, *Cell* ***100***, 57–70.
2. Fidler, I. J. (2001) Seed and soil revisited: contribution of the organ microenvironment to cancer metastasis, *Surg Oncol Clin N Am* ***10***, 257–269, vii–viiii.
3. Chambers, A. F., Groom, A. C., and MacDonald, I. C. (2002) Dissemination and growth of cancer cells in metastatic sites, *Nat Rev Cancer* ***2***, 563–572.
4. Kauffman, E. C., Robinson, V. L., Stadler, W. M., Sokoloff, M. H., and Rinker-Schaeffer, C. W. (2003) Metastasis suppression: the evolving role of metastasis suppressor genes for regulating cancer cell growth at the secondary site, *J Urol* ***169***, 1122–1133.
5. Pantel, K., and Brakenhoff, R. H. (2004) Dissecting the metastatic cascade, *Nat Rev Cancer* ***4***, 448–456.
6. Minn, A. J., Gupta, G. P., Siegel, P. M., Bos, P. D., Shu, W., Giri, D. D., Viale, A., Olshen, A. B., Gerald, W. L., and Massague, J. (2005) Genes that mediate breast cancer metastasis to lung, *Nature* ***436***, 518–524.
7. Hoon, D. S., Kitago, M., Kim, J., Mori, T., Piris, A., Szyfelbein, K., Mihm, M. C., Jr., Nathanson, S. D., Padera, T. P., Chambers, A. F., Vantyghem, S. A., MacDonald, I. C., Shivers, S. C., Alsarraj, M., Reintgen, D. S., Passlick, B., Sienel, W., and Pantel, K. (2006) Molecular mechanisms of metastasis, *Cancer Metastasis Rev* ***25***, 203–220.
8. Suzuki, M., Mose, E. S., Montel, V., and Tarin, D. (2006) Dormant cancer cells retrieved from metastasis-free organs regain tumorigenic and metastatic potency, *Am J Pathol* ***169***, 673–681.
9. Eccles, S. A., and Welch, D. R. (2007) Metastasis: recent discoveries and novel treatment strategies, *Lancet* ***369***, 1742–1757.
10. Barkan, D., Kleinman, H., Simmons, J. L., Asmussen, H., Kamaraju, A. K., Hoenorhoff, M. J., Liu, Z. Y., Costes, S. V., Cho, E. H., Lockett, S., Khanna, C., Chambers, A. F., and

Green, J. E. (2008) Inhibition of metastatic outgrowth from single dormant tumor cells by targeting the cytoskeleton, *Cancer Res* ***68***, 6241–6250.

11. Hunter, K. W., Crawford, N. P., and Alsarraj, J. (2008) Mechanisms of metastasis, *Breast Cancer Res* ***10*** Suppl 1, S2.
12. Taylor, J., Hickson, J., Lotan, T., Yamada, D. S., and Rinker-Schaeffer, C. (2008) Using metastasis suppressor proteins to dissect interactions among cancer cells and their microenvironment, *Cancer Metastasis Rev* ***27***, 67–73.
13. Weber, G. F. (2008) Molecular mechanisms of metastasis, *Cancer Lett* ***270***, 181–190.
14. Albini, A. (1998) Tumor and endothelial cell invasion of basement membranes. The matrigel chemoinvasion assay as a tool for dissecting molecular mechanisms, *Pathol Oncol Res* ***4***, 230–241.
15. Eccles, S. A., Box, C., and Court, W. (2005) Cell migration/invasion assays and their application in cancer drug discovery, *Biotechnol Annu Rev* ***11***, 391–421.
16. Leighton, J. (1964) Invasion and Metastasis of Heterologous Tumors in the Chick Embryo, *Prog Exp Tumor Res* ***4***, 98–125.
17. Locker, J., Goldblatt, P. J., and Leighton, J. (1969) Hematogenous metastasis of Yoshida ascites Hepatoma in the chick embryo liver: ultrastructural changes in tumor cells, *Cancer Res* ***29***, 1245–1253.
18. McAllister, R. M., Peer, M., Gilden, R. V., Klement, V., and Landing, B. H. (1974) Tumors formed by human rhabdomyosarcoma cells in chorioallantoic membrane of embryonated hens' eggs, *Int J Cancer* ***13***, 886–890.
19. Chambers, A. F., Shafir, R., and Ling, V. (1982) A model system for studying metastasis using the embryonic chick, *Cancer Res* ***42***, 4018–4025.
20. Chambers, A. F., and Wilson, S. (1985) Cells transformed with a ts viral src mutant are temperature sensitive for in vivo growth, *Mol Cell Biol* ***5***, 728–733.
21. Gordon, J. R., and Quigley, J. P. (1986) Early spontaneous metastasis in the human epidermoid carcinoma HEp3/chick embryo model: contribution of incidental colonization, *Int J Cancer* ***38***, 437–444.
22. Chambers, A. F., Schmidt, E. E., MacDonald, I. C., Morris, V. L., and Groom, A. C. (1992) Early steps in hematogenous metastasis of B16F1 melanoma cells in chick embryos studied by high-resolution intravital videomicroscopy, *J Natl Cancer Inst* ***84***, 797–803.
23. MacDonald, I. C., Schmidt, E. E., Morris, V. L., Chambers, A. F., and Groom, A. C. (1992) Intravital videomicroscopy of the chorioallantoic microcirculation: a model system for studying metastasis, *Microvasc Res* ***44***, 185–199.
24. Koop, S., Khokha, R., Schmidt, E. E., MacDonald, I. C., Morris, V. L., Chambers, A. F., and Groom, A. C. (1994) Overexpression of metalloproteinase inhibitor in B16F10 cells does not affect extravasation but reduces tumor growth, *Cancer Res* ***54***, 4791–4797.
25. Quigley, J. P., and Armstrong, P. B. (1998) Tumor cell intravasation alu-cidated: the chick embryo opens the window, *Cell* ***94***, 281–284.
26. Ossowski, L., Aguirre Ghiso, J., Liu, D., Yu, W., and Kovalski, K. (1999) The role of plasminogen activator receptor in cancer invasion and dormancy, *Medicina (B Aires)* ***59***, 547–552.
27. Aguirre-Ghiso, J. A., Ossowski, L., and Rosenbaum, S. K. (2004) Green fluorescent protein tagging of extracellular signal-regulated kinase and p38 pathways reveals novel dynamics of pathway activation during primary and metastatic growth, *Cancer Res* ***64***, 7336–7345.
28. Wilson, S. M., and Chambers, A. F. (2004) Experimental metastasis assays in the chick embryo, *Curr Protoc Cell Biol* Chapter ***19***, Unit 19.6.
29. Zijlstra, A., Lewis, J., Degryse, B., Stuhlmann, H., and Quigley, J. P. (2008) The inhibition of tumor cell intravasation and subsequent metastasis via regulation of in vivo tumor cell motility by the tetraspanin CD151, *Cancer Cell* ***13***, 221–234.
30. Deryugina, E. I., and Quigley, J. P. (2008) Chick embryo chorioallantoic membrane model systems to study and visualize human tumor cell metastasis, *Histochem Cell Biol* ***130***, 1119–1130.
31. Jilani, S. M., Murphy, T. J., Thai, S. N., Eichmann, A., Alva, J. A., and Iruela-Arispe, M. L. (2003) Selective binding of lectins to embryonic chicken vasculature, *J Histochem Cytochem* ***51***, 597–604.
32. Welch, D. R. (1997) Technical considerations for studying cancer metastasis in vivo, *Clin Exp Metastasis* ***15***, 272–306.
33. Chishima, T., Miyagi, Y., Wang, X., Yamaoka, H., Shimada, H., Moossa, A. R., and Hoffman, R. M. (1997) Cancer invasion and micrometastasis visualized in live tissue by green fluorescent protein expression, *Cancer Res* ***57***, 2042–2047.
34. Naumov, G. N., Wilson, S. M., MacDonald, I. C., Schmidt, E. E., Morris, V. L., Groom, A. C., Hoffman, R. M., and Chambers, A. F.

(1999) Cellular expression of green fluorescent protein, coupled with high-resolution in vivo videomicroscopy, to monitor steps in tumor metastasis, *J Cell Sci* ***112*** (Pt 12), 1835–1842.

35. Lewis, J. D., Destito, G., Zijlstra, A., Gonzalez, M. J., Quigley, J. P., Manchester, M., and Stuhlmann, H. (2006) Viral nanoparticles as tools for intravital vascular imaging, *Nat Med* ***12***, 354–360.
36. Sahai, E. (2007) Illuminating the metastatic process, *Nat Rev Cancer* 7, 737–749.
37. Hoffman, R. M. (2009) Imaging cancer dynamics in vivo at the tumor and cellular level with fluorescent proteins, *Clin Exp Metastasis* ***26***, 345–355.
38. Hoffman, R.M. (2005) The multiple uses of fluorescent proteins to visualize cancer in vivo. *Nat Rev Cancer 5*, 796–806.
39. Hoffman, R. M., and Yang, M. (2006) Subcellular imaging in the live mouse. *Nat Protoc 1*, 775–782.
40. Hoffman, R. M., and Yang, M. (2006) Color-coded fluorescence imaging of tumor-host interactions. *Nat Protoc 1*, 928–935.
41. Hoffman, R. M., and Yang, M. (2006) Whole-body imaging with fluorescent proteins. *Nat Protoc 1*, 1429–1438.

Chapter 2

The Use of Fluorescent Proteins for Intravital Imaging of Cancer Cell Invasion

James Hulit, Dmitriy Kedrin, Bojana Gligorijevic, David Entenberg, Jeffrey Wyckoff, John Condeelis, and Jeffrey E. Segall

Abstract

The analysis of cancer cell behavior in the primary tumor in living animals provides an opportunity to explore the process of invasion and intravasation in the complex microenvironment that is present in vivo. In this chapter, we describe the methods that we have developed for performing intravital imaging of mammary tumors. We provide procedures for generating tumors through injection of tumor cell lines, and multiphoton imaging using a skin-flap tumor dissection and a mammary imaging window.

Key words: Intravital imaging, Multiphoton imaging, Live animal imaging

1. Introduction

The malignancy of cancer is due in large part to local and distant spread (1). In tumors whose locations make it difficult for surgery to completely remove them, such as glioblastoma (2) and head and neck cancer (3), local invasion can lead to local recurrence. For other cancers which quite often can be fully removed, such as breast cancer, cells which have spread to distant organs prior to surgery can form metastases in multiple sites, making further treatment very difficult. Currently most therapeutics are based on blocking tumor cell growth. Thus, an understanding of the processes of local tumor cell invasion and intravasation (which enables distant metastasis) may allow the development of new treatments which can reduce the malignancy of cancer.

Although 2D and 3D in vitro studies of cell motility and invasion are useful to understand basic mechanisms that regulate

Robert M. Hoffman (ed.), *In Vivo Cellular Imaging Using Fluorescent Proteins: Methods and Protocols*, Methods in Molecular Biology, vol. 872, DOI 10.1007/978-1-61779-797-2_2, © Springer Science+Business Media, LLC 2012

tumor cell movement, given the complexity of the tumor microenvironment, it is important to directly evaluate cancer cell motility in vivo. Specific features of the in vivo tumor microenvironment that are difficult to mimic in vitro include the presence of a variety of cell types, a range of extracellular molecules and fibers which are poorly defined, and blood vessels which have varying diameter and cell types associated with them.

Important stromal cell types that can affect tumor cell behavior include tumor-associated macrophages (4), neutrophils (5), fibroblasts (6, 7), and bone marrow-derived stem cells (8).

Extracellular fibers of varying diameter and composition run through the tumor and can potentially provide pathways for invasion (9, 10), hinder tumor cell movement, or provide a level of stiffness that can stimulate signaling for growth and invasion (11).

Blood vessels and lymphatics provide opportunities for transport to distant sites (12, 13). The arrangement of endothelial and smooth muscle cells in tumors is disorganized and more permeable compared to the vasculature of normal tissue (14, 15).

Mouse models of cancer provide an accessible mammalian system for studying cancer invasion in vivo. Using immunocompromised mice, human cancers can be transplanted either direct from human tumors as fragments or from cell lines (16). The advantage is that the focus is on the most clinically relevant (human) cancer, but the stromal effects cannot include potential contributions from T and B cells or other immune cells depending on the precise immunodeficiency model that is utilized (17). Conversely, transgenic models make use of specific oncogenes to induce tumors in the organ of interest in fully immunocompetent animals (18), with the limitations of using highly promoted oncogenes that may be unusual for human tumors as well as taking much longer and more sporadically to form a tumor than many xenograft models.

Both immunocompetent and immunodeficient mouse models have been utilized for intravital imaging of tumor cell invasion. In both cases, a key component is the expression of fluorescent proteins in the cells of interest (16). For xenograft transplants into immunodeficient mice, expression of fluorescent proteins is quite straightforward for cell lines, enabling relatively rapid testing of various proteins with specific functional or localization properties. However, imaging of stromal cells is challenging, with macrophages being the major cell type that has been imaged due to their uptake of i.v. injected fluorescent dextran (19). With transgenic animals, labeling of both tumor cells and stromal cells with fluorescent proteins is possible. As the appropriate transgenic strains become available, the label can be crossed into immunodeficient lines as well (20–23).

Some of our studies of in vivo tumor properties using intravital imaging have used confocal microscopy. Single-photon confocal microscopy can provide greater flexibility in terms of excitation of multiple fluorophores (due to the relatively lower cost of using multiple lasers and a wider range of wavelengths available) but provides more limited depth penetration (24). Multiphoton microscopy allows deeper penetration, as well as the detection of some (but not all) matrix fibers using second harmonic generation (SHG) (9, 25, 26). We currently use multiphoton microscopy to allow relatively deep (>100 μm) imaging into tumors in animals with relatively little bleaching or damage. We originally used a skin-flap method to expose the tumor surface for imaging, which is difficult to use for multiple imaging sessions. We then developed a mammary imaging window to evaluate tumor cell motility over multiple sessions, in combination with the photoconvertible protein Dendra2 (27). In this chapter, we provide technical details for generating and observing mammary tumors using intravital imaging, including injection of cancer cells, skin flap dissection, and utilization of a mammary imaging window.

2. Materials

We describe below materials required for orthotopic (i.e., mammary gland) growth (see Subheading 2.1), cancer cell imaging by the skin-flap method (see Subheading 2.2), generating mammary imaging windows (see Subheading 2.3), window implantation (see Subheading 2.4), and microscope setup (see Subheading 2.5).

2.1. Generating the Tumors

In our work, MTLn3 rat mammary adenocarcinoma and MDA-MB-231 human mammary adenocarcinoma cell lines are frequently used. These well-characterized cell lines have been transfected with vectors driving constitutive expression of GFP for visualization of organelles of individual cells. It is, however, possible to restrict expression of fluorescent markers to a specific organelle (nucleus, Golgi, etc.) depending on the study design.

Materials required for orthotopic (i.e., mammary gland) tumor growth and imaging (for both skin flap methods as well as mammary imaging window methods) are listed below:

1. MTLn3 (rat) or MDA-MB-231 (human) adenocarcinoma cell lines (available from authors) or an alternative.
2. Solution 1: For cell detachment. Phosphate-buffered saline (PBS) without Ca^{2+} or Mg^{2+} containing 2 mM ethylenediamine tetraacetic acid (EDTA) pH 8. Autoclaved or filter sterilized using 0.22 μm pore size.
3. Solution 2: For resuspension and injection of MTLn3 cells. PBS with Ca^{2+} and Mg^{2+}, containing 0.2% (w/v) bovine serum albumin (BSA). Filter sterilized using 0.22 μm pore size.

4. Solution 3: For resuspension and injection of MDA-MB-231 (4173) cells. PBS with Ca^{2+} and Mg^{2+}, containing 0.2% (w/v) BSA and 20% (v/v) collagen (type 1). Filter sterilized using 0.22 μm pore size.
5. Immunodeficient mice, i.e., female 5–6-week-old BALB/c SCID/Ncr mice (National Cancer Institute). MTLn3 cells can also be injected into immunocompetent Fisher 344 rats since they are derived from this strain.
6. 70% (v/v) ethanol.
7. 15-cm tissue-culture dishes.
8. 15-mL conical polypropylene centrifuge tubes, sterile.
9. Cell scrapers.
10. Centrifuge.
11. 1-mL syringes with 25-G needles for cell injection.

2.2. Animal Preparation for Imaging via Skin Flap

An oxygen/anesthesia system and equipment are used when imaging with the mammary imaging window in conjunction with the custom-imaging box described below.

1. Oxygen/anesthesia vaporizer apparatus (we use Forane Vaporizer, model 100, from SurgiVet).
2. Rodent-sized anesthesia box.
3. Compressed oxygen (100% pure, for anesthesia delivery/mixture).
4. Isoflurane anesthesia, USP.
5. Anesthesia delivery/breathing circuit (tubing, nose cone, scavenging system).
6. Heated stage or heater/blower system for microscope box.
7. 70% ethanol solution in dH_2O.
8. Surgical gloves.
9. Rodent-sized surgical scissors (sterilized).
10. Rodent-sized surgical forceps (sterilized).
11. Sterile cotton swabs (or forceps to reposition fat-pad covering tumor imaging surface).
12. Optional: Magnification glass with light.
13. Heating pad for surgical procedure.
14. Sterile tape for securing animal and anesthesia equipment to microscope stage.

2.3. Production of Imaging Window

The mammary imaging window consists of a glass coverslip mounted atop a tissue-culture grade plastic platform designed for implantation over a murine mammary gland or mammary tumor, located proximal to the hind leg. Cancer cells can be injected under the window to monitor the initial steps of tumor growth. Alternatively, the window can be inserted over spontaneously grown mammary tumors either in transgenic animals, such as the MMTV-PyMT strain, or xenograft tumors formed by orthotopic implantation. See refs. 26–28 for details.

1. Tissue-culture grade plastic (tissue-culture dishes can be utilized).
2. Cyanoacrylate glue.
3. Bunsen burner.
4. 23-G needle.
5. Dremel tool for sanding and shaping of the plastic parts.
6. 8-mm diameter, circular glass cover-slips.
7. 70% Ethanol.

2.4. Mammary Imaging Window Implantation

1. Anesthesia (Avertin).
2. Betadine.
3. Ethanol solution (70%) in dH_2O.
4. Hair removal cream and small animal shaver.
5. Sterile dissecting microscissors.
6. Sterile dissection microforceps.
7. Suturing thread (nonwicking) and needle.
8. Suturing forceps.
9. Sterile Q-tips.
10. Sterile cloth.
11. Heating pad (90°F).
12. Cyanoacrylate glue.
13. Sterile gauze.
14. Ophthalmic ointment.
15. TMP-SMX antibiotic mix: sulfamethoxazole 0.6 mg/mL, trimethoprim 0.12 mg/mL.
16. Analgesia, Flunixin (Banamine): Used at 2.0–2.5 mg/kg bodyweight.
17. Surgical gown, hairnet, gloves, and mask.

2.5. Imaging Microscope Setup

We use several different multiphoton microscopy systems for intravital imaging in mice and rats. These include one of the first available multiphoton turnkey systems, the BioRad Radiance 2000 (manual laser tuning, two PMT detectors), the Olympus FV1000-MPE (computer-tuned laser, four PMT detectors), equipped both with confocal and multiphoton capabilities, which has a 15-W Tsunami (Spectra-Physics) Ti Sapphire laser. With these two microscopy systems we are able to image second harmonic generated excitation, CFP and Texas Red simultaneously. We have found that imaging at 880 nm allows for the best simultaneous imaging of these fluorophores. We also use a custom-built dual-laser multiphoton system with an optical parametric oscillator (OPO). With this microscope, we are able to image the above-mentioned chromophores along with Red Dendra and all other red fluorescent proteins, including those fluorescing into the far red (26).

3. Methods

The ultimate goal of intravital imaging is to capture the behavior of cells (in our case cancer cells) in their native microenvironment. In our group, two methods are used to access the tumor for imaging. Both methods make use of either tumors of transgenic origin or tumors generated from orthotopic implantation. The first method makes use of a surgically-created skin flap to expose the tumor surface. This method allows for imaging sessions lasting up to several hours (see Note 1). The second method described utilizes a custom-built mammary imaging window that is subcutaneously implanted over a growing tumor, but can also be inserted first over the mammary gland, with cancer cells injected afterward. The window allows multiple imaging sessions using the same animal and tumor over a period of several days (see Note 2). For detailed video instructions on window implantation and its use for mammary imaging in vivo, see (28). Both the skin flap and window techniques have their pros and cons in terms of preparation time, types of data acquired, number of animals per experiment, and flexibility in experimental design. All are important considerations when choosing the appropriate approach for testing new hypotheses (see Note 3).

3.1. Growth and Generation of Tumors for In Vivo Imaging

In our work, we have utilized fluorescent derivatives of both MTLn3 cells—a highly metastatic rat mammary adenocarcinoma cell line and the human MDA-MB-231 breast cancer cell line. Here, we describe the method for preparing and injecting MTLn3 and MDA-MB-231 cells. Our group utilizes 5–6-week-old BALB/c SCID mice or 7–8-week-old Fischer 344 rats (used for MTLn3 cells). The animals are injected with a cell suspension into the mammary fat pad under the fourth (abdominal) nipple. MTLn3 tumors typically reach suitable imaging size within 3–4 weeks, while MDA-MB-231 require approximately 8 weeks of growth.

1. MTLn3 or MDA-MB-231 cells are grown to 80% confluence in a 15-cm tissue-culture dish.
2. The dish is rinsed three times with 5–7 mL of prewarmed (37°C) solution 1. After rinsing, add another 7 mL of solution 1

and incubate the cells at 37°C until detachment, which takes 10–30 min depending on cell density. Cells can be dislodged by gently tapping the side of the plate. Alternatively, trypsin/EDTA can be used to detach cells.

3. Once the cells have detached, harvest and transfer them to a sterile 15-mL conical tube. Scrape the plate to collect the cell matrix and transfer the mixture to the same 15-mL conical tube.
4. Add 7 mL of solution 2 to the plate to wash and collect the remaining matrix. At this point an aliquot of cell suspension can be taken for counting.
5. Pellet the cells by centrifugation for 5 min at 180–200 × *g* at room temperature. Upon completion, aspirate the supernatant.
6. The cells are resuspended in solution 2 to a concentration of not more than 10^7 cells/mL (10^6 cells/100 μL). If injecting MDA-MB-231 cells, resuspend in solution 3, which contains 20% collagen (the collagen is an important component in the initial stages of tumor formation for this cell line). This suspension is kept on ice until injection into the mammary fat pad, which should not be more than 30 min after resuspending them. Alternatively, a 50% (v/v) Matrigel-cell suspension mixture can be injected.
7. Keep the cells suspended and load the mixture into a 1-mL syringe fitted with a 25-G needle.
8. The area around the fourth abdominal nipple is sprayed with 70% ethanol. Inject 100 μL of the cell suspension under the skin into the mammary fat pad (for rats the volume can be increased to 200 μL).
9. To inject conscious animals, it is best to have two researchers working as a team. One person restrains the animal by simultaneously grasping the base of the tail and the scruff of the neck with one hand, while holding and extending the hind leg with their other hand to expose the nipple area. The second researcher performs the injection by slowly inserting the needle into the skin under the nipple. A proper injection into the mammary fat pad should induce the formation of a small raised area under the nipple. This raised area will not appear if the injection misses the fat pad. Alternatively, the animal can be anesthetized and a single researcher can perform the injection.
10. The animals are kept in a barrier-room facility until the tumors are large enough to image. Typically, this is 5–7 mm in diameter when using the mammary imaging window and 5–20 mm for the skin-flap technique.

3.2. Preparation of the Animal for Imaging Using the Skin Flap Method

Animals with tumors of diameter 5–20 mm can be used for imaging using the skin-flap method.

1. The animal is placed under anesthesia with 5% isoflurane via the oxygen/anesthesia apparatus. For mice, isoflurane levels are immediately reduced to 2.5% upon achieving unconsciousness. Importantly, animals with heavier lung metastasis burdens will require lower isoflurane levels than animals with lungs free of metastasis. Labored or erratic breathing is a danger sign.
2. Anesthesia is maintained throughout surgery and the imaging session. If vessel labeling is used, 70 kDa Texas Red- or FITC-dextran can be introduced via tail vein injection prior to or immediately after the skin-flap surgery.
3. Prior to making the surgical incision, thoroughly swab and clean the skin at the incision site with 70% ethanol. Allow the solution to dry before making the incision in order to prevent seepage into the wound.
4. Skin is then incised medially to the tumor using an incision long enough to permit exposure of the tumor surface facing the medial line of the animal.
5. Since fat cells can alter the path of photons coming into and exiting the imaging plane, the area over the exposed imaging surface of the tumor should be cleared of fascia and fat. However, care should be taken to maintain the vasculature and minimize bleeding, which can potentially alter cellular behavior within the tumor microenvironment. An attempt should first be made to carefully push aside the fascia to expose a region on the tumor surface. If fat and fascia obscure the image quality, delicately microdissect away the tissue.
6. The animal is subsequently transferred to a prewarmed and preferably enclosed microscope stage. As air-conditioning and contact with metal stages can cause extensive loss of body heat, external heaters should be used to maintain an environment between 25°C and 30°C. The exposed tumor surface is placed against the upper surface of the coverslip directly over the objective lens. It is necessary at this point to secure the tumor to the stage in order to minimize breathing artifacts and drift caused by animal movement. We have found that the most efficient method is to isolate the tumor with surgical or laboratory tape by simultaneously securing both the skin flap/tumor and lower tail region of the animal to the stage. By providing the lungs room to expand and the upper torso freedom to absorb the movement, we can minimize the breathing artifacts. Care should be taken to minimize pressure on the tumor so as not to block the blood flow or compress the tumor as these will cause image drift during

time-lapse acquisition. However, as pressure from the tape will introduce a small amount of drift along the *Z*-axis during imaging, the setup should be allowed to stabilize before the start of image collection.

7. Lower breathing rates minimize image drift and distortions, which can interfere with data analysis. As the imaging session progresses, the level of anesthesia deepens in the animal. As a result, during the imaging session it is advisable to progressively lower the isoflurane levels. By periodically lowering the isoflurane in steps of 0.25%, a constant breathing rate can be maintained while providing adequate tissue oxygenation and preventing suffocation. A proper breathing rate is defined by one or fewer distortions (breathing artifacts) per image scan when scanning at 166 lines/s.
8. After each *z*-stack time series is captured on a specific field, the tumor (microscope stage) can be repositioned to image additional fields. Multiple series can be captured as long as the animal's breathing and pulse are stable.

3.3. Mammary Imaging Window Construction (See Fig. 1)

1. The bottom of a tissue-culture dish is softened by heating with a flame (we typically use an alcohol lamp). Subsequently, a curved surface (2 cm diameter, 8 mm center height) is created by using a mold (e.g., an appropriately sized rounded Dremel bit) to press in to the softened plastic. The centerpiece of the curved surface, 1 cm in diameter, is then cut out using a hot scalpel.
2. Using a small, cone-shaped Dremel bit, a 6-mm hole is formed in the center of the curved plastic surface. Edges of the plastic base are smoothened by sanding with the Dremel and further filing.
3. The top of the plastic base is filed to make a flat surface (7 mm in diameter) for the glass coverslip.
4. A circular glass coverslip (8 mm) is glued using cyanoacrylate adhesive and the glue is allowed to dry for 15 min.
5. Eight suturing holes are made in the plastic by using a flame-heated 23-G needle.
6. The imaging window is cleaned with ethanol. A Q-tip can be used to clean the glass. If there are foggy spots present from the glue vapor, carefully use acetone with a Q-tip to remove them.
7. The window can be sterilized by overnight UV exposure in a tissue-culture hood. Ensure that both sides of the window are exposed to UV light.

In order to position and secure the animal on the microscope stage, a custom-designed 1/8-in. thick plexiglass-imaging box was constructed and used in combination with the mammary imaging

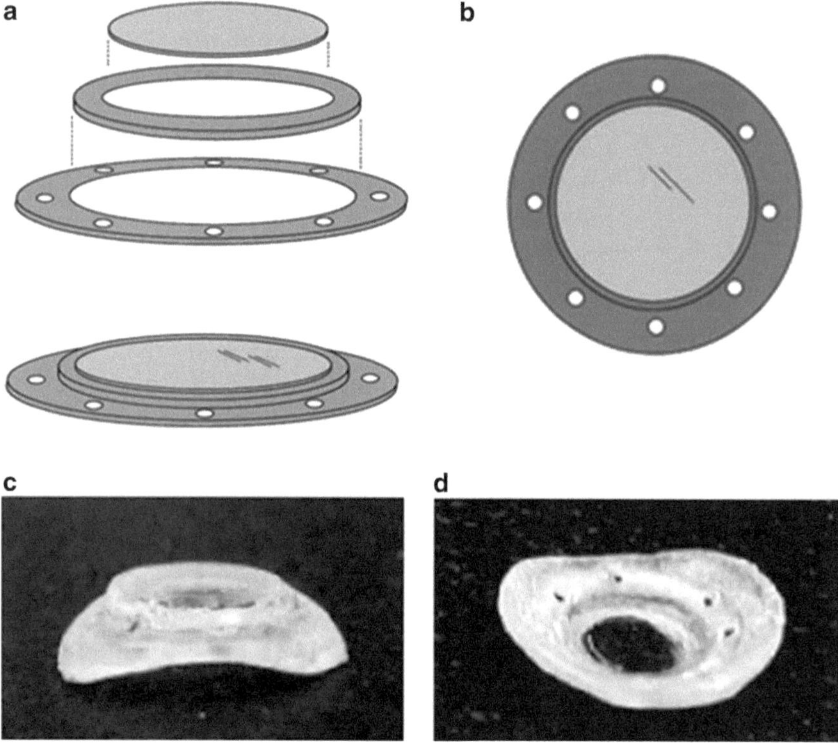

Fig. 1. Mammary imaging window design. (**a**) Overview of the construction with two spacer rings attached, with a coverslip mounted on the top. (**b**) Front view of the coverslip. (**c**) Alternative design of the window, where the bottom ring is curved to better accommodate a growing tumor. (**d**) Bottom view of a completed window ready for implantation. Adapted from *(27)*.

window (see Fig. 2). The mouse is placed inside of the box with the imaging windows glass coverslip projecting out from the underside. Precise placement of the window and coverslip against the microscope objective is achieved by adjusting two sliding doors. Vents have been constructed in the sides of the box to allow anesthesia to enter and exit while maintaining a constant flow. Finally, depending on the specific microscope and stage to be used, the design of the bottom of the imaging box can be altered to accommodate the different set-ups.

3.4. Mammary Imaging Window Implantation Method

1. As mentioned, the mammary imaging window can either be implanted over a normal mammary gland or a growing tumor. The optimal tumor size for window implantation is 5–7 mm, which assures that the tumor will be in contact with the coverslip during imaging and which permits space under the window for further tumor growth. It is also important to note the appearance of the skin over the tumor surface. In order to maximize the number of imaging sessions and data acquisition, tumors selected for the procedure should not be protruding or invading through the skin.

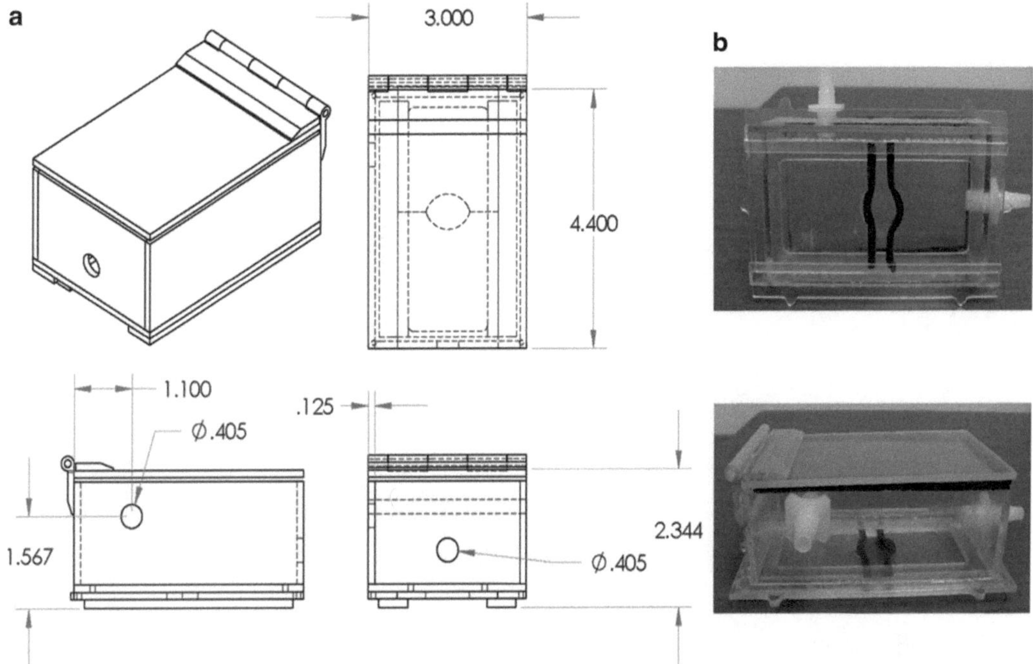

Fig. 2. Imaging box used in combination with the mammary imaging window. (**a**) This general imaging box schematic can be optimized to fit any microscope stage. (**b**) A photograph of a currently-used version of the imaging box.

2. Sterile technique should be used throughout the surgery—optimally inside a positive airflow sterile hood. All instruments used should be sterilized prior to surgery. Full gown, mask, and gloves should be worn throughout the procedure.
3. The mouse is anesthetized by intraperitoneal injection of 2.5% Avertin (20 μL/g body weight) in HBSS.
4. Hair should be removed by shaving the area above the tumor using a small animal shaver. The rest of the hair can be removed using hair removal cream.
5. The skin is disinfected using betadine and ethanol-dipped Q-tips. Ophthalmic ointment is applied to the eyes to protect from infection and drying.
6. First a 2-mm incision is made medial to the fourth (abdominal) nipple; this is where the tumor tends to form closest to the surface of the skin. The initial incision can be expanded, while simultaneously the underlying mammary fat pad is separated from the skin using blunt dissection with scissors and forceps until it is large enough to accommodate the imaging window. Care should be taken to avoid severing of major blood vessels supplying the tumor or surrounding area. This step is necessary

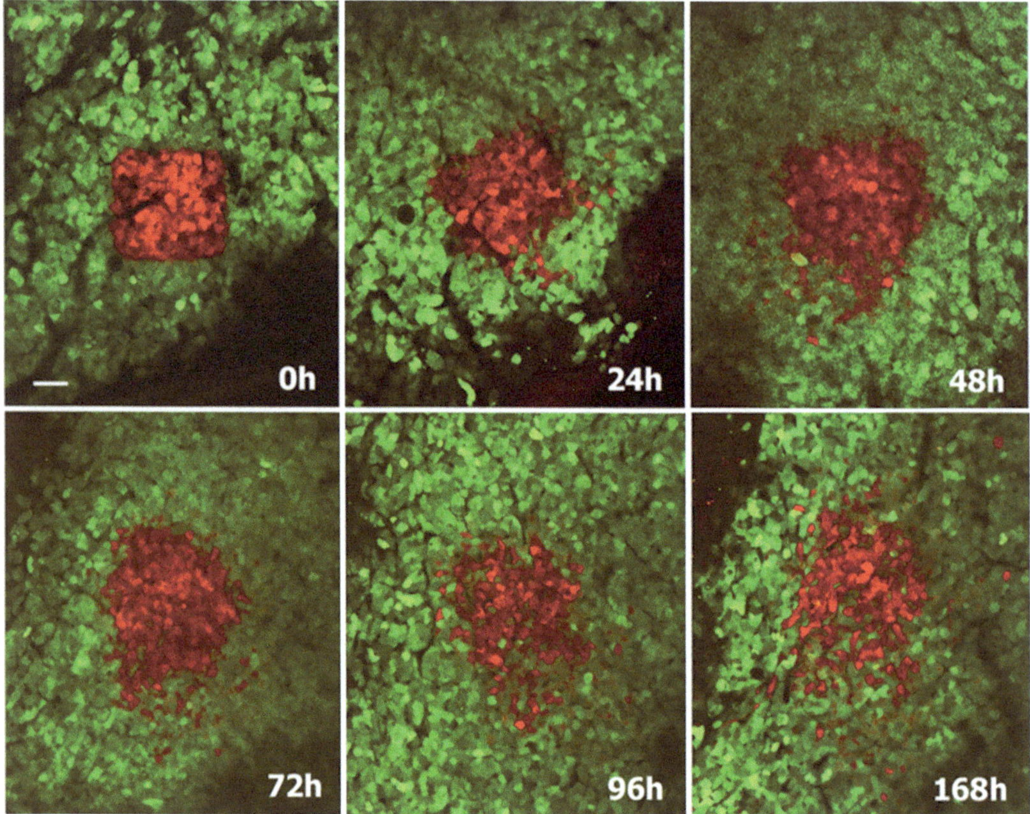

Fig. 3. Example of use of the mammary imaging window. When using the mammary imaging window and imaging box, multiple imaging sessions can be done using the same animal. Using expression of a photoconvertible protein, Dendra2, one can monitor the same cell population over 7 days (168 h).

to allow adequate space for the insertion of the imaging window. The imaging window is inserted such that there is skin on top of the imaging window base and sutured in place using nonabsorbable thread and a reverse cutting needle.

7. Tissue adhesive is used to fill in the suturing holes and secure the imaging window to the skin.
8. An analgesic should be administered in the immediate postoperative period for pain control.
9. TMP-SMX antibiotic mix is added into the cage water bottle for 3 days before and after the surgery.
10. The animal is allowed to recover over the next 3–4 days before the first imaging session is attempted. This allows for healing of the surgery area and thus minimizes inflammatory/tissue-remodeling artifacts. Imaging can subsequently be carried out up to 7 days (see Fig. 3).

3.5. Image Acquisition

While performing intravital microscopy, the data are collected as a 4D-data set (*x*, *y*, *z* series at each time point) of images at various scan speeds, adjusted depending on information to be obtained from analysis. The exact microscope parameters are highly dependent on the specific fluorophores used to obtain the data. Most commonly, we use a 20× water or glycerol immersion objective, which presents a field of view 1 mm in diameter in epifluorescence mode. Depending on the microscope, this becomes an imaging field of 500–775 μm in confocal or multiphoton mode. In confocal (single photon) microscopes, we use a laser power of 30–60 μW at the sample, 4.9 μs pixel dwell time, 1 Airy disk pinhole, and 50–100% PMT setting. For multiphoton microscopy, we use a laser with a 300-fs pulse width, 80-MHz repetition rate and <30 mW average laser power at the sample, 7.6 μs pixel dwell time, and 50–100% PMT setting. The imaging procedure is as follows:

1. Initial field selection and focusing are performed in epifluorescence mode through the microscope eyepiece. Select the channel from which the brightest signal is emitted and focus on the tumor surface. We typically use GFP-expressing cancer cells and introduce labeled dextrans into the circulation. Either the GFP fluorescence of the cancer cells or an intravenously-injected fluorescent label can be used for focusing and field selection. The eyepiece allows for a larger field of view than the PMT detectors, and epifluorescence mode allows quick illumination of the whole field and several depths at once. The brightest cancer cells are typically those closer to the objective lens and will yield the best quality images.
2. After selecting and focusing on the initial field, the microscope is switched from visualization to data collection in multiphoton mode. Ideal areas for our experiments are bright fluorescent fields of view populated with vessels containing active blood flow. Red blood cells absorb excitation light due to their high hemoglobin content. Therefore, flowing red blood cells can be observed indirectly in epifluorescence as moving shadows, or directly if labeled dextran has been injected into the vasculature. Areas that should be avoided are those densely covered by adipocytes (the mammary fat pad), which heavily refract and scatter light, and those regions where red blood cells have leaked from damaged vessels onto the tumor surface. To capture cell motility, we typically average 30 min to capture a 4D data set. However, depending on the specific cellular manipulation, animal model, or experimental aims, this time can be extended further.
3. While in multiphoton mode, we select new fields by scanning at higher speeds while adjusting the stage to find the best area to image. After selecting a new field, the scan speed is decreased to allow a longer dwell time per pixel (which results in higher

signal-to-noise images). In our experience, the areas with motile cells are found at the tumor–matrix borders and which possess blood vessels with active flow. We try to select the edge of a tumor outgrowth with well-defined cell boundaries, and which includes extracellular matrix fibers as cancer cells typically migrate along their length.

4. Optimally, the beginning of a *z*-stack should be set on the outer tumor surface, one or two slices above the tumor–matrix border, so that the local tumor microenvironment is intact. If the system used for imaging allows for varying the laser power in *z*, more power can be used to image at greater depths (though it is important to confirm that this does not affect the tissue), thus allowing for a greater number of slices recorded in a given *z*-stack.
5. Set up the time-lapse *z*-series (generally we use 10-μm steps for a 100-μm scan and 1-min intervals for 30 min of imaging time) and start collecting images.
6. At the end of the imaging session (or sessions for experiments using the mammary imaging window), the animal should be euthanized using protocols approved by the user's institute animal care committee.

3.6. Processing Intravital 4D Data

Until recently, most available image-processing tools offered were not optimized for intravital or 3D imaging. They were commonly designed for 2D images or time-lapse sequences without measurement of axial movements or ability to follow movement of the same cell throughout a stack. This task is somewhat complicated by difficulty in the display of 4D (3D information over time), as well as difficulty in analysis of 3D data due to decreased resolution in *z* as compared to *x*–*y* dimensions. Also, results of intravital imaging often include slight drift in one of the *x*, *y*, *z* dimensions as well as irregular breathing artifacts. However, as the number of groups imaging in vivo or in 3D cultures is growing, solving this issue becomes increasingly important. All techniques reported so far are based on the ability of the software to automatically segment and threshold images based on fluorescence intensity. Unfortunately, this is a task that cannot be fully automated for cells in solid tissues, such as tumors, where the cells are packed closely to each other in three dimensions. Our typical quantitation procedure involves visually counting the number of moving cancer cells in all the slices in a 4D stack. Additional parameters that can be evaluated include movement of cancer cells on visible matrix fibers, movement of host cells (nonfluorescent cells seen as shadows against the fluorescent cancer-cell background), orientation around blood vessels, and matrix density.

4. Notes and Expected Results

1. The skin flap technique is very flexible in terms of tumor size (29). Furthermore, it does not require a lengthy surgery or postoperative recovery time in preparation for imaging. However, data are acquired over an imaging session during which the animal is under anesthesia and its physiological processes may be somewhat impaired. Conclusions are made by combining data not only from several animals, but also from different tumor microenvironments from within the same tumor.
2. The flexible nature of the imaging-window design provides the researcher with several options for introducing fluorescent protein-expressing cells for imaging. One important advantage of the window system is that cancer cells (or other cell types), with stable or transient expression of fluorescent proteins, can be delivered to the imaging environment during any stage (i.e., into either a mammary fat pad or growing tumor) and imaged immediately. Furthermore, through the use of photo-convertible proteins, the imaging window allows the researcher to follow cell fate over days from within the same tumor microenvironment.
3. It is important to note that the skin flap and mammary imaging window methods do not image the same side of the tumor—the skin flap exposes the side of the tumor adjacent to the peritoneum, while the window exposes the side oriented towards the skin. Therefore, differences exist in the amount of collagen and fat in the outer layer of the tumor, hence the microenvironment will also differ.

Acknowledgments

The work described herein was supported by CA100324, CA113395, and CA126511 (B.G., D.E., J.W., and J.C.) and CA100324 and CA77522 (J.H., D.K., and J.E.S.).

References

1. Chambers, A.F., Groom, A.C., MacDonald, I.C. (2002) Dissemination and growth of cancer cells in metastatic sites. *Nat Rev Cancer.* **2**, 563–72.
2. Tuettenberg, J., Grobholz, R., Seiz, M., et al. (2009) Recurrence pattern in glioblastoma multiforme patients treated with anti-angiogenic chemotherapy. *J Cancer Res Clin Oncol.* **135**, 1239–44.
3. Arons, M.S., Smith, R.R. (1961) Distant metastases and local recurrence in head and neck cancer. *Annals of Surgery.* **154**, 235–40.
4. Condeelis, J., Pollard, J.W. (2006) Macrophages: obligate partners for tumor cell migration, invasion, and metastasis. *Cell.* **124**, 263–6.
5. Queen, M.M., Ryan, R.E., Holzer, R.G., Keller-Peck, C.R., Jorcyk, C.L. (2005) Breast

cancer cells stimulate neutrophils to produce oncostatin M: potential implications for tumor progression. *Cancer Res.* **65**, 8896–904.

6. Orimo, A., Gupta, P.B., Sgroi, D.C., et al. (2005) Stromal fibroblasts present in invasive human breast carcinomas promote tumor growth and angiogenesis through elevated SDF-1/CXCL12 secretion. *Cell.* **121**, 335–48.
7. Gaggioli, C., Hooper, S., Hidalgo-Carcedo, C., et al. (2007) Fibroblast-led collective invasion of carcinoma cells with differing roles for RhoGTPases in leading and following cells. *Nature Cell Biol.* **9**, 1392–400.
8. Karnoub, A.E., Dash, A.B., Vo, A.P., et al. (2007) Mesenchymal stem cells within tumour stroma promote breast cancer metastasis. *Nature.* **449**, 557–63.
9. Wang, W., Wyckoff, J.B., Frohlich, V.C., et al. (2002) Single cell behavior in metastatic primary mammary tumors correlated with gene expression patterns revealed by molecular profiling. *Cancer Res.* **62**, 6278–88.
10. Provenzano, P.P., Eliceiri, K.W., Campbell, J.M., Inman, D.R., White, J.G., Keely, P.J. (2006) Collagen reorganization at the tumor-stromal interface facilitates local invasion. *BMC Medicine.* **4**, 38.
11. Levental, K.R., Yu, H., Kass, L., et al. (2009) Matrix crosslinking forces tumor progression by enhancing integrin signaling. *Cell.* **139**, 891–906.
12. Pepper, M.S., Tille, J.C., Nisato, R., Skobe, M. (2003) Lymphangiogenesis and tumor metastasis. *Cell Tissue Res.* **314**, 167–77.
13. Alexander, S., Koehl, G.E., Hirschberg, M., Geissler, E.K., Friedl, P. (2008) Dynamic imaging of cancer growth and invasion: a modified skin-fold chamber model. *Histochem Cell Biol.* **130**, 1147–54.
14. Dvorak, H.F. (2003) Rous-Whipple Award Lecture. How tumors make bad blood vessels and stroma. *Am J Pathol.* **162**, 1747–57.
15. Hashizume, H., Baluk, P., Morikawa, S., et al. (2000) Openings between defective endothelial cells explain tumor vessel leakiness. *Am J Pathol.* **156**, 1363–80.
16. Hoffman, R.M. (2005) Orthotopic metastatic (MetaMouse) models for discovery and development of novel chemotherapy. *Methods in Molecular Medicine.* **111**, 297–322.
17. Lohela, M., Werb, Z. (2009) Intravital imaging of stromal cell dynamics in tumors. *Curr Opin Genet Dev.* **20**, 72–8.
18. Frese, K.K., Tuveson, D.A. (2007) Maximizing mouse cancer models. *Nat Rev Cancer* **7**, 645–58.
19. Wyckoff, J.B., Wang, Y., Lin, E.Y., et al. (2007) Direct visualization of macrophage-assisted tumor cell intravasation in mammary tumors. *Cancer Res.* **67**, 2649–56.
20. Hillen, F., Kaijzel, E.L., Castermans, K., oude Egbrink, M.G., Lowik, C.W., Griffioen, A.W. (2008) A transgenic Tie2-GFP athymic mouse model; a tool for vascular biology in xenograft tumors. *Biochem Biophys Res Commun.* **368**, 364–7
21. Yang, M., Reynoso, J., Jiang, P., Li, L., Moossa, A.R., and Hoffman, R.M. (2004) Transgenic nude mouse with ubiquitous green fluorescent protein expression as a host for human tumors. *Cancer Res.* **64**, 8651–6.
22. Yang, M., Reynoso, J., Bouvet, M., and Hoffman, R.M. (2009) A transgenic red fluorescent protein-expressing nude mouse for color-coded imaging of the tumor microenvironment. *J. Cell. Biochem.* **106**, 279–84.
23. Tran Cao, H.S., Reynoso, J., Yang M., Kimura, H., Kaushal, S., Snyder, C.S., Hoffman, R.M., and Bouvet M. (2009) Development of the transgenic cyan fluorescent protein (CFP)-expressing nude mouse for "Technicolor" cancer imaging. *J. Cell. Biochem.* **107**, 328–34.
24. Centonze, V.E., White, J.G. (1998) Multiphoton excitation provides optical sections from deeper within scattering specimens than confocal imaging. *Biophys J.* **75**, 2015–24.
25. Brown, E., McKee, T., diTomaso, E., et al. (2003) Dynamic imaging of collagen and its modulation in tumors in vivo using second-harmonic generation. *Nature Med.* **9**, 796–800.
26. Wyckoff, J., Gligorijevic, B., Entenberg, J., Segall, J.E., Condeelis, J. (2010) "High-Resolution Multiphoton Imaging of Tumors in Vivo", *Live Cell Imaging: A Laboratory Manual.* **2nd Edition ed**, CSHL Press.
27. Kedrin, D., Gligorijevic, B., Wyckoff, J., et al. (2008) Intravital imaging of metastatic behavior through a mammary imaging window. *Nature Methods.* **5**, 1019–21.
28. Gligorijevic, B., Kedrin, D., Segall, J.E., Condeelis, J., van Rheenen, J. (2009) Dendra2 photoswitching through the Mammary Imaging Window. *J Vis Exp.* (http://www.jove.com/index/Details.stp?ID=1278).
29. Yang, M., Baranov, E., Wang, J-W., et al. (2002) Direct external imaging of nascent cancer, tumor progression, angiogenesis, and metastasis on internal organs in the fluorescent orthotopic model. *Proc Natl Acad Sci USA* **99**, 3824–3829.

Chapter 3

High-Resolution In Vivo Imaging of Fluorescent Proteins Using Window Chamber Models

Gregory M. Palmer, Andrew N. Fontanella, Siqing Shan, and Mark W. Dewhirst

Abstract

Fluorescent proteins enable in vivo characterization of a wide and growing array of morphological and functional biomarkers. To fully capitalize on the spatial and temporal information afforded by these reporter proteins, a method for imaging these proteins at high resolution longitudinally is required. This chapter describes the use of window chamber models as a means of imaging fluorescent proteins and other optical parameters. Such models essentially involve surgically implanting a window through which tumor or normal tissue can be imaged using existing microscopy techniques. This enables acquisition of high-quality images down to the cellular or subcellular scale, exploiting the diverse array of optical contrast mechanisms, while also maintaining the native microenvironment of the tissue of interest. This makes these techniques applicable to a wide array of problems in the biomedical sciences.

Key words: Window chamber, Intravital imaging, Fluorescence, Fluorescent protein, GFP, Microscopy, In vivo imaging

1. Introduction

1.1. History and Background of Fluorescent Proteins and Application of the Window Chamber Model

Green fluorescent protein (GFP) was discovered by Shimomura et al. while attempting to discern the mechanism of autofluorescence in the *Aequorea* jellyfish (1). The mechanism was unlike any others known at that time in that a chemiluminescent calcium-binding protein, aequorin, emits a blue photon upon binding to calcium. This blue photon is then transferred to GFP via fluorescence resonance energy transfer (FRET) to provide the green bioluminescence characteristic of the animal. After much painstaking effort, the GFP protein was sequenced and cloned, and fortuitously found to autocatalytically form its active chromophore, making it suitable for use in virtually any species as a molecular marker.

Robert M. Hoffman (ed.), *In Vivo Cellular Imaging Using Fluorescent Proteins: Methods and Protocols*,
Methods in Molecular Biology, vol. 872, DOI 10.1007/978-1-61779-797-2_3,

Subsequently, mutants and alternative fluorescent proteins have provided a whole palette of colors spanning the visible and, recently, near infrared wavelengths (2, 3). It has by now become an indispensible tool of biology, useful for investigating protein localization, protein interactions (via FRET), cell localization, interrogation of signaling pathways, and more (4). It has also been adapted for use as a genetically-encoded biosensor for such diverse measurands as calcium concentration, pH, proximity, and protease activity (5).

Given the diverse and wide ranging tool set that fluorescent proteins can potentially provide, it is desirable to have an effective means of imaging them in vivo. A variety of techniques exist, including invasive imaging, diffuse optical (deep tissue) imaging, or simply using confocal or multiphoton imaging transcutaneously or directly. In this chapter, we focus on window chamber techniques, which have the advantages of providing direct access for high-resolution microscopy, while also enabling longitudinal monitoring of the same site.

The development of the window chamber model for the investigation of cancer in vivo has proven an invaluable resource in the elucidation of real-time tumor inception, growth, adaptation, and treatment response. A critical advancement in the optical interrogation of living tissue came in 1928 with J.C. Sandison's observations of blood vessel growth through a transparent chamber implanted in a rabbit ear (6). In 1939, this technique was later adapted to carcinoma studies by Ide and Warren (7) in one of the first direct observations of angiogenesis and its influence on tumor growth. The window chamber model was further refined in 1943 by Glenn Algire, with his publication of a number of novel window chamber designs (8)—among them, the dorsal window chamber model, which is widely used in tumor xenograft and allograft animal studies to this day (9). As an example, Fig. 1 illustrates the use of fluorescent protein reporters of hypoxia-inducible factor-1 (HIF-1) upregulation in response to radiotherapy (RT) (10). The use of the window chamber in conjunction with fluorescent protein reporters enabled a detailed understanding of the mechanism and temporal dynamics with which HIF-1 is upregulated post-therapy and has spurred several studies investigating the use of HIF-1 inhibitors as a means of improving the therapeutic efficacy of RT (11, 12).

1.1.1. Animal and Tumor Models

A variety of genetically-engineered animals and tumor models exist that incorporate fluorescent proteins for imaging. This includes transgenic animals that express GFP in their germ line, which enables in vivo imaging of host tissue. Another common approach is to use standard animal strains, with a genetically-engineered cell line introduced into the mature animal in the form of tumor-cell xenografts, or other transplants.

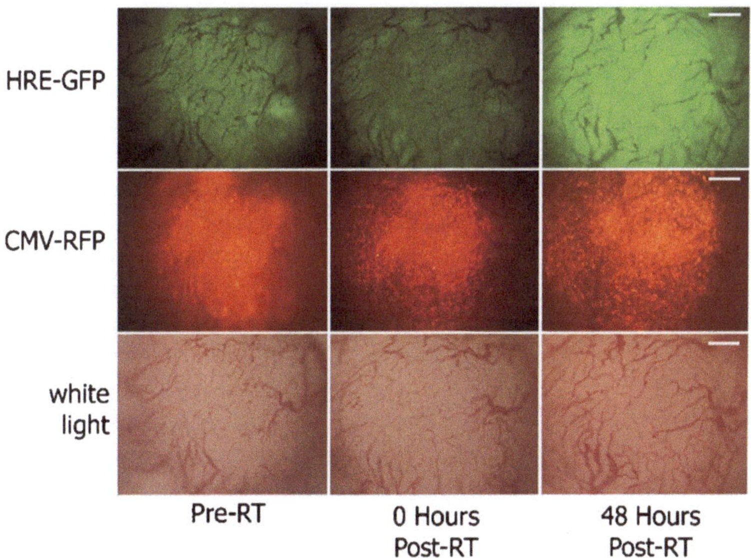

Fig. 1. Fluorescent proteins are used to study the induction of HIF-1 after radiotherapy. A GFP construct with an hypoxia responsive element is used to report HIF-1 activity, while a constitutively-active RFP construct with a cytomegalovirus (CMV) promoter reports overall tumor extent. White light imaging shows the vascular morphology and reformation over time. It can be seen that HIF-1 activity increases significantly post-RT, while the vasculature becomes dilated. Reprinted from (10) with permission from Elsevier.

1.2. Orthotopic Mammary Window Chamber Model

Orthotopic and ectopic organ environments differentially influence cancer cell gene expression, tumor growth, invasiveness, angiogenesis, metastasis, drug delivery, and sensitivity to therapeutic agents in many tumor types (13–15). Orthotopic breast cancer models with rodent syngeneic tumors or human xenografts have been widely used in the studies of hormone dependency, novel metastasis models, molecular targets, growth factors, angiogenesis, tumor growth, and gene therapy (16–21).

For intravital imaging of orthotopic breast cancer, we have established a rodent mammary window model, which combines the unique features of a tissue window with growth in an orthotopic organ environment, which allows, continuous, and noninvasive monitoring of tumor growth and angiogenesis (22). There is significant overlap with the dorsal skin-fold model, so we focus on an overview of the procedures, while highlighting the differences in technique.

1.3. Imaging

Once the window chamber has been implanted, imaging can be carried out using any of the established techniques for fluorescence microscopy, including wide field, confocal, or multiphoton fluorescence. The primary advantages of confocal and multiphoton being better resolution and depth-resolved imaging. The following protocol describes wide-field fluorescence imaging, but any of these techniques can be used with similar approaches.

2. Materials

2.1. Materials for Dorsal Flank Model and Common Materials for Both Window Types

2.1.1. Cell Culture

The cancer cells to be used for inoculation should be screened for pathogens before they are introduced into animals. If using cells from cryo-storage, the cells should be plated out and passed at least once prior to inoculation. The number of cells needed for injection may vary significantly depending upon the particular cell line and animal strain. Nude mice, for example, will readily take a wide variety of cancer-cell lines. 10,000 cells per animal often achieves a near-100% tumor rate with highly tumorigenic cell lines such as the 4T1 mammary carcinoma line. For less tumorigenic cell lines, or when using animals strains with a competent immune system, however, larger quantities of cells may be needed (up to $>10^7$). Some cell lines will not grow well in certain strains, or may require special conditions (e.g., matrigel cell suspensions, implantable estrogen pellets). It is important to research the cells and animals intended for use in order to avoid wasting time and animals on failed surgeries.

2.1.2. Anesthetics and Analgesics

The concentrations and doses listed here are for mice. These doses provide a recommendation for animal protocol development, but it is important to ensure that all drug administration adheres to your IACUC guidelines (see Note 1).

1. Ketamine/xylazine is recommended for window chamber surgeries. Pentobarbital can also be used for surgeries (for hairy mice that need shaving and depilatory creams), although it is prone to a higher incidence of accidental overdose. Ketamine/xylazine should be diluted in sterile saline solution to a concentration of 10 mg/mL ketamine and 1 mg/mL xylazine. The administered dose should be 80–120 mg/kg ketamine, and 8–12 mg/kg xylazine i.p. for mice. At this concentration and dose, this results in approximately 0.01 mL/kg body weight, or 0.2 mL for a 20-g mouse. If using pentobarbital, it should be diluted in sterile saline to a concentration of 10 mg/mL, and the administered dose should be 75–90 mg/kg i.p. for mice.
2. Buprenorphine is used for postsurgical pain management. It is diluted to a concentration of 15 µg/mL and administered at a dose of 100 µg/kg subcutaneously (s.c.) for mice, or 50 µg/kg for rats. A second dose can be given after 8–12 h, if necessary.

2.1.3. Surgical Equipment and Accessories

Figure 2 shows the surgical equipment, numbered as described below. These items should be placed in a surgical tray, wrapped in

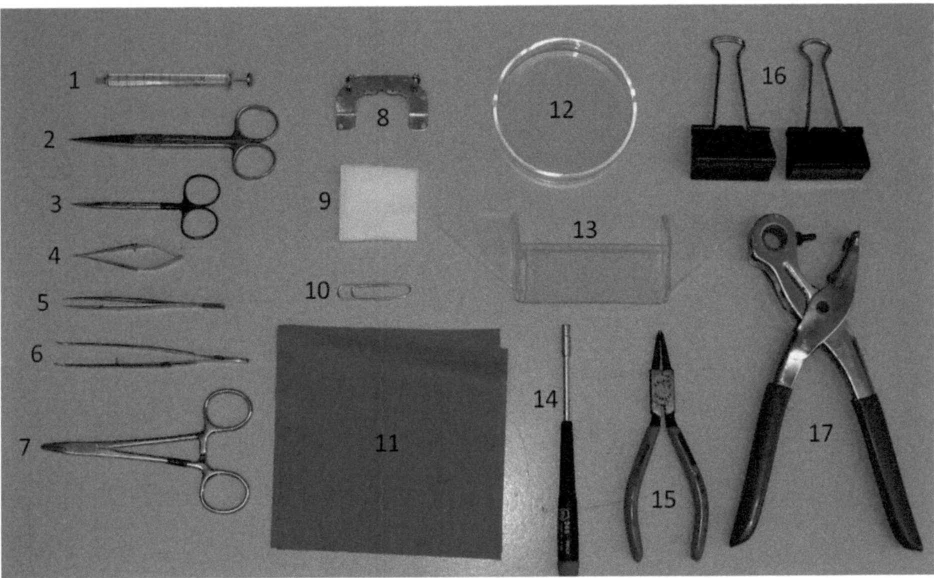

Fig. 2. Surgical equipment used in implanting dorsal skin-fold window.

40 in. × 40 in. sterilization wrap, and steam or dry-heat sterilized prior to surgery:

1. Glass microsyringe (25–100 μL volume).
2. Mayo scissors.
3. Iris scissors.
4. Conjunctival scissors.
5. Jeweler type forceps.
6. Mosquito forceps.
7. Needle holder.
8. C-clamp.
9. Cotton gauze.
10. Metal paper clip.
11. 6 in. × 6 in. squares cut from a piece of spare sterilization wrap (one square per animal).

These items should also be packaged and sterilized prior to surgery. Items with plastic and nonstainless components should be gas sterilized separately:

12. Petri dish.
13. Plexiglass viewing stage.
14. Nut driver.
15. Retaining-ring pliers.
16. Two large metal binder clips.
17. Leather hole puncher with 1/8 in. hollow punch (or a 16-G needle can also be used).

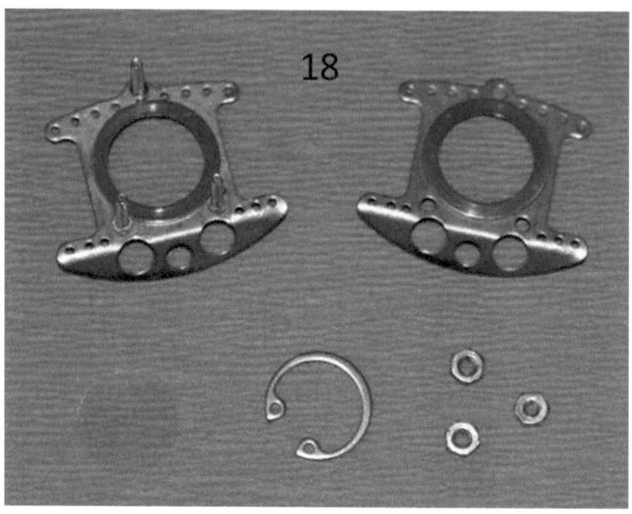

Fig. 3. Dorsal window chamber frames.

18. Window chambers with nuts, glass cover slips, and retaining rings, seen in Fig. 3.

The following items can be purchased as prepackaged sterile units:

19. 18 in. × 26 in. sterile field.
20. Skin marker.
21. 4-0 Monosof or silk suture.
22. Disposable surgical blade.
23. 1-mL syringe.
24. Two 30-G needles.
25. Surgical gloves.

The following items do not need to be presterilized:

26. Water-circulating heating blanket.
27. Fiber-optic lamp.
28. Hot plate.
29. Glass-bead sterilizer.
30. Plexiglass surgical platform.
31. Paraffin heating pad.
32. Absorbent paper.
33. Exidine solution.
34. 70% Ethanol solution.
35. Vortexer.
36. Antibiotic ointment.
37. Ophthalmic ointment.

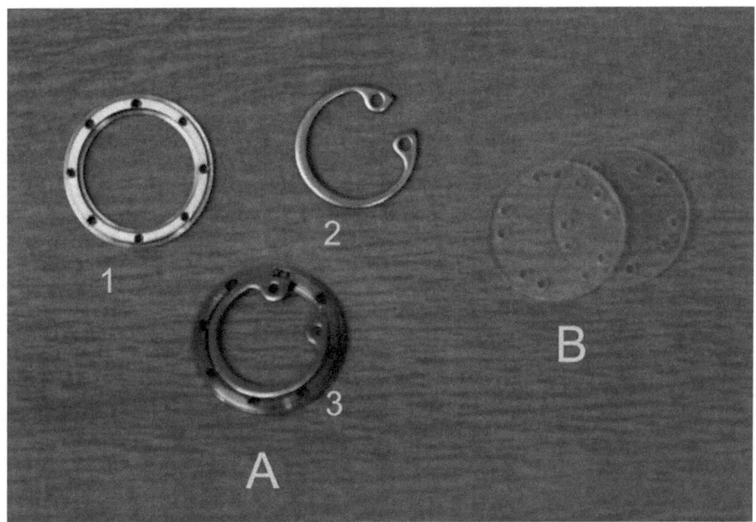

Fig. 4. Window choices: Metal/glass or polycarbonate disk can be used. (**a**) Metal ring with glass cover slide (*1*) stainless steel ring with holes, (*2*) internal retaining ring to fix cover glass into the ring, and (*3*) assembled window. (**b**) Acrylic or polycarbonate disk with holes for suturing.

38. Surgical mask.
39. Hair removal lotion (Nair) and electric shaver (for non-nude mice only).

2.2. Mammary Window Chamber Materials

2.2.1. Mammary Window Materials

Most surgical equipment and supplies are shared in common with the dorsal skin-fold window chamber model, with the notable exception of the window itself. This is shown in Fig. 4.

Glass windows have better optic quality and scratch resistance but are a bit thicker and heavier. Acrylic or polycarbonate windows are lighter, but optical quality and scratch resistance are not as good as glass. The window sizes for rats and mice are 12 and 8 mm in diameter, respectively.

2.2.2. Animals and Tumors

Depending on tumor types and study designs, different female rats or mouse strains can be used. For the most commonly used syngenic rat mammary-adenocarcinoma line, R3230Ac, the Fischer 344 rat is desired. Since this tumor grows best with tumor-fragment implants from donor rats rather than cancer-cell inoculation, retired female breeders or lactation-weaned rats are preferable because of a well-developed nipple sinus where tumor fragments are implanted. However, any female rat or mouse can be used depending on tumor types.

2.2.3. Anatomical Consideration

Rats and mice have six pairs of mammary glands, and they are referred to by location as cervical, cranial-thoracic, caudal-thoracic, abdominal, cranial-inguinal, and caudal-inguinal, or by numbers, anterior to posterior as L1, R1, L2, R2, etc. (23). The mammary glands are compound tubuloalveolar glands comprised of a highly branched system of ducts and terminal secretary alveoli arranged in lobules. Each gland in the rat and mouse has a single lactiferous duct entering the nipple. The duct widens to form the nipple sinus, which then opens onto the surface by way of the nipple canal. The nipple, nipple canal, and nipple sinus are lined by squamous epithelium continuous with the epidermis. The second to fifth pairs of mamma on either side can be used for window surgery, but for convenience of intravital microscopy, mammary glands R4 or R5 are most commonly used.

2.3. Imaging Materials

2.3.1. Equipment

- *Fluorescence microscope and camera.* This can be any type of fluorescence microscope, inverted or upright. For the mammary window, breathing artifacts are more of a concern. A stage with an adjustable restraining device is preferable (22), although an inverted microscope is better for the mammary window to minimize breathing artifacts and enable the mouse to remain upright. Ideal objectives depend on the application, but can range from 2.5× for a broad field of view suitable for imaging large segments of the window, up to 60× or higher, suitable for imaging at a subcellular resolution. Filter cubes with the appropriate filter sets for each fluorophore of interest should be in place, as well as a sensitive CCD camera. Standard filter sets are available from several manufacturers that are capable of imaging overall intensity spanning the typical excitation/emission wavelength range of common fluorescent protein variants. A more ideal solution is to acquire hyperspectral data sets using a liquid-crystal tunable filter (LCTF), acousto-optic tunable filter (AOTF), or similar system. This enables pixel-by-pixel measurement of the emission spectrum as a function of wavelength, which facilitates quantitative discrimination of multiple fluorophores using spectral discrimination techniques. For bright light epi-illumination, a cold flexible fiber optic lamp can be used. A ring lamp attached to the lens will provide more even illumination (see Note 7).
- *Wavelengths.* Filters used for wavelength selection will depend on the source of contrast used. GFP is compatible with fluorescein or FITC filters, which are common on fluorescence microscopes. RFP is compatible with rhodamine or TRITC filter sets. Most manufacturers sell filter sets optimized for a variety of fluorescent protein variants.
- *Imaging mount.* This is used to secure the window chamber in place during imaging. For the dorsal window, this can be as simple as a metal sheet with three holes drilled in to allow placement of the three window frame bolts to be fed through and secured. A mammary window requires use of a strap to secure the animal in place over the objective.

- *Heating pad.* A thermally-regulated heating pad designed for veterinary use is ideal. Alternatives include heated paraffin pads designed to maintain a constant suitable temperature as they undergo a phase transition from liquid to solid.
- *Anesthesia equipment.* This depends on the method of anesthesia, but for isoflurane anesthesia a vaporizer, scavenger, gas tank, and regulator are needed, as well as tubing and a nose cone to deliver the gas to the animal.

3. Methods

3.1. Dorsal Skin-Fold Window Chamber and Common Procedures

Note. Surgeries are most efficiently performed as a two-person team. While one person prepares the cells, the other can prepare the surgical area. During surgeries, one person can anesthetize and prepare the animal and handle the cells while the other person remains sterile and performs the surgeries.

3.1.1. Preparation

Cells

1. Immediately prior to surgery, remove the media from plates containing cancer cells and incubate them in trypsin until the majority of cells detach.
2. Dilute the trypsinized cells in standard media, pipetting repeatedly to break apart large clumps.
3. Pipet the cell suspension into a sterile conical tube and centrifuge at $680 \times g$ for 5 min.
4. Aspirate the supernatant and resuspend the cells in a minimal quantity of phenol red-free media.
5. Count the cells and dilute the cell suspension to a concentration such that a 10–20-μL volume contains the desired number of cells for inoculation.
6. Pipet approximately 500 μL of the cell suspension into a sterile microcentrifuge tube, and place the tube on ice.

Surgical Area

Surgery should be performed in a laminar-flow HEPA-filtered hood or other isolated environment with an accessible electrical outlet. Once a sterile environment has been set up, care should be taken not to unintentionally introduce any nonsterile objects into the area. Any sterile item or surface that accidentally comes into contact with a nonsterile object should be considered contaminated and must be resterilized or replaced. A complete setup is shown in Fig. 5.

1. Place the fiber-optic lamp to the far end of the surgical area and turn the lamp on to its highest setting.
2. Place the hot plate in an accessible corner of the surgical area.

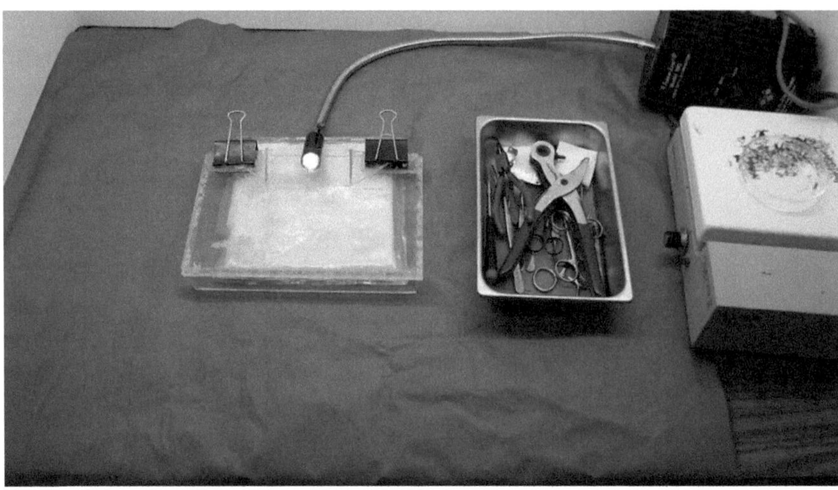

Fig. 5. The prepared surgical workstation.

3. Place the glass-bead sterilizer in an accessible location immediately adjacent to the surgical area and turn it on.
4. Thoroughly clean and disinfect all exposed surfaces within the surgical area using Sporicidin or similar hospital grade antiseptic. Clean and disinfect the Plexiglass surgical platform and place it to the side.
5. Place the wrapped and sterilized surgical tray in the middle of the surgical area. Fully unwrap the tray and use the inner surface of the sterilization wrap to set up a broad, sterile working surface. Do not touch the inner surface of the wrap or the tray as you are preparing your field. The wrap should not cover the hot plate or obstruct the fiber lamp.
6. Placing your hands underneath the wrap (in contact with only the nonsterile outer surface), position the surgical tray to the side of the surgical area corresponding to the surgeon's dominant hand.
7. While handling only the outer packaging, hold items 13–24 over the surgical tray and open them (without touching the sterile inner contents), allowing the items to fall into the tray.
8. Open the Petri dish packaging, and without touching the inner portion of the dish, place it on the hot plate. Using a syringe, fill the Petri dish with a few milliliters of sterile saline, and set the temperature of the hot plate to heat the saline to approximately 37°C.
9. Microwave the paraffin pad for a few minutes until the wax is partially liquefied. Knead the pad to distribute the heat evenly across the surface.
10. Place the heated pad directly on the sterilization wrap within your surgical area. It should be placed towards the center of

the working surface and about half a forearm's length from the front edge.

11. Place the Plexiglass surgical platform on top of the paraffin wax pad. You may need to place an item beneath the wax pad to elevate it to a level such that it is in contact with the inner surface of the platform. This will ensure that the platform remains heated during surgery. Remember that although the surgical platform is disinfected, it is not aseptic. Do not allow any sterile items to come into contact with it.
12. Using binder clips, attach the Plexiglass viewing stage to the far end of the platform such that it slopes downward towards the user. The clips and the outer portion of the viewing stage will become nonsterile as you set this up. Ensure that the forward-facing, sloped portion of the viewing stage remains sterile.
13. Position the fiber lamp's illumination source directly behind the Plexiglass surgical platform, facing forward and slightly upwards.

Animals

It is important to remember that the animal's core body temperature will drop while it is under anesthesia. Steps must be taken to ensure that the animal does not become dangerously hypothermic. Throughout the entire course of anesthesia, the animal should remain in contact with a heated surface, and body temperature should be regularly monitored (see Note 2).

1. Power-on the water-circulating heating blanket and cover it with the absorbent paper. Allow it to warm to approximately 37°C, which should be calibrated by measuring the actual temperature on the metal surface. Placing a thin metal surface on the heating blanket will improve thermal conductance and help maintain the anesthetized animal's body temperature more efficiently (see Note 5a).
2. Prepare a working solution of either ketamine/xylazine or pentobarbital anesthesia at the concentrations listed in Subheading 2. You will need approximately 1 mL of either anesthetic for every three animals (allowing for redosing).
3. Prepare a working solution of buprenorphine for postsurgical pain management. At the concentration listed in Subheading 2, you will need approximately 0.15 mL per animal.
4. Weigh the animal and administer the proper dose of anesthesia with an i.p. injection (see Note 1).
5. Isolate the animal until the anesthesia takes effect. Once the animal is down, place it on the heating pad in a sternally recumbent position with limbs spread.
6. Apply ophthalmic ointment to the eyes to prevent corneal desiccation during surgery.

7. If using a non-nude mouse, use the electric shaver to shave the hair from the base of the tail to the nape and from limb to limb on both sides. Apply hair-removal lotion and follow the product instructions to remove remnant hair. Skip this step if using nude mice.
8. Using cotton gauze, apply 2% chlorhexidine disinfectant solution to the animal's visible torso and tail. Wipe away with cotton gauze soaked with 70% ethanol. Repeat this process a total of three times. Avoid recontaminating the skin afterwards.

3.1.2. Surgery

1. Don sterile surgical gloves and surgical mask. Throughout the surgery, your gloved hands should not come into contact with any nonsterile object. In the case of contamination, replace the gloves.
2. Using the Mayo scissors, cut out a rectangular portion in one of the folds of the sterile field, approximately the size of the forward-sloping portion of the Plexiglass viewing stage. Drape the sterile field over the surgical platform so that only the sterile portion of the viewing stage is visible and protruding through the hole in the sterile field.
3. Take one of the 6 in. × 6 in. squares cut from the sterilization wrap and fold it diagonally. Cut a slit, approximately 1.5 in. long, along the diagonal axis.
4. With the skin marker, lightly trace a line along the length of the animal's spine. On either side of this line, make a small mark to note the highest-rising point on the animal's back. This will serve as a guide for the correct placement of the window chamber.
5. Take the 6 in. × 6 in. square and drape it over the animal such that the slit you cut runs parallel to the spine and the highest point on the back is at the center of the slit. Touching only the wrapping, maintain the animal in this position while rolling up the loose corners of the square until the animal is held snugly in place. Secure the rolled up corners with a sterilized paper clip.
6. Place the animal on the sterile field within the surgical area, and gently pull the lose skin on the back through the slit in the wrapping, producing a skin fold. Ensure the skin is evenly stretched up from either side and folded along the marked line running parallel to the spine. The marked area corresponding to the peak of the back should be centered.
7. Using the needle driver, suture the skin fold to the c-clamp in four places, stretching the skin evenly and creating a taut area for window placement. Hang the c-clamp from the top of the viewing stage so that the surgical area is well illuminated.
8. Hold the screw-less half of a window chamber flush against the skin fold in a position such that the window fully fits within

the area of the skin fold. Use the marker to mark the positions of the screw holes on the skin flap, and mark the area for the window by tracing a circle along the circumference of the window hole.

9. Remove the c-clamp from the viewing stage. Using the marks for the screw holes as a guide, punch a hole through both sides of the skin flap at each one of the marked locations using a leather hole puncher or a 16-G needle.
10. Return the c-clamp to the viewing stage. With the iris scissors in your dominant hand, cut away the forward-facing half of the skin flap along the circumference of the circle marked on the skin. Use a mosquito forceps in your other hand to hold the skin in place as you cut. You will need to cut away the connective tissue beneath the skin to remove the skin disk.
11. Holding a conjunctival scissors in your dominant hand and a jeweler forceps in the other, pull up and cut away any residual connective tissue in the area, leaving a thin layer of translucent fascia covering the visible dermis of the back-skin fold. Take care not to puncture this dermal layer, as this will lead to visible lesions within the window (see Notes 3 and 4).
12. Throughout the rest of the surgery until the cover slip is secured, control bleeding by flushing the exposed wound with saline (using the warmed saline on the hot plate and the 1-mL syringe) and absorbing fluids with a sterile cotton gauze.
13. Remove the c-clamp from the viewing stage. Take the half of the window chamber with the attached screws and insert the screws through the holes you punched in the skin fold. You may need to clear away connective tissue with the iris scissors to fully insert the screws through the skin fold.
14. Manipulate the forward facing skin such that the dermal layer of the far side of the skin fold is fully visible within the window area.
15. Insert the other half of the window chamber through the protruding screws to clamp the skin fold in place. Securely fasten the two halves together with the nuts and nut driver, but ensure that the window is not fastened so tightly as to cut off circulation to any region of tissue.
16. Holding the skin taut, suture the window chamber in place using the suture holes in the window chamber frame. Use multiple knots to ensure that the suture remains in place for the duration of the window chamber study. (If the suture comes off, it can be replaced at a later point in the study using aseptic technique).
17. Once the window is secured, use the disposable surgical blade to cut away the sutures holding the mouse to the c-clamp.

18. Position the window chamber such that the exposed surface is level. Use the syringe and gauze to flush out and remove any remaining fluid within the window.
19. Attach one of the 30-G needles to the glass microsyringe.
20. Retrieve the microcentrifuge vial containing the cancer cells, and use the vortexer to break apart any large cell aggregates (in order to maintain sterility, this step is best performed by an assistant).
21. Draw the cell suspension into the glass microsyringe, noting the volume (ideally 10–20 μL) needed for your calculated cell-inoculation number.
22. Inserting the needle at the shallowest possible angle with the bevel up, inject the appropriate volume of the cell suspension between the dermal layer and the overlying fascia while avoiding major blood vessels. Bending the needle tip at a 45° angle with the needle driver helps achieve a shallower angle of entry. With a successful injection, a small bubble should appear. If no bubble is visible, attempt the injection again.
23. Attach the other 30-G needle to a 1-mL syringe, and draw it full with warm saline.
24. With the very end of the needle placed level with the window surface and slightly within the window area, place one of the glass cover slips on top such that it is positioned directly over the window, but with one edge supported by the needle tip.
25. Inject saline into the window until it begins to spill over. Remove the needle tip, allowing the cover slip to fall in place over the window. The cover slip should now seal off the saline-covered dermis, and no air bubble should be visible within the window. If an air bubble remains, push up on the window opening on the far skin fold to remove the cover slip, and repeat the process.
26. Once the window is in place with no air bubbles, use the retaining-ring pliers to flex one of the retaining rings, place it flush above the cover slip, and release the pliers leaving the retaining-ring in place to secure the cover slip.
27. Remove the animal from its wrapping and allow it to recover on the heating blanket.
28. As the anesthesia wears off, inject the appropriate dose of buprenorphine subcutaneously for post-surgical pain management. The mouse can be returned to its cage once it regains mobility.
29. If performing additional surgeries, replace the sterile field. Metal surgical tools can be sterilized with the glass-bead sterilizer, although it is important to remember to allow the tools time to cool in order to avoid burning the animal.

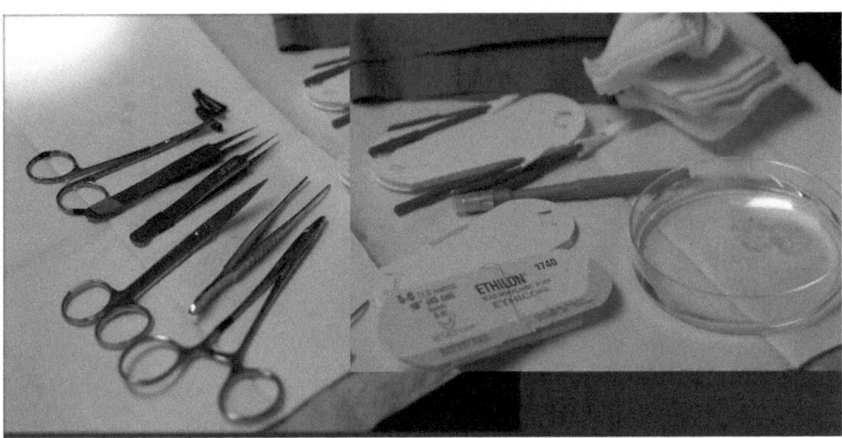

Fig. 6. Sterile surgical instrument, suture, skin puncher, and acrylic disks in sterile saline.

30. After all surgeries have been performed return the animals to the appropriate animal facility.
31. Observe the animals daily for the next few days for signs of distress.

3.2. Mammary Window Procedures

3.2.1. Surgical Procedure

The basic procedures in rats and mice are similar, except for the window size. The surgical tools specific to this protocol are shown in Fig. 6.

1. Animals are anesthetized with sodium pentobarbital given i.p. at 45 mg/kg body weight for rats and 80 mg/kg for mice. Animals are kept warm using a circulating water blanket. Set temperature at 37°C.
2. For hairy rodents, the anterior aspect of lower thorax and abdomen is shaved and depilated with Nair (Carter-Wallace, Inc., New York, NY). Skin is wiped with Chlorhexidine (Baxter-Healthcare, Co., Deerfield, IL) followed by alcohol, alternately for three times.
3. Surgery is performed with aseptic technique with the aid of a dissecting microscope. A circular incision with a diameter of 8 mm for rats or 5 mm for mice, is made on the skin around the nipple. For better marking, a clinically-used skin-biopsy puncher, with an 8 or 5 mm diameter, can be used for rats or mice, respectively.
4. The thin layer of skin around the base of the nipple is removed within the circular incision.
5. The nipple is cut at its base and the nipple sinus is exposed. The lining epithelium is carefully removed.
6. Cancer cells or tumor fragments are implanted into the nipple sinus (see below). Cancer-cell suspensions from tissue culture

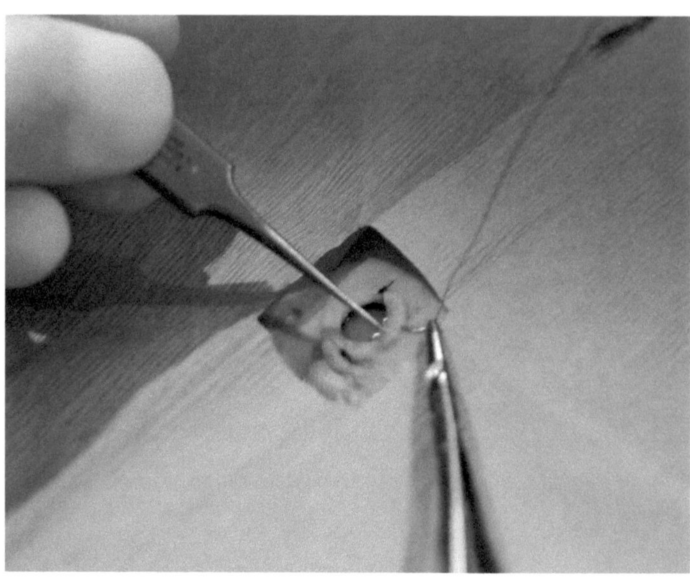

Fig. 7. The window disk has been inserted into the circular wound and is being sutured with the skin edge.

or tumor fragments from a donor rat should be prepared before the surgery starts.

7. Pre-gas sterilized window disk (see Subheading 2) is placed into the wound, with the tumor implant located at the center, as shown in Fig. 7.
8. The skin edge should cover the periphery of the disk. The disk is sutured to the skin edge using modified subepidermal sutures (5-0 Monosoft® nylon suture). Neosporin ointment is applied around the wound.
9. The analgesic buprenorphine HCl can be injected after surgery, at 0.05 or 0.01 mg/kg s.c., for mice and rats, respectively.
10. Animals should be monitored and kept on the warm blanket until fully recovered from anesthesia.

3.2.2. Tumor Transplantation into the Window

Different techniques can be used for tumor implantation. For better monitoring of cancer-cell morphological change and tumor–host interactions, single-cell suspensions of fluorescent protein-expressing cancer cells is preferable. The cancer cells can be injected into the mammary tissues before the window disk is mounted. Tumor fragments of R3230 Ac, derived from subcutaneously-implanted tumors in a donor animal can be transplanted into the nipple sinus of Fischer 344 rats, which yields faster tumor growth. For cancer-cell inoculation, half-confluent cells are trypsonized and washed with PBS twice, immediately prior to surgery. Viable cell numbers are counted, using trypan blue exclusion, with a hemocytometer. Defined concentrations of viable cell suspensions are made by resuspending cell pellets with PBS. A cell-suspension tube can be stored on ice prior to transplantation (see Note 5b).

11. Cancer-cell injection: 10–20 μL of cancer cells are injected into the mouse mammary gland at an appropriate cell density depending on cancer-cell type and mouse strain. For the murine mammary carcinoma cell line 4T1 used in Balb/C mice, the usual cell density is 1×10^6/mL.
12. Cancer cells in Gelfoam transplants: GFP-R32330Ac cells can be used for cancer-cell transplantation. Three microliters of GFP-R32330Ac cancer cells, at 1×10^7/mL, are soaked into a 1-mm^3 piece of absorbable gelatin sponge (Gelfoam, Pharmacia & Upjohn, Kalamazoo, MI), which is placed into the nipple sinus.
13. Tumor fragments of R3230Ac: A 0.5-mm^3 piece of tumor tissue from a donor animal, with tumor growing subcutaneously on the thigh, is placed into the nipple sinus for the first generation of the mammary window. Fragments from orthotopically-transplanted R3230Ac tumors in donor rats are used for later surgeries. The methods for tumors fragment implantation, from donor rats bearing subcutaneous tumors, have been reported previously (24). Briefly, donor animals are anesthetized with Nembutal (Abbott Laboratories, North Chicago, IL) at 50 mg/kg i.p. and tumor tissue is removed aseptically. After removal, tumor tissue is rinsed with sterile saline, cut to 0.5 mm^3 fragments in filtered DMEM, and maintained in DMEM on ice, not longer than 2 h before being implanted into the nipple sinus.

3.2.3. Postoperative Care

To protect the window disk from being damaged by animals after surgery, a lightweight rat Elizabethan collar (Harvard Apparatus, Holliston, MA) is applied before animals emerged from anesthesia. For mice, an 18-mm wide adhesive bandage (Band-Aid by Johnson & Johnson Medical Inc., Arlington, TX) can be applied for the first 2 days. Animals should be kept in a cage with a wire floor to minimize wound contamination for the first 5 days. The collars and bandages can be removed 2 days after surgery (see Note 6).

3.3. Imaging Methods

1. *Prepare the imaging equipment.* A fluorescent lamp should be turned on and warmed up for the recommended time, typically 5–20 min. A warming pad is required to maintain the animal's body temperature during imaging under anesthesia. The camera should also be turned on and allowed to stabilize at its operating temperature over a similar time frame. A thermally-regulated heating pad, designed for veterinary use, should be allowed to warm up prior to anesthetizing the animal.

2. *Anesthetize the animal.* Perhaps the best option is to use 1–2.5% isoflurane + oxygen, which enables a safe and easily controllable depth of anesthesia. The isoflurane dose can be adjusted based on the animals breathing rate and motion. Alternatively, medical air can be used, if tissue oxygenation/hypoxia is being studied, to avoid perturbing the oxygen tension of the tissue. A well-ventilated room and a scavenging system are required to prevent gas build up. Alternatively, injectable anesthesia may be used following the anesthesia protocols described above (see Subheading 2.1.2) (see Note 1).
3. *Prepare the animal for imaging.* The animal should be placed on an imaging mount designed to hold the window securely in place. The window cover glass should be placed directly facing the objective. The warming pad should be placed such that the animal is in direct contact during imaging. Rectal temperature should be monitored to ensure the animal is not hypothermic. Eye drops should be placed on the animal's eyes to prevent drying. A cotton swab wetted with ethanol should be used to clean the window prior to imaging.
4. *Acquire images.* Images are acquired, as with any fluorescence sample, adjusting exposure time, light intensity, gain, aperture size, and other settings specific to each microscope, to ensure adequate signal levels. Important points to observe include: (1) quantitative analysis requires that the signal is not saturated, (2) reducing the illumination power levels minimizes photobleaching, (3) appropriate controls are needed to distinguish autofluorescence or signal bleed-through from true signal, and (4) the animal's temperature and breathing rate should be carefully monitored during imaging.
5. *Post-imaging.* The animal is removed from anesthesia and monitored while body temperature is maintained until it is alert and able to stand.

4. Notes

1. Inadequate depth of anesthesia is one of the most common problems associated with this procedure. It is important to remember that sensitivity to anesthesia can vary across mouse strains, and even individual mice of the same strain can have significantly different sensitivities. The doses listed here are adequate in most cases, but they may require adjustment based on experience or prior knowledge of sensitivities. Be sure that your doses are in compliance with all approved animal protocols.

2. It is important to ensure that all animals are large enough to accommodate the window chamber prior to beginning surgeries. Mice generally need to be greater than 20 g.
3. When performing incisions, avoid severing major vessels. The placement of the window can be altered somewhat to avoid large vessels. If severing a major vessel is unavoidable, allow a few minutes for clotting to occur before continuing with the surgery. Flush the window area with sterile saline to remove any blood which may inhibit visibility through the window.
4. When cutting away the excess fascia, it is important to leave a layer in place so that the cell solution can be injected between the fascia and the dermis. If too much fascia has been removed, no bubble will form, and the cell solution will not be held in place. If this is the case, try injecting in an area where the fascia remains intact.

5a. Mice that are inadequately heated throughout the surgery may become dangerously hypothermic. Make sure to periodically check the animal's body temperature.

5b. Different cell lines have different capacities for tumor formation, and this also varies greatly among mouse strains. Failure of a tumor to grow may be due to an inherent incompatibility between a cell line and mouse strain, although other cell lines may simply require higher cell concentrations. It is important to research the cell line of interest to ensure it is compatible with the proposed strain. If a tumor still fails to grow despite known compatibility, try increasing the number of injected cells.

6. Continue to monitor the animals at frequent intervals post-surgery. If the window chamber becomes grayish or translucent, it is likely that the tissue being held by the frame is suffering from lack of circulation. If it becomes cloudy or whitish, it may be infected. During future surgeries, ensure that the window chamber is not being tightened to the point of inhibiting circulation. Maintain proper aseptic technique.
7. Inverted or upright microscopes are suitable for imaging the dorsal skin-fold model, but an inverted microscope is preferred for the mammary fat pad to minimize motion artifact.

Acknowledgments

We would like to acknowledge Katherine Hansen who assisted with the technical details of the surgical procedures. We would also like to acknowledge funding from the Department of Defense Breast Cancer Research Program (grant number W81XWH-07-1-0355) and the National Institutes of Health (grant number R01 - CA40355-26).

References

1. Shimomura, O., Johnson, F., and Saiga, Y. (1962) Extraction, purification and properties of aequorin, a bioluminescent protein from the luminous hydromedusan, Aequorea., *J Cell Comp Physiol* **59**, 223–239.
2. Zimmer, M. (2009) GFP: from jellyfish to the Nobel prize and beyond., *Chem Soc Rev* **38**, 2823–2832.
3. Shcherbo, D., Merzlyak, E., Chepurnykh, T., Fradkov, A., Ermakova, G., Solovieva, E., Lukyanov, K., Bogdanova, E., Zaraisky, A., Lukyanov, S., and Chudakov, D. (2007) Bright far-red fluorescent protein for whole-body imaging., *Nat Methods* **4**, 741–746.
4. Sullivan, K. F. (2008) *Fluorescent proteins*, 1st ed., Academic Press, London; San Diego, CA.
5. Frommer, W., Davidson, M., and Campbell, R. (2009) Genetically encoded biosensors based on engineered fluorescent proteins., *Chem Soc Rev* **38**, 2833–2841.
6. Sandison, J. (1928) Observations on growth of blood vessels as seen in transparent chamber introduced into rabbit's ear, *Am J Anat* **41**, 475–496.
7. Ide, A., and Warren, S. (1939) Vascularization of the Brown Pearce rabbit epithelioma transplant as seen in the transparent ear chamber, *Am J Roentgenol* **42**, 891–889.
8. Algire, G. (1939) An adaptation of the transparent chamber technique to the mouse, *J Natl Cancer Inst* 4.
9. Huang, Q., Shan, S., Braun, R. D., Lanzen, J., Anyrhambatla, G., Kong, G., Borelli, M., Corry, P., Dewhirst, M. W., and Li, C. Y. (1999) Noninvasive visualization of tumors in rodent dorsal skin window chambers, *Nat Biotechnol* **17**, 1033–1035.
10. Moeller, B. J., Cao, Y., Li, C. Y., and Dewhirst, M. W. (2004) Radiation activates HIF-1 to regulate vascular radiosensitivity in tumors: role of reoxygenation, free radicals, and stress granules, *Cancer Cell* **5**, 429–441.
11. Moeller, B. J., Dreher, M. R., Rabbani, Z. N., Schroeder, T., Cao, Y., Li, C. Y., and Dewhirst, M. W. (2005) Pleiotropic effects of HIF-1 blockade on tumor radiosensitivity, *Cancer Cell* **8**, 99–110.
12. Dewhirst, M. W., Cao, Y., Li, C. Y., and Moeller, B. (2007) Exploring the role of HIF-1 in early angiogenesis and response to radiotherapy, *Radiother Oncol* **83**, 249–255.
13. Fidler, I. J., Yano, S., Zhang, R. D., Fujimaki, T., and Bucana, C. D. (2002) The seed and soil hypothesis: vascularisation and brain metastases, *Lancet Oncology.* **3**, 53–57.
14. Tsuzuki, Y., Carreira, C. M., Bockhorn, M., Xu, L., Jain, R. K., and Fukumura, D. (2001) Pancreas microenvironment promotes VEGF expression and tumor growth: novel window models for pancreatic tumor angiogenesis and microcirculation., *Lab Invest* **81**, 1439–1451.
15. Hoffman, R. M. (1998–1999) Orthotopic transplant mouse models with green fluorescent protein-expressing cancer cells to visualize metastasis and angiogenesis, *Cancer Metastasis Rev* **17**, 271–277.
16. Yoneda, T., Michigami, T., Yi, B., Williams, P. J., Niewolna, M., and Hiraga, T. (2000) Actions of bisphosphonate on bone metastasis in animal models of breast carcinoma (Review), *Cancer* **88** (12 Suppl.), 2979–2988.
17. Skobe, M., Hawighorst, T., Jackson, D. G., Prevo, R., Janes, L., Velasco, P., Riccardi, L., Alitalo, K., Claffey, K., and M., D. (2001) Induction of tumor lymphangiogenesis by VEGF-C promotes breast cancer metastasis., *Nature Med* **7**, 192–198.
18. Brandt, R., Wong, A. M., and Hynes, N. E. (2001) Mammary glands reconstituted with Neu/ErbB2 transformed HC11 cells provide a novel orthotopic tumor model for testing anti-cancer agents, *Oncogene* **20**, 5459–5465.
19. Chatzistamou, L., Schally, A. V., Nagy, A., Armatis, P., Szepeshazi, K., and Halmos, G. (2000) Effective treatment of metastatic MDA-MB-435 human estrogen-independent breast carcinomas with a targeted cytotoxic analogue of luteinizing hormone-releasing hormone AN-207, *Clinical Cancer Research.* **6**, 4158–4165.
20. Nakagawa, H., Tsuta, K., Kiuchi, K., Senzaki, H., Tanaka, K., Hioki, K., and Tsubura, A. (2001) Growth inhibitory effects of diallyl disulfide on human breast cancer cell lines., *Carcinogenesis* **22**, 891–897.
21. Lebedeva, S., Bagdasarova, S., Tyler, T., Mu, X., Wilson, D. R., and Gjerset, R. A. (2001) Tumor suppression and therapy sensitization of localized and metastatic breast cancer by adenovirus p53, *Human Gene Ther* **12**, 763–772.
22. Shan, S., Sorg, B., and Dewhirst, M. W. (2003) A novel rodent mammary window of orthotopic breast cancer for intravital microscopy, *Microvascular Research* **65**, 109–117.
23. Boorman, G. A. (1990) *Pathology of the Fischer rat: reference and atlas*, Academic Press, San Diego.
24. Shan, S., Lockhart, A., Saito, W., Knapp, A., Laderoute, K., and Dewhirst, M. (2001) The novel tubulin-binding drug BTO-956 inhibits R3230AC mammary carcinoma growth and angiogenesis in Fischer 344 rats., *Clin Cancer Res* **7**, 2590–2596.

Chapter 4

In Vivo Imaging of Pancreatic Cancer with Fluorescent Proteins in Mouse Models

Michael Bouvet and Robert M. Hoffman

Abstract

In this chapter, we describe protocols for clinically-relevant, metastatic orthotopic mouse models of pancreatic cancer, made imageable with genetic reporters. These models utilize human pancreatic-cancer cell lines which have been genetically engineered to selectively express high levels of green fluorescent protein (GFP) or red fluorescent protein (RFP). Tumors with fluorescent genetic reporters are established subcutaneously in nude mice by injection of the GFP- or RFP-expressing pancreatic cancer cell lines, and fragments of the subcutaneous tumors are then surgically transplanted onto the pancreas of additional nude mice. Loco-regional tumor growth and distant metastasis of these orthotopic tumors occurs spontaneously and rapidly throughout the abdomen in a manner consistent with clinical human disease. Highly-specific, high-resolution, real-time quantitative fluorescence imaging of tumor growth, and metastasis is achieved in vivo without the need for contrast agents, invasive techniques, or expensive imaging equipment. Transplantation of RFP-expressing tumor fragments onto the pancreas of GFP- or cyan fluorescent protein (CFP)-expressing transgenic nude mice was used to facilitate visualization of tumor–host interaction between the pancreatic cancer cells and host-derived stroma and vasculature. Such in vivo models have enabled us to visualize in real time and acquire images of the progression of pancreatic cancer in the live animal. These models can demonstrate the real-time antitumor and antimetastatic effects of novel therapeutic strategies on pancreatic malignancy. These fluorescent models are therefore powerful and reliable tools with which to investigate metastatic human pancreatic cancer and novel therapeutic strategies directed against it.

Key words: GFP, RFP, CFP, Pancreas, Pancreatic cancer, Mouse models, Non-invasive imaging

1. Introduction

Pancreatic cancer is almost always a fatal disease with 5-year survival rates of only 1–4% (1, 2). It is the fourth leading cause of cancer-related mortality in the USA. Reasons for low survival in this disease include aggressive tumor biology, high metastatic

Robert M. Hoffman (ed.), *In Vivo Cellular Imaging Using Fluorescent Proteins: Methods and Protocols*, Methods in Molecular Biology, vol. 872, DOI 10.1007/978-1-61779-797-2_4, © Springer Science+Business Media, LLC 2012

potential and late presentation at the time of diagnosis (3, 4). The symptoms of pancreatic cancer may include jaundice, pain, weight loss, digestive problems, and new-onset diabetes (5). By the time an individual with pancreatic cancer develops these symptoms, the tumor has often reached a large size and metastasized to other organs including liver, lung, and peritoneum (4). Although chemotherapy can offer some palliation, it is not curative and the median survival is <6 months. For the few patients who have localized disease, surgical resection offers the only chance for cure. However, even with potentially curative surgery the 5-year survival rates are only 15–20% (1). Clearly, techniques for earlier more effective diagnosis and new treatment modalities need to be explored if progress is to be made.

In order to discover more effective treatment modalities for pancreatic cancer and improve detection, we have developed orthotopic models of human pancreatic cancer in the nude mouse that have patient-like tumor growth, progression, and metastasis and allow for testing of novel treatment strategies (6–25). The models take advantage of green fluorescent protein (GFP)- and red fluorescent protein (RFP)-expressing cancer cells that require no preparative procedures, contrast agents, substrates, anesthesia, or light-tight boxes for imaging, as do other imaging techniques (26). We have developed technology that has enabled the stable transduction of the GFP and RFP genes into a large series of human cancer cell lines (6, 27–39). The cancer cell lines are able to stably express GFP and/or RFP at high levels both in vitro and in vivo. As pioneered by our laboratory, GFP and RFP are uniquely suited for whole-body imaging of tumor growth and metastasis in live animals (40–42) (see Figs. 1 and 2). We have previously demonstrated the important parameter that GFP- and RFP-expressing cancer cells could be directly visualized in transplanted animals at very high resolution down to the subcellular level (29–33, 43).

Transplantation of RFP-expressing tumor fragments onto the pancreas of GFP- or cyan fluorescent protein (CFP)-expressing transgenic mice was used to facilitate visualization of tumor–host interaction between the pancreatic cancer cells and host-derived stroma and vasculature (21, 25, 44, 45) (see Figs. 3 and 4). Such in vivo models have enabled us to visualize in real time and acquire images of the progression of pancreatic cancer in the live animal and to demonstrate the real-time antitumor and antimetastatic efficacy of several novel therapeutic strategies on pancreatic malignancy (23). These fluorescent models are therefore powerful and reliable tools with which to investigate metastatic human pancreatic cancer and novel therapeutic strategies directed against it.

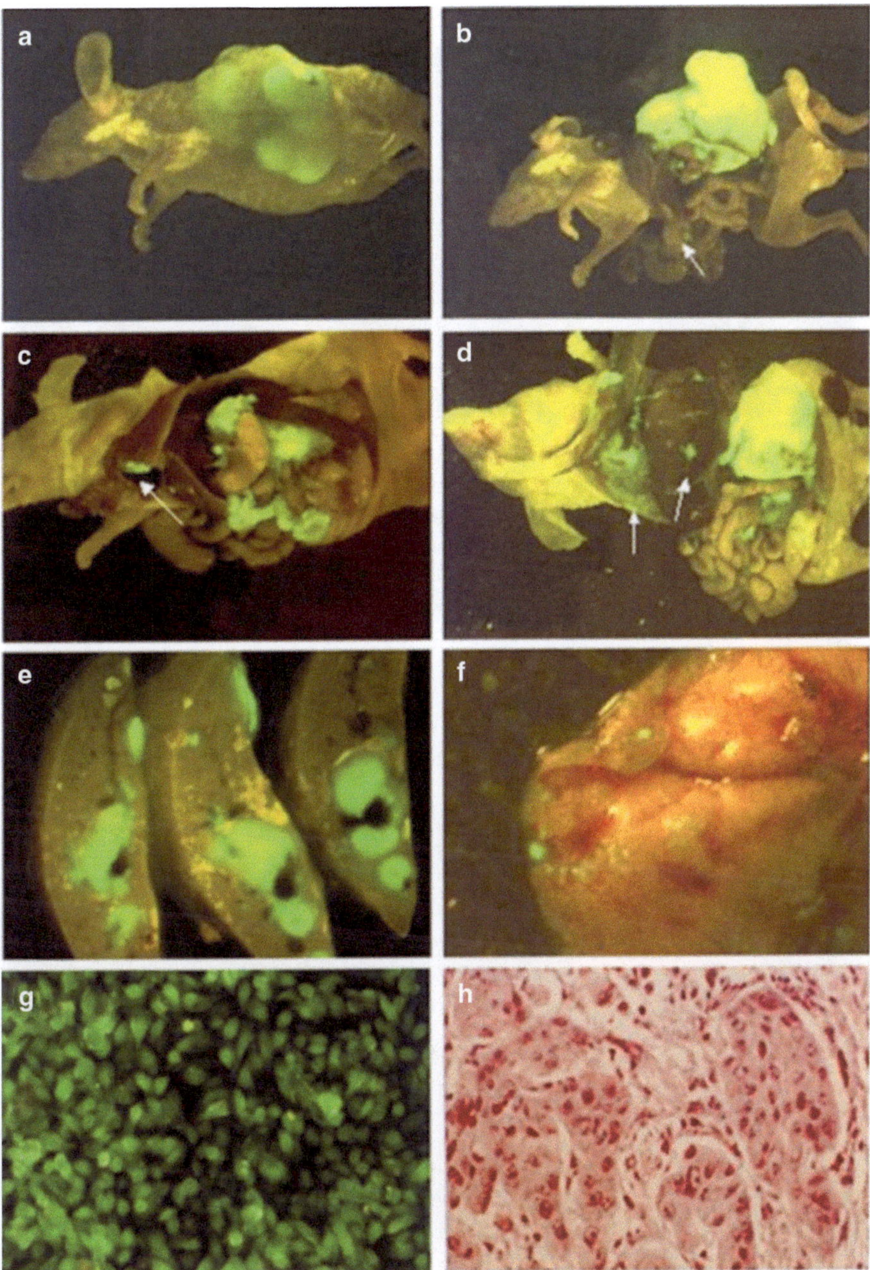

Fig. 1. (**a**) The human BxPC-3-GFP pancreatic tumor, transplanted by surgical orthotopic implantation (SOI), is externally visualized, with fluorescence imaging, through the skin of the nude mouse. (**b**) Laparotomy of the same mouse in (**a**) showing locally-advanced BxPC-3-GFP tumor with portal lymph node metastases. (**c**) The primary tumor formed in the pancreas at 12 weeks after SOI is visualized under bright-field microscopy. Numerous metastases and micrometastases can be visualized by GFP under fluorescence microscopy to the stomach, spleen, periportal nodes (*arrow*), liver, and mediastinum (*arrow*). (**d**) MIA PaCa-2-GFP tumor at week 10 post-SOI. *Left arrow* shows diaphragm metastases and *right arrow* shows liver metastases. (**e**) Multiple high-expressing GFP liver metastases are observed under fluorescence microscopy. (**f**) GFP-expressing lung metastases are observed. (**g**) The human pancreatic cancer cell line MIA PaCa-2 was transduced with the RetroXpress vector pLEIN that expresses enhanced GFP and the neomycin resistance gene on the same bicistronic message. The stable high expression clone was selected in 800 μg/ml G418. (**h**) H&E section of MIA PaCa-2-GFP tumor (46, 47).

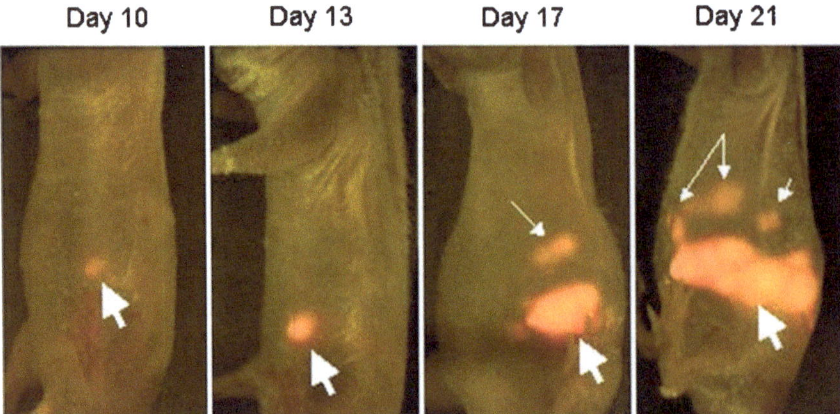

Fig. 2. RFP tumor fluorescence enabled real-time, whole-body imaging of tumor growth and metastasis of the human MIA PaCa-2-RFP tumor after surgical orthotopic implantation (SOI). *Panels* represent sequential fluorescence imaging of a single mouse on days 10, 13, 17, and 21 after SOI. Progressive primary tumor growth and the development of metastases are clearly visualized through the skin in the live animal. *Heavy arrow* primary tumor, *thin arrows* metastatic tumors (13–15).

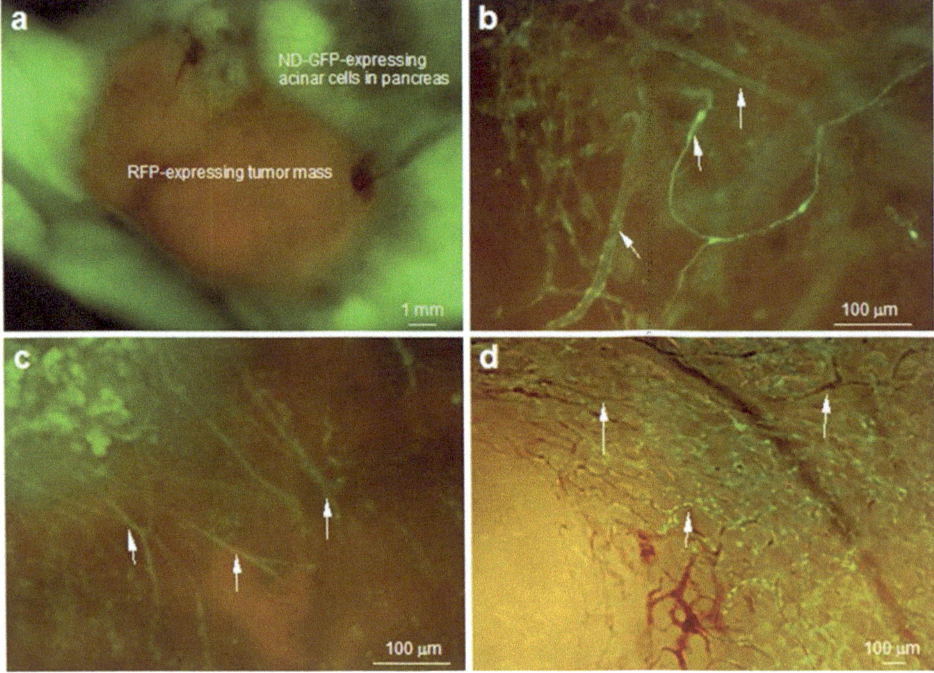

Fig. 3. Dual-color imaging of nascent blood vessels in the orthotopically-growing MIA PaCa-2 pancreatic tumor. (**a–d**) Day 14 after orthotopic implantation of RFP-expressing MIA PaCa-2 human pancreatic cancer cells to nestin-expressing green fluorescent protein (ND-GFP) transgenic nude mice. (**a**) RFP-expressing MIA PaCa-2 human pancreas tumor growing in the ND-GFP-expressing pancreas in an ND-GFP transgenic nude mouse. (**b–d**) ND-GFP-expressing nascent blood vessels (*white arrows*) growing in the RFP-expressing tumor mass. (**d**) Newly formed ND-GFP-expressing blood vessels with blood flow (44).

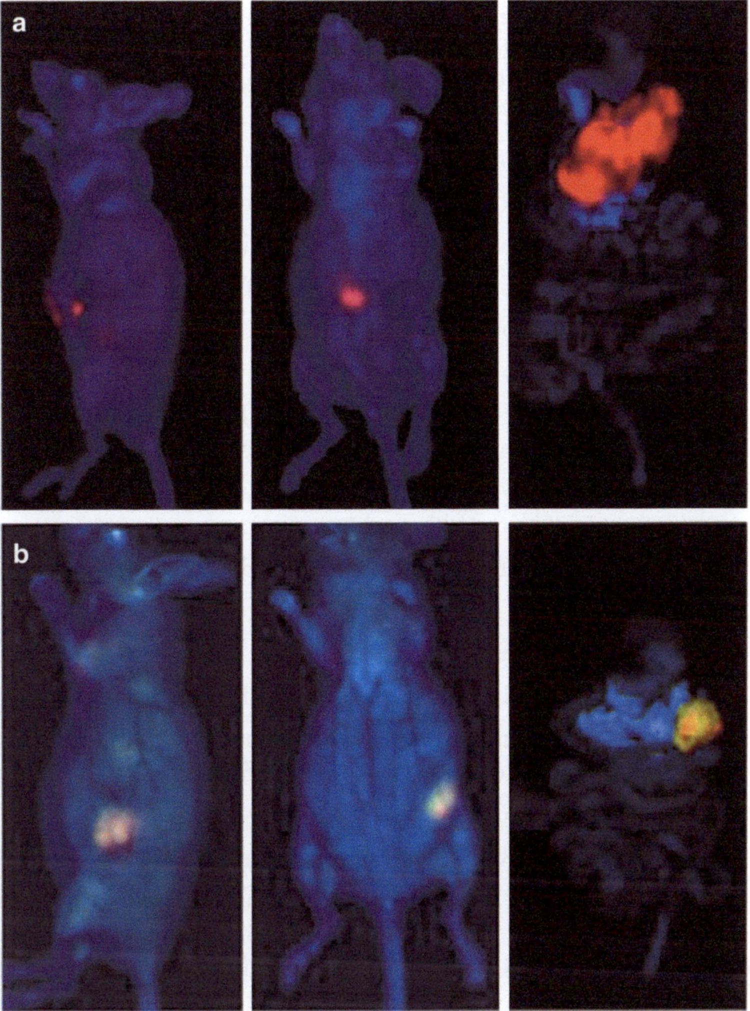

Fig. 4. Imaging the orthotopic human pancreatic cancer CFP nude mouse model with XPA-1-RFP human pancreatic cancer cells (21). Orthotopic injection of human pancreatic cancer cells was performed in CFP nude mice. Images were obtained at week-6 post-implantation. Exposure time for CFP signal acquisition was 1 s. (**a**) In this series, whole-body dual-color imaging permits identification of the tumor in the whole animal. This brightly-red fluorescent tumor had grown extensively into the brightly-blue pancreas. This could be seen after isolation of the GI tract, when the blue pancreas could be seen around the red tumor. (**b**) The dual-colored XPA-1-GFP-RFP tumor was grown from human pancreatic cancer cells engineered to express GFP in the nucleus and RFP in the cytoplasm. Whole body tri-color imaging permitted identification of the tumor within the mouse. After isolation of the GI tract, one can see a dual-colored tumor growing on a blue-fluorescent pancreas (21).

2. Materials

2.1. Reagents

1. Immunocompetent and immunodeficient mice (Charles River; Taconic; Harlan Teklad; see REAGENT SETUP).
2. Nude mice expressing GFP ("GFP mice"; AntiCancer Inc., San Diego, CA, USA) or CFP ("CFP mice"; AntiCancer Inc., San Diego, CA, USA).

3. Cell lines, such as those derived from pancreatic human cancer, transfected with genes coding for fluorescent proteins (AntiCancer Inc.; AntiCancer Japan, Osaka, Japan; Wako Pure Chemical Industries, Osaka, Japan).
4. Enhanced GFP gene (Clontech Laboratories, Inc.).
5. RFP (DsRed2) gene (Clontech Laboratories, Inc.).
6. Retroviral vector pLEIN (Clontech Laboratories, Inc.).
7. Retroviral vector pLNCX (Clontech Laboratories, Inc.).
8. Retroviral pLACX vector (Clontech Laboratories, Inc.).
9. Retroviral pFB-GFP vector (Clontech Laboratories, Inc.).
10. PT67-GFP cells, PT67-RFP cells, and PT67 H2B-GFP cells (Clontech Laboratories, Inc.).
11. Growth medium (normal and selective), appropriate for cell culture (Life Technologies).
12. HEPES buffer (20 mM, pH 7.2; Sigma).
13. DOTAB reagent (Boehringer).
14. LipofectAMINE Plus (Life Technologies).
15. Cloning cylinders (Bel-Art Products).
16. Anesthetic reagents (ketamine, xylazine, acepromazine maleate, isofluorane) (Butler Animal Health Supply).
17. Nair (Carter-Wallance).
18. G418 neomycin (Life Technologies).
19. PT67 packaging cells (Clontech Laboratories, Inc.).

2.2. Equipment

1. OV100 Small Animal Imaging System (Olympus Corp.).
2. FluorVivo Small Animal Imaging System (Indec Biosystems, Santa Clara, CA, USA).
3. iBox Small Animal Imaging System (UVP, Upland, CA, USA).
4. Sony VCR model SLV-R1000 (Sony Corp.).
5. Image Pro Plus 4.0 software (Media Cybernetics).
6. 1-ml 27G2 latex-free syringe (Becton Dickinson).
7. 25-ml Hamilton syringe (Fisher Scientific).
8. Humidified incubator with an atmosphere of 5% CO_2 (Hotpack).
9. Blue LED flashlight (LDP LLC).
10. D470/40 excitation filter (Chroma Technology).
11. GG475 emission filter (Chroma Technology).
12. Culture dishes, including 6-well and 96-well dishes (Fisher Scientific).

13. Cloning cylinders (Bel-Art Products).
14. Lipofectamine Plus transfection kit (Life Science).

2.3. Mice

1. Immunocompetent and immunodeficient mice. All animal studies are conducted in accordance with the principles and procedures outlined in the National Institutes of Health National Research Council's Guide for the Care and Use of Laboratory Animals (available at http://www.nap.edu/readingroom/) under assurance number A3873-1. Animals are kept in a barrier facility under high-efficiency particulate air (HEPA) filtration. Mice are fed with autoclaved laboratory rodent diet (Tecklad LM-485, Western Research Products).
2. Use a gene encoding GFP or CFP under the control of either the β-actin promoter, resulting in ubiquitous GFP or CFP expression, or the promoter of the gene encoding nestin, resulting in selective GFP expression, including expression in nascent blood vessels in tumors. To obtain nude GFP or CFP nude mice, first cross 6-week-old female C57BL/6 GFP or C57BL/6 CFP mice with 6- to 8-week-old BALB/c homozygous nude (nu/nu) or NCR nu/nu male mice, then cross male F_1 mice with female F_1 C57BL/6 GFP or CFP mice (17, 21, 27).
3. Retroviral pLACX vector. This requires the maintenance of a Biosafety Level 2 facility, including performing the work in a limited-access area, posting biohazard warning signs, minimizing the production of aerosols, decontaminating potentially infectious waste before disposal and taking precautions with sharps. For more information on Biosafety Level 2, see the following reference: Biosafety in Microbiological and Biomedical Laboratories, 3rd edn HHS Pub. #(CDC) 93-8395 (U.S. Department of Health and Human Services, PHS, CDC, NIH, 1993).

2.4. Equipment Setup

1. Whole-body imaging equipment use an Olympus OV100 Small Animal Imaging System, containing an MT-20 light source (Olympus Biosystems) and DP70 CCD camera (Olympus), for whole-body imaging in live mice at variable magnification. The optics of the OV100 fluorescence imaging system have been specially developed for macroimaging as well as microimaging, with high light-gathering capacity. The instrument incorporates a unique combination of high numerical aperture and long working distance. Four individually optimized objective lenses, parcentered and parfocal, provide a 105-fold magnification range for seamless imaging of the entire body down to the subcellular level without disturbing the animal. The OV100 has the lenses mounted on an automated turret with a high magnification range of 1.6× to 16× and a field-of-view ranging from 6.9 to 0.69 mm. The optics and

antireflective coatings ensure optimal imaging of multiplexed fluorescent reporters in small animals. High-resolution images are captured directly on a PC (Fujitsu Siemens, Celsius W340, Model MTR-D2156). An equivalent model can also be used. Images are processed for contrast and brightness and analyzed with the use of Paint Shop Pro 8 (Corel Corp.) and Cell (Olympus Biosystems).

Many other fluorescence imaging systems can also be used for dual-color tumor–host imaging: the iBox Small Animal Imaging System (UVP, Upland, CA, USA) or the FluorVivo Small Animal Imaging System (Indec Biosystems, Santa Clara, CA, USA). The iBox imaging system is used by our laboratory to image the CFP mice (21, 25).

3. Methods

3.1. GFP Retrovirus Production

1. For a GFP expression vector, use the pLEIN or equivalent retroviral vector expressing GFP and the neomycin resistance gene on the same bicistronic message (27, 28, 43).
2. Use PT67, an NIH3T3-derived packaging cell line, expressing the 10 A1 viral envelope to produce retrovirus. Culture PT67 cells in DMEM medium supplemented with 10% (vol/vol) heat-inactivated FCS. It takes approximately 3 days for the cells to reach approximately 70% confluence after seeding of 3×10^5 PT67 cells in a 25-mm^2 flask with DMEM medium containing 10% FCS.
3. For vector production, use PT67 packaging cells, at 70% confluence. Plate PT67 cells on a 60-mm dish at 60–80% confluence 12 h before transfection. Use Lipofectamine Plus transfection kit. Add 7 ml of precomplexed GFP DNA vector in 87 ml of serum-free medium and then add 6 ml lipofectamine reagent in a tube; mix and incubate at room temperature (22–26°C) for 15 min.
4. Dilute 4 ml lipofectamine in 96 ml serum-free medium in a second tube. Mix and incubate at room temperature for 15 min.
5. Combine precomplexed DNA and diluted lipofectamine reagent; then mix and incubate at room temperature for 15 min.
6. While the complexes are forming, replace medium on the cells with 800 ml serum-free DMEM. Add the DNA–lipofectamine reagent complex to the dish with cells containing fresh DMEM. Mix the complexes into the medium gently; incubate at 37.1°C, 5% CO_2 for 4 h.

7. After 4-h incubation, increase the volume of medium to 5 ml. Incubate at 37.1°C, 5% CO_2 for 24 h.
8. After 24-h incubation, clone the packaging cells by limit dilution in 96-well plates where cells are plated to a density of less than one cell per well.
9. Examine the cells by fluorescence microscopy 48 h posttransduction.
10. For selection, culture the cells in the presence of 300, 400, 500–2,000 μg/ml G418, increased in a stepwise manner, to select for a clone producing high amounts of a GFP retroviral vector (PT67-GFP). Culture the cells for 1–2 days in each concentration of G418. High-viral-titer production clones of GFP-PT67 cells are identified with NIH 3T3 cells used for virus tittering. Clones with titer higher than 10^6 plaque-forming units (pfu) are used for GFP vector production. Increasing the level of G418 in a stepwise manner is very important to induce the expression of the transgene. This procedure assures high-level production of GFP retrovirus (27, 28) (see Note 1).

3.2. RFP Retrovirus Production

1. Insert the *Hind*III/*Not*I fragment from pDsRed2, containing the full-length RFP cDNA, into the *Hind*III/*Not*I site of pLNCX2 that has the neomycin resistance gene to establish the pLNCX2-DsRed2 plasmid (27, 28, 43).
2. Incubate PT67 cells at 70% confluence for 24 h at 37°C. Lipofectamine reagent is used as described in steps 3–8 to transfect the pLNCX2-DsRed2 vector into PT67 cells.
3. Culture the cells in the presence of 200–1,000 μg/ml G418 to select a clone producing high amounts of RFP retroviral vector (PT67-DsRed2). Culture the cells for 1–2 days in each concentration of G418 in stepwise increases of 100 μg/ml. This procedure assures high-level production of RFP retrovirus as described in Subheading 3.1, step 10 (27, 28).

3.3. Production of Histone H2B-GFP Vector

1. Insert the histone H2B-GFP fusion gene at the *Hind*III/*Cla*I site of the pLHCX retrovirus that has the hygromycin resistance gene (43).
2. To establish a packaging cell clone producing high amounts of the histone H2B-GFP retroviral vector, transfect the pLHCX histone H2B-GFP plasmid in PT67 cells, using the same methods described above for PT67-DsRed2.
3. Culture the transfected cells in the presence of 200–400 μg/ml hygromycin to establish stable PT67 H2B-GFP packaging cells. The amount of hygromycin is increased stepwise as described above for G418. This procedure assures high-level production of histone H2B-GFP retrovirus.

3.4. RFP or GFP Gene Transduction of Cancer Cell Lines

1. For RFP or GFP gene transduction, use 20% confluent cancer cells. 12–18 h before infection with GFP or RFP retrovirus, plate the target cells at a density of $1–2 \times 10^5$ per 60-mm plate.
2. For retroviral infection, collect conditioned medium from packaging cells (PT67/pFB GFP or PT67/pLNCX2-DsRed2) and filter medium through a 0.45-mm polysulfonic filter. Add virus-containing filtered medium to target cells. Add polybrene to a final concentration of 8 mg/ml. Incubate cells for 24 h at 37°C.
3. Replace medium with DMEM and 10% (vol/vol) FCS after 24-h incubation.
4. Check for GFP- or RFP-expressing cells by fluorescence microscopy.
5. Harvest cancer cells with trypsin/EDTA and subculture them at a ratio of 1:15 in a selective medium that contains 50 μg/ml G418.
6. To select brightly fluorescent cells, increase the level of G418 to 800 μg/ml in a stepwise manner. Culture the cells for 1–2 days in each concentration of G418, using at least four different concentrations. Increasing the level of G418 in a stepwise manner is very important to induce the expression of the transgene. This procedure ensures high-level production of GFP or RFP in the cells.
7. Isolate clones expressing GFP or RFP with cloning cylinders with trypsin/EDTA and amplify them in DMEM and 10% FCS in the absence of selective agent. Further select cells for brightness and stability. This step ensures that the cells will stably express GFP or RFP in the absence of antibiotic selection, which is the case in vivo (27, 28).

3.5. Double RFP and Histone H2B-GFP Gene Transduction of Cancer Cells

1. To establish dual-color cancer cells, use clones of cancer cells expressing RFP in the cytoplasm at 70% confluence. Incubate RFP cancer cells, produced as described above (see Subheading 3.4 steps 1–7), with the retroviral-containing medium supernatants of PT67 H2B-GFP cells and the above culture medium for 48 h at 37°C. To obtain the double transformants, incubate cells with hygromycin for 48 h after transfection and select as described above (43).

3.6. Establishment of Imageable Tumor Models

A number of options are available for establishing tumor models using fluorescent protein-expressing tumor cells, including cell injection, subcutaneous injection of cancer cells, and surgical orthotopic implantation.

1. Cell injection to establish experimental metastasis model
   (a) Harvest fluorescent protein-expressing cancer cells by trypsinization using 0.25% (wt/vol) trypsin for 3 min at 37°C.

(b) Wash cells three times with cold serum-free medium using a tabletop centrifuge at 2,000 rpm (670 × *g*) for 5 min at room temperature.

(c) Resuspend the cells in approximately 0.2 ml of serum-free medium.

(d) Within 30 min of harvesting, inject 6-week-old GFP-C57CL/6 or nude (nu/nu) GFP- or nude CFP-mice with 10^6 cancer cells into the lateral tail vein or subcutaneously, in a total volume of 0.2 ml using a 1-ml 27G2 latex-free syringe. Cells in suspension can lose viability over time and therefore should be injected as soon as possible (see Note 2).

(e) For liver colonization, inject fluorescent protein-expressing cells directly into the spleen under anesthesia (please see below for details on inducing anesthesia).

2. Subcutaneous injection of cancer cells

(a) Detach cancer cells from the culture dish using 0.25% trypsin for 3 min at 37°C.

(b) Wash cells three times with cold serum-containing medium using a tabletop centrifuge at 500 × *g* for 5 min at room temperature, and then keep on ice.

(c) Inject 6-week-old GFP male C57BL/6, GFP-, or CFP-nude mice subcutaneously with 1×10^6 cancer cells that were collected and washed. This is done by inoculating cells subcutaneously on the dorsal skin of the mouse in a total volume of 50 μl cell culture medium, within 40 min of collection.

3. Surgical orthotopic implantation (SOI) to establish spontaneous metastasis model (11, 41, 42) (see Note 3)

(a) Induce anesthesia with ketamine mixture (10 ml ketamine HCl, 7.6 ml xylazine, 2.4 ml acepromazine maleate and inject subcutaneously).

(b) Perform all procedures of the operation under a 7× magnification microscope (see Note 4).

(c) Isolate fluorescent protein-expressing tumor fragments (1 mm³) from subcutaneously-growing tumors, which were formed from injected RFP- or GFP-expressing tumor cells by mincing tumor tissue. After proper exposure of the pancreas, implant three tumor fragments per mouse pancreas.

(d) Using an 8-0 surgical suture, penetrate the tumor fragments and suture the fragments onto the pancreas. Orthotopic implantation of tumor fragments results in higher spontaneous metastatic rates than injection of a cell suspension.

(e) Keep animals in a barrier facility under HEPA filtration.

3.7. Whole Body Imaging of Mice

A number of methods are available for whole-body imaging of mice, including chamber, microscopy, flashlight imaging, and light-box imaging.

1. Whole-body imaging with a microscope (27)
   (a) Use a fluorescence stereo microscope equipped with a 50 W mercury lamp power supply or equivalent.
   (b) Produce selective excitation of GFP through a D425/60 band-pass filter and 470 DCXR dichroic mirror (see Note 5).
   (c) Collect emitted fluorescence through a long-pass filter (GG475) on a cooled color CCD camera or equivalent.
   (d) Process images for contrast and brightness with the use of Image Pro Plus 4.0 software or equivalent.
   (e) Capture high-resolution images of 1,024×724 pixels directly on an IBM PC or continuously through video output on a high-resolution Sony VCR, model SLV-R1000 or equivalent.
   (f) In the case of C57BL/6 mice, remove hair with depilatory cream (e.g., Nair) or by shaving. Hair is highly autofluorescent, so improper removal of hair will result in low-quality images (see Note 6).
2. Whole-body imaging with a flashlight (27)
   (a) Use a blue LED flashlight with an excitation filter (mid-point wavelength peak of 470 nm) and a D470/40 emission filter for whole-body imaging of RFP- or GFP-expressing mice with RFP-expressing tumors growing in or on internal organs. Correct filters are necessary to eliminate tissue autofluorescence.
   (b) Acquire images with a digital camera such as a Nikon Cool-PIX or a simple CCD camera and store on a PC as described above.
   (c) In the case of C57BL/6 mice, remove hair with depilatory cream or by shaving. Hair is highly autofluorescent, so improper removal of hair will result in low-quality images.
3. Whole-body imaging in a light box (27)
   (a) Perform whole-body imaging in a fluorescence light box illuminated by fiberoptic lighting at 470 nm.
   (b) Collect emitted fluorescence through a GG475 long-pass filter on a cooled color CCD camera or equivalent (use of separate band-pass filters for RFP or GFP emission allows a monochrome camera to be used).
   (c) Capture high-resolution images of 1,024×724 pixels directly on an IBM PC or equivalent.
   (d) Process images for contrast and brightness with the use of Image Pro Plus 4.0 software or equivalent.

(e) In the case of C57BL/6 mice, remove hair with depilatory cream or by shaving. Hair is highly autofluorescent, so improper removal of hair will result in low-quality images.

4. Whole-body imaging in a chamber

(a) Perform whole-body imaging in an Olympus OV100 imaging system with 470 nm excitation light originating from an MT-20 light source. For CFP imaging, use an iBox Small Animal Imaging System (UVP, Upland, CA, USA). The FluorVivo (Indec Biosystems, Santa Clara, CA, USA) and the Maestro (Caliper Life Sciences, Hopkinton, MA, USA) can be used as well.

(b) Collect emitted fluorescence through appropriate filters configured on a filter wheel with a DP70 CCD camera. Variable magnification imaging can be performed with a series of four objective lenses.

(c) Capture images on a PC, and process images for contrast and brightness with Paint Shop Pro 8 and CellR.

(d) In the case of C57BL/6 mice, remove hair with depilatory cream or by shaving. Hair is highly autofluorescent, so improper removal of hair will result in low-quality images.

3.8. Image Analysis

Measuring the intensity of GFP images of internally-growing tumors is carried out as described below. Image size can be analyzed by using Image Pro Plus or Adobe Photo shop. The pixel area of the image is determined and converted to mm^2. It should be noted that image area and actual volume have been shown to correlate for tumors (13).

Steps for quantification of image size

1. Open the image file with Image Pro Plus, use the manual measurement function and then mark the periphery around the image.
2. Make measurements by clicking on the area icon or double-click on indicated length.
3. Export the data to Microsoft Excel for statistical analysis.

Measuring the intensity of GFP expression is described below. The ratio of GFP to intrinsic red fluorescence intensity of the skin is a reliable and convenient index for analyzing and quantifying the intensity of a GFP image. A ratio (γ) of green to red channels in a color CCD camera can be determined for each pixel in the image with and without GFP. Values of γ for mouse skin in the absence of GFP vary between 0.7 and 1.0. The intensity of GFP fluorescence can then be determined by subtracting the intrinsic value of γ from that obtained from the GFP image. The integral of GFP fluorescence intensity above the intrinsic maximum value of γ for skin without GFP can then be obtained by making the determination for each pixel in the image (6).

Steps for quantification of intensity

1. Open the image file with Image Pro Plus software and use the icon to outline the area of interest and then mark the area of interest either automatically or manually.
2. Open image histogram and then read the intensity of green or red channels to obtain the values for γ as outlined above for each pixel, and multiply by the number of pixels to obtain the total intensity.
3. Export the data to Microsoft Excel for statistical analysis.

It should be noted that image attenuation will occur as a function of depth. Such attenuation can be estimated by comparing the area of the whole-body image to that of the intravital image in the opened animal.

4. Notes

1. Increasing the level of G418 in a stepwise manner is very important to induce the expression of the transgene. This procedure assures high-level production of GFP retrovirus.
2. Cells in suspension can lose viability over time and therefore should be injected as soon as possible.
3. Orthotopic implantation of tumor fragments results in higher spontaneous metastatic rates than injection of a cell suspension.
4. Orthotopic surgical implantation should be performed with a 7× magnification stereo microscope.
5. Correct filters are necessary to eliminate tissue autofluorescence.
6. Hair is highly autofluorescent, so improper removal of hair will result in low-quality images.

Acknowledgments

This work was supported in part by grants from the National Cancer Institute CA109949 (to M.B.) and CA132971 (to M.B. and AntiCancer, Inc.) and American Cancer Society RSG-05-037-01-CCE (to M.B.), and National Cancer Institute grant CA103563 (to AntiCancer, Inc.).

References

1. Bouvet, M., Gamagami, R. A., Gilpin, E. A., Romeo, O., Sasson, A., Easter, D. W., and Moossa, A. R. (2000) Factors influencing survival after resection for periampullary neoplasms, *Am J Surg* ***180***, 13–17.
2. Katz, M. H., Bouvet, M., Al-Refaie, W., Gilpin, E. A., and Moossa, A. R. (2004) Non-pancreatic periampullary adenocarcinomas: an explanation for favorable prognosis, *Hepatogastroenterology* ***51***, 842–846.
3. Bouvet, M., Binmoeller, K. F., and Moossa, A. R. (2001) Diagnosis of adenocarcinoma of the pancreas, in *American Cancer Society Atlas of Clinical Oncology: Pancreatic Cancer* (Cameron, J. L., Ed.), BC Decker, Hamilton, Ontario, Canada.
4. Moossa, A. R., Bouvet, M., and Gamagami, R. (2002) The pancreas, in Essential surgical practice (Cuschieri, A., Steele, R. J. C., and Moossa, A. R., Eds.) 4th ed., pp 477–525, Arnold Publishing, London.
5. Katz, M. H., Savides, T. J., Moossa, A. R., and Bouvet, M. (2005) An evidence-based approach to the diagnosis and staging of pancreatic cancer, *Pancreatology* ***5***, 576–590.
6. Yang, M., Baranov, E., Moossa, A. R., Penman, S., and Hoffman, R. M. (2000) Visualizing gene expression by whole-body fluorescence imaging, *Proc Natl Acad Sci USA* ***97***, 12278–12282.
7. Kiguchi, K., Kubota, T., Aoki, D., Udagawa, Y., Yamanouchi, S., Saga, M., Amemiya, A., Sun, F. X., Nozawa, S., Moossa, A. R., and Hoffman, R. M. (1998) A patient-like orthotopic implantation nude mouse model of highly metastatic human ovarian cancer, *Clin Exp Metastasis* ***16***, 751–756.
8. Tomikawa, M., Kubota, T., Matsuzaki, S. W., Takahasi, S., Kitajima, M., Moossa, A. R., and Hoffman, R. M. (1997) Mitomycin C and cisplatin increase survival in a human pancreatic cancer metastatic model, *Anticancer Res* ***17***, 3623–3625.
9. An, Z., Wang, X., Kubota, T., Moossa, A. R., and Hoffman, R. M. (1996) A clinical nude mouse metastatic model for highly malignant human pancreatic cancer, *Anticancer Res* ***16***, 627–631.
10. Furukawa, T., Kubota, T., Watanabe, M., Kitajima, M., and Hoffman, R. M. (1993) A novel patient-like treatment model of human pancreatic cancer constructed using orthotopic transplantation of histologically intact human tumor tissue in nude mice, *Cancer Res* ***53***, 3070–3072.
11. Fu, X., Guadagni, F., and Hoffman, R. M. (1992) A metastatic nude-mouse model of human pancreatic cancer constructed orthotopically with histologically intact patient specimens, *Proc Natl Acad Sci USA* ***89***, 5645–5649.
12. Bouvet, M., Nardin, S. R., Burton, D. W., Lee, N. C., Yang, M., Wang, X., Baranov, E., Behling, C., Moossa, A. R., Hoffman, R. M., and Deftos, L. J. (2002) Parathyroid hormone-related protein as a novel tumor marker in pancreatic adenocarcinoma, *Pancreas* ***24***, 284–290.
13. Katz, M. H., Takimoto, S., Spivack, D., Moossa, A. R., Hoffman, R. M., and Bouvet, M. (2003) A novel red fluorescent protein orthotopic pancreatic cancer model for the preclinical evaluation of chemotherapeutics, *J Surg Res* ***113***, 151–160.
14. Katz, M. H., Bouvet, M., Takimoto, S., Spivack, D., Moossa, A. R., and Hoffman, R. M. (2003) Selective Antimetastatic Activity of Cytosine Analog CS-682 in a Red Fluorescent Protein Orthotopic Model of Pancreatic Cancer, *Cancer Res* ***63***, 5521–5525.
15. Katz, M. H., Takimoto, S., Spivack, D., Moossa, A. R., Hoffman, R. M., and Bouvet, M. (2004) An imageable highly metastatic orthotopic red fluorescent protein model of pancreatic cancer, *Clin Exp Metastasis* ***21***, 7–12.
16. Amoh, Y., Yang, M., Li, L., Reynoso, J., Bouvet, M., Moossa, A. R., Katsuoka, K., and Hoffman, R. M. (2005) Nestin-linked green fluorescent protein transgenic nude mouse for imaging human tumor angiogenesis, *Cancer Res* ***65***, 5352–5357.
17. Yang, M., Reynoso, J., Jiang, P., Li, L., Moossa, A. R., and Hoffman, R. M. (2004) Transgenic nude mouse with ubiquitous green fluorescent protein expression as a host for human tumors, *Cancer Res* ***64***, 8651–8656.
18. Bouvet, M., Spernyak, J., Katz, M. H., Mazurchuk, R. V., Takimoto, S., Bernacki, R., Rustum, Y. M., Moossa, A. R., and Hoffman, R. M. (2005) High correlation of whole-body red fluorescent protein imaging and magnetic resonance imaging on an orthotopic model of pancreatic cancer, *Cancer Res* ***65***, 9829–9833.
19. Tsuji, K., Yang, M., Jiang, P., Maitra, A., Kaushal, S., Yamauchi, K., Katz, M. H., Moossa, A. R., Hoffman, R. M., and Bouvet, M. (2006) Common bile duct injection as a novel method for establishing red fluorescent protein (RFP)-expressing human pancreatic cancer in nude mice, *JOP* ***7***, 193–199.

20. Snyder, C. S., Kaushal, S., Kono, Y., Tran Cao, H. S., Hoffman, R. M., and Bouvet, M. (2009) Complementarity of ultrasound and fluorescence imaging in an orthotopic mouse model of pancreatic cancer, *BMC Cancer* ***9***, 106.
21. Tran Cao, H. S., Reynoso, J., Yang, M., Kimura, H., Kaushal, S., Snyder, C. S., Hoffman, R. M., and Bouvet, M. (2009) Development of the transgenic cyan fluorescent protein (CFP)-expressing nude mouse for "Technicolor" cancer imaging, *J Cell Biochem* ***107***, 328–334.
22. Yang, M., Reynoso, J., Bouvet, M., and Hoffman, R. M. (2009) A transgenic red fluorescent protein-expressing nude mouse for color-coded imaging of the tumor microenvironment, *J Cell Biochem* ***106***, 279–284.
23. Bouvet, M., and Hoffman, R. M. (2009) Clinically-relevent orthotopic metastatic models of pancreatic cancer imageable with fluorescent genetic reporters, *Minerva Chir* ***64***, 521–539.
24. McElroy, M. K., Kaushal, S., Tran Cao, H. S., Moossa, A. R., Talamini, M. A., Hoffman, R. M., and Bouvet, M. (2009) Upregulation of thrombospondin-1 and angiogenesis in an aggressive human pancreatic cancer cell line selected for high metastasis, *Mol Cancer Ther* ***8***, 1779–1786.
25. Tran Cao, H. S., Kimura, H., Kaushal, S., Snyder, C. S., Reynoso, J., Hoffman, R. M., and Bouvet, M. (2009) The cyan fluorescent protein (CFP) transgenic mouse as a model for imaging pancreatic exocrine cells, *JOP* ***10***, 152–156.
26. Budinger, T. F., Benaron, D. A., and Koretsky, A. P. (1999) Imaging transgenic animals, *Annual Review of Biomedical Engineering* ***1***, 611–648.
27. Hoffman, R. M., and Yang, M. (2006) Whole-body imaging with fluorescent proteins, *Nat Protoc* ***1***, 1429–1438.
28. Hoffman, R. M., and Yang, M. (2006) Color-coded fluorescence imaging of tumor-host interactions, *Nat Protoc* ***1***, 928–935.
29. Chishima, T., Miyagi, Y., Wang, X., Yamaoka, H., Shimada, H., Moossa, A. R., and Hoffman, R. M. (1997) Cancer invasion and micrometastasis visualized in live tissue by green fluorescent protein expression, *Cancer Res* ***57***, 2042–2047.
30. Chishima, T., Miyagi, Y., Wang, X., Baranov, E., Tan, Y., Shimada, H., Moossa, A. R., and Hoffman, R. M. (1997) Metastatic patterns of lung cancer visualized live and in process by green fluorescence protein expression, *Clin Exp Metastasis* ***15***, 547–552.
31. Chishima, T., Miyagi, Y., Wang, X., Tan, Y., Shimada, H., Moossa, A., and Hoffman, R. M. (1997) Visualization of the metastatic process by green fluorescent protein expression, *Anticancer Res* ***17***, 2377–2384.
32. Chishima, T., Miyagi, Y., Li, L., Tan, Y., Baranov, E., Yang, M., Shimada, H., Moossa, A. R., and Hoffman, R. M. (1997) Use of histoculture and green fluorescent protein to visualize tumor cell host interaction [letter], *In Vitro Cell Dev Biol Anim* ***33***, 745–747.
33. Chishima, T., Yang, M., Miyagi, Y., Li, L., Tan, Y., Baranov, E., Shimada, H., Moossa, A. R., Penman, S., and Hoffman, R. M. (1997) Governing step of metastasis visualized in vitro, *Proc Natl Acad Sci USA* ***94***, 11573–11576.
34. Yang, M., Hasegawa, S., Jiang, P., Wang, X., Tan, Y., Chishima, T., Shimada, H., Moossa, A. R., and Hoffman, R. M. (1998) Widespread skeletal metastatic potential of human lung cancer revealed by green fluorescent protein expression, *Cancer Res* ***58***, 4217–4221.
35. Yang, M., Jiang, P., Sun, F. X., Hasegawa, S., Baranov, E., Chishima, T., Shimada, H., Moossa, A. R., and Hoffman, R. M. (1999) A fluorescent orthotopic bone metastasis model of human prostate cancer, *Cancer Res* ***59***, 781–786.
36. Yang, M., Jiang, P., An, Z., Baranov, E., Li, L., Hasegawa, S., Al-Tuwaijri, M., Chishima, T., Shimada, H., Moossa, A. R., and Hoffman, R. M. (1999) Genetically fluorescent melanoma bone and organ metastasis models, *Clin Cancer Res* ***5***, 3549–3559.
37. Yang, M., Baranov, E., Jiang, P., Sun, F. X., Li, X. M., Li, L., Hasegawa, S., Bouvet, M., Al-Tuwaijri, M., Chishima, T., Shimada, H., Moossa, A. R., Penman, S., and Hoffman, R. M. (2000) Whole-body optical imaging of green fluorescent protein-expressing tumors and metastases, *Proc Natl Acad Sci USA* ***97***, 1206–1211.
38. Yang, M., Baranov, E., Li, X. M., Wang, J. W., Jiang, P., Li, L., Moossa, A. R., Penman, S., and Hoffman, R. M. (2001) Whole-body and intravital optical imaging of angiogenesis in orthotopically implanted tumors, *Proc Natl Acad Sci USA* ***98***, 2616–2621.
39. Yang, M., Baranov, E., Wang, J. W., Jiang, P., Wang, X., Sun, F. X., Bouvet, M., Moossa, A. R., Penman, S., and Hoffman, R. M. (2002) Direct external imaging of nascent cancer, tumor progression, angiogenesis, and metastasis on internal organs in the fluorescent orthotopic model, *Proc Natl Acad Sci USA* ***99***, 3824–3829.
40. Flotte, T. R., Beck, S. E., Chesnut, K., Potter, M., Poirier, A., and Zolotukhin, S. (1998) A fluorescence video-endoscopy technique for detection of gene transfer and expression, *Gene Ther* ***5***, 166–173.

41. Fu, X. Y., Besterman, J. M., Monosov, A., and Hoffman, R. M. (1991) Models of human metastatic colon cancer in nude mice orthotopically constructed by using histologically intact patient specimens, *Proc Natl Acad Sci USA* ***88***, 9345–9349.

42. Sun, F. X., Sasson, A. R., Jiang, P., An, Z., Gamagami, R., Li, L., Moossa, A. R., and Hoffman, R. M. (1999) An ultra-metastatic model of human colon cancer in nude mice, *Clin Exp Metastasis* ***17***, 41–48.

43. Hoffman, R. M., and Yang, M. (2006) Subcellular imaging in the live mouse, *Nat Protoc* ***1***, 775–782.

44. Amoh, Y., Li, L., Tsuji, K., Moossa, A. R., Katsuoka, K., Hoffman, R. M., and Bouvet, M. (2006) Dual-color imaging of nascent blood vessels vascularizing pancreatic cancer in an orthotopic model demonstrates antiangiogenesis efficacy of gemcitabine, *J Surg Res* ***132***, 164–169.

45. Amoh, Y., Nagakura, C., Maitra, A., Moossa, A. R., Katsuoka, K., Hoffman, R. M., and Bouvet, M. (2006) Dual-color imaging of nascent angiogenesis and its inhibition in liver metastases of pancreatic cancer, *Anticancer Res* ***26***, 3237–3242.

46. Bouvet, M., Yang, M., Nardin, S., Wang, X., Jiang, P., Baranov, E., Moossa, A. R., and Hoffman, R. M. (2000) Chronologically-specific metastatic targeting of human pancreatic tumors in orthotopic models, *Clinical & Experimental Metastasis* ***18***, 213–218.

47. Bouvet, M., Wang, J., Nardin, S. R., Nassirpour, R., Yang, M., Baranov, E., Jiang, P., Moossa, A. R., and Hoffman, R. M. (2002) Real-time optical imaging of primary tumor growth and multiple metastatic events in a pancreatic cancer orthotopic model, *Cancer Res* ***62***, 1534–1540.

Chapter 5

Lentivirus-Based DsRed-2-Transfected Pancreatic Cancer Cells for Deep In Vivo Imaging of Metastatic Disease

Zeqian Yu, Jiahua Zhou, and Robert M. Hoffman

Abstract

Pancreatic cancer is one of the most aggressive human malignancies. One of the leading causes of pancreatic cancer death is metastasis. The early stages of tumor progression and micrometastasis formation have been difficult to analyze and have been hampered by the inability to identify small numbers of tumor cells against a background of many host cells. The intrinsic brightness of fluorescent proteins has been taken advantage of to develop a technology of whole-body imaging of tumors and gene expression in mouse internal organs. Stable transformation with fluorescent protein genes can be effected using lentiviral vectors containing a selectable marker such as neomycin resistance. The cells that stably express fluorescent proteins can then be transplanted into appropriate mouse models. For whole-body imaging, nude mice are very appropriate, the hair does not need to be removed by shaving or depilation. The instruments used are very simple, they need appropriate excitation and emission filters. It is crucial that proper filters be used such that background autofluorescence is minimal. Fluorescent protein-based imaging technology can be used for whole-body imaging of fluorescent cells on essentially all organs.

Key words: Pancreatic cancer, Metastasis, Micrometastasis, Green fluorescent protein, Red fluorescent protein

1. Introduction

This protocol enables the user to develop a model of pancreatic cancer in which a lentivirus is used to transform the cancer cells with red fluorescent protein (RFP). This model can be used for imaging tumors and metastasis at deep levels. The technique has many applications, including visualizing cancer growth and metastasis, gene expression, infectious disease, stem cells, immunology, neurobiology, and numerous others. This protocol focuses on pancreatic cancer growth, progression, and metastasis.

Robert M. Hoffman (ed.), *In Vivo Cellular Imaging Using Fluorescent Proteins: Methods and Protocols*, Methods in Molecular Biology, vol. 872, DOI 10.1007/978-1-61779-797-2_5, © Springer Science+Business Media, LLC 2012

The early stages of tumor progression and micrometastasis formation have been difficult to analyze and have been hampered by the inability to identify small numbers of cancer cells against a background of many host cells. The visualization of tumor cell emboli, micrometastases, and their progression over real-time during the course of the disease models has not been easy to study in current models of metastasis. Previous studies used transfection of tumor cells with the *Escherichia coli* β-*galactosidase (lacZ)* gene to detect micrometastases (1). However, detection of lacZ requires extensive histological preparation; therefore, it is impossible to detect and visualize cancer cells in viable fresh tissue or the living animal at the microscopic level.

Another approach for visualizing tumor cells in vivo involves insertion of the luciferase gene into tumor cells causing them to emit light (2). However, once transferred to mammalian cells, luciferase enzymes require exogenous delivery of their luciferin substrate, which is an impractical procedure in an intact animal. Because of the low resolution and signal achieved with this approach, it takes a substantial amount of time to collect sufficient photons to construct a pseudo-image from an anesthetized animal. Also, whether luciferase genes can function stably in tumors over long periods and in the metastases derived from them is not known.

The use of the green fluorescent protein (GFP) gene cloned from the bioluminescent jellyfish *Aequorea victoria* (3), as a marker for localization of tumor cells in host tissues, has previously been described (4–6). The tracking of cancer cells that stably express the GFP gene has been shown to facilitate in vivo identification of single-cell metastases and permits real-time imaging of the growth and dissemination of tumors in live animals without the use of anesthesia, contrast agents, laparotomy, or invasive procedures (7).

RFP from the *Discosoma* coral possesses enhanced optical qualities that enable it to be used instead of, or as an adjunct to, GFP-based systems (8–10).

A crucial point is the stable expression of fluorescent proteins in the cells to be imaged. Such stable expression can be effected by appropriate vectors containing the fluorescent protein gene and a selectable marker such as antibiotic resistance. Lentiviral vectors have the advantages of being able to incorporate into genomic DNA with high efficiency, especially in cells that are not actively dividing. Lentiviral vector-mediated transgene expression can also be maintained for long periods of time. RFP-expression remained stable for 5 months in transducted SW1990-RFP cells. The cells could easily be monitored by fluorescence microscopy and FACS analysis (see Fig. 1a, b), suggesting a stable chromosomal integration of the amplified RFP gene. In addition, RFP fluorescence is relatively unaffected by the external environment. The expression of RFP in tumor cells does not interfere with

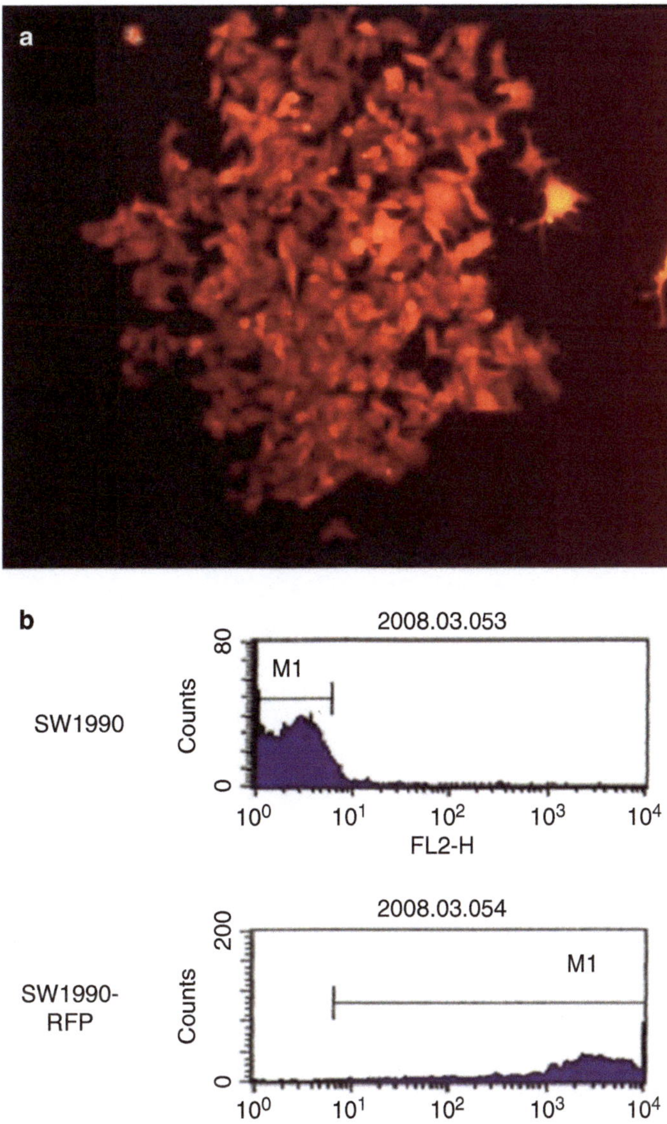

Fig. 1. Human pancreatic cancer line SW1990 was transduced by lentivirus which resulted in high-level RFP expression. (**a**) Strong red fluorescence was detected under fluorescence microscopy after 5 months culture without G418 selection. (**b**) Strong red fluorescence was also detected under fluorescence microscopy (200×). FACS analysis showed 99.62% RFP-expression of SW1990-RFP after 5 months culture without G418.

tumor growth (see Fig. 2a). These advantages of the lentiviral vector system make it promising for establishment of transgenic animals and the manipulation of the mammalian genome.

A wide range of instrumentation can be used for fluorescent protein-based imaging, ranging from an LED flashlight and appropriate excitation and emission filters to highly specific instrumentation such as the Olympus OV100, a small animal imaging apparatus with variable magnification from macro to micro. The technology

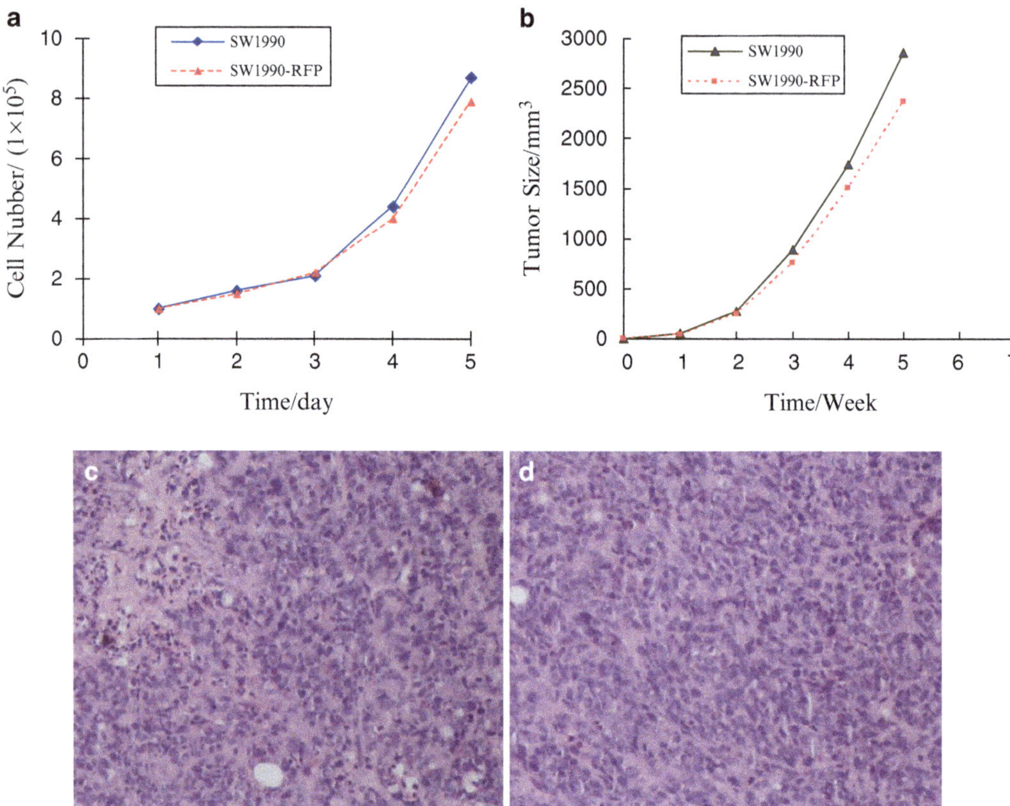

Fig. 2. Biological properties of SW1990-RFP compared to non-transfected SW1990 cells. (**a**) No difference in vitro growth between SW1990 and SW1990-RFP was observed over 5 days. (**b**) No difference in vivo growth between SW1990 and SW1990-RFP was observed over 2 months. (**c**) H&E slide of SW1990 subcutaneous tumor (200×). (**d**) H&E slide of SW1990-RFP subcutaneous tumor (200×).

requires incident excitation light as well as detection of fluorescence emission. Internally-growing SW1990-RFP orthotopic tumors become visible through the skin in the live animal within 10 days after implantation (see Fig. 3a). The red fluorescence emitted by the tumor enables real-time, sequential whole-body imaging, and quantification of tumor burden (see Fig. 4) without the need for laparotomy, contrast agents, or invasive procedures. Orthotopically-transplanted SW1990-RFP tumors produced both extensive locoregional and disseminated disease. At the time of autopsy, most of the mice had micrometastasis and metastasis in the greater omentum, periportal lymph nodes, liver, and lung (see Fig. 5). Fluorescence imaging enabled identification of micrometastases. Solid tumors were found to have histological features consistent with moderately differentiated pancreatic adenocarcinoma upon examination of hematoxylin- and eosin-stained tissue sections (see Fig. 4g).

Metastatic patterns of SW1990-RFP cells were also imaged after tail vein and spleen injection of cancer cells into nude mice.

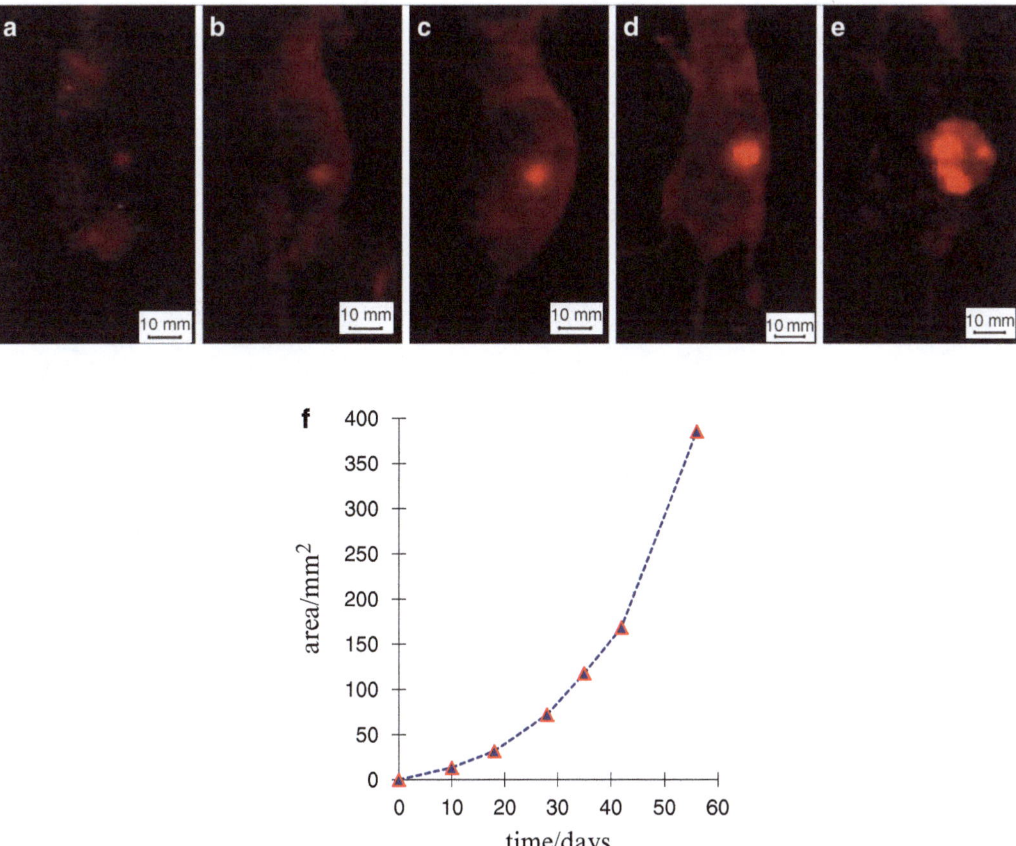

Fig. 3. RFP fluorescence enabled real-time, noninvasive in vivo imaging of tumor growth of SW1990-RFP tumors after surgical orthotopic implantation (SOI). (**a–e**) Represent sequential fluorescence imaging of a single mouse taken on days 10, 18, 28, 35, and 56 after SOI. Progressive primary tumor growth was clearly visualized through the skin in the live animal. (**f**) Quantification of RFP tumor fluorescence enabled real-time quantification of total tumor burden in vivo after SOI.

One month following tail vein injection, we detected metastasis and micrometastasis in the lungs (see Fig. 5), which were confirmed by H&E staining (see Fig. 5c). Liver micrometastasis was also detected, which was also confirmed by H&E staining. Forty-five days after spleen injection, mice were observed with many small red fluorescent tumor nodules on the surface of the liver (see Fig. 6). Frozen sections expressing RFP could distinguish cancer cells and normal cells by their bright fluorescence, often down to the single-cell level (see Fig. 6c–e). We did not detect metastasis in other organs besides liver.

In this study, RFP gene-transfected pancreatic cancer cells were successfully used to noninvasively image micrometastases of liver, omentum, and peritoneum in nude mice, following surgical orthotopic implantation (SOI) in the pancreas of subcutaneous-growth tumor fragments. These models provided the opportunity to demonstrate the capability of RFP for detection and visualization of

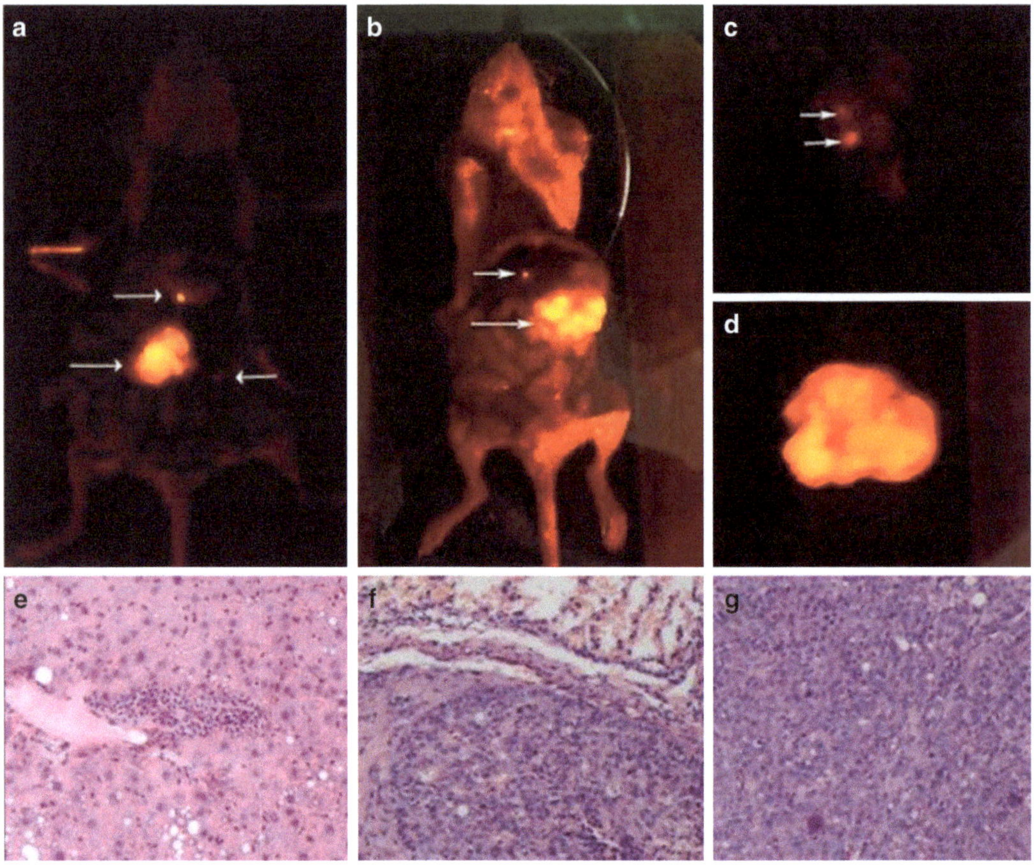

Fig. 4. Imaging of primary tumor and micrometastasis of SW1990-RFP after SOI. (**a**) The *top arrow* shows a micrometastasis of greater omentum; the *middle arrows* show the primary tumor; and the *bottom arrows* show a micrometastasis of the peritoneum on day 45 after SOI. (**b**) The *top arrows* show the micrometastasis of liver and the *bottom arrow* shows the primary tumor taken on day 37 after SOI. (**c**) The two *arrows* show micrometastases of the lung taken on day 60 after SOI. Note the lungs have been removed from the mouse. (**d**) Primary tumor emitted strong red fluorescence. (**e**) H&E slide of micrometastasis of liver (200×). (**f**) H&E slide of micrometastasis of lung (200×). (**g**) H&E slide of primary tumor (200×).

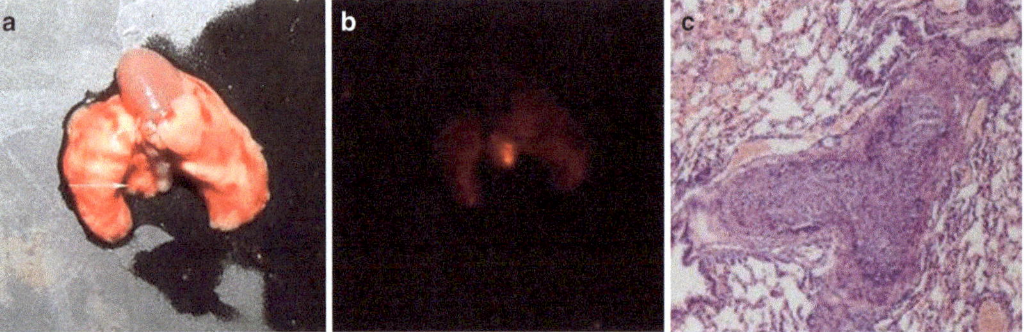

Fig. 5. Metastasis of the lung 30 days after tail vein injection. (**a**) Brightfield image. (**b**) Fluorescence image of lung metastasis. (**c**) H&E stain of lung metastasis (100×).

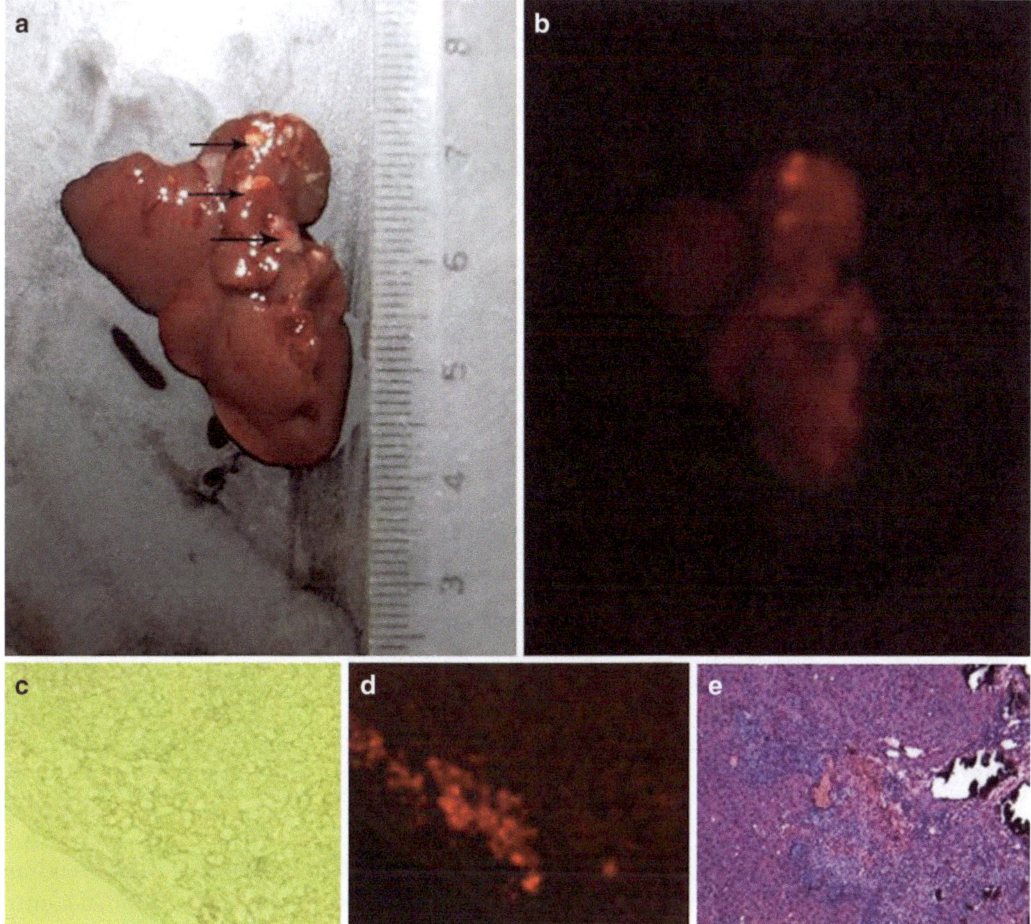

Fig. 6. Metastasis of the liver 45 days after spleen injection. (**a**) Brightfield image. (**b**) Fluorescence image. (**c**) Brightfield image of frozen section of liver metastasis (100×). (**d**) The same section as (**c**) observed under fluorescence microscopy (100×). (**e**) H&E slide of section in (**c**) (100×).

metastases and micrometastases even in deep organs. Some limits might occur when only a few cells are present in or on a deep organ. Excitation light is scattered and attenuated as it goes through skin, body tissues, and fluids. To overcome limits on fluorescent protein imaging imposed by the skin, reversible skin-flap window models have been developed that allow single-cell imaging on most organs of the mouse. In addition, red-shifted RFPs are now being developed (11). Katushka combines far-red emission and brightness features of a fluorescent protein necessary for visualizing very deep within living tissues (12). The spectral characteristics make fluorescent proteins powerful tools for cancer research, developmental biology, gene expression, and other whole-body imaging studies even in deep organs.

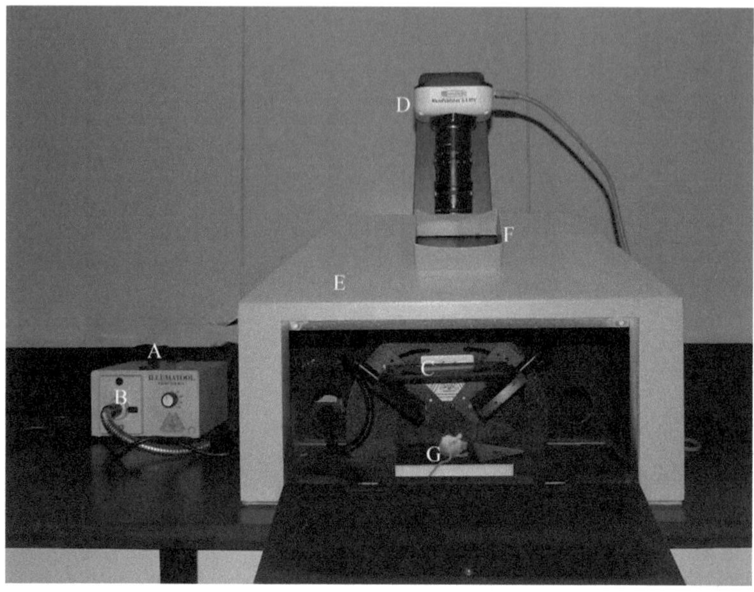

Fig. 7. Lightools Research imaging system. (**a**) Light source. (**b**) D540/40-nm band-pass filter. (**c**) D580 nm long-pass filter. (**d**) High-sensitivity CCD. (**e**) Black-box system. (**f**) Observation Window. (**g**) The mouse.

2. Materials

2.1. Reagents

- Immunodeficient mice. BALB/C nu/nu nude mice weighing 20–22 g, between 4 and 6 weeks (Experimental Animal Centre of Shanghai Slaccas).
- Cell lines. The SW1990 human pancreatic cancer cell line, to be transfected with genes coding for fluorescent proteins and TAT-free packaging cells 293T (HEK) cells, to produce recombinant lentivirus (American Type Culture Collection [ATCC]).
- RFP. DsRed2 gene (Clontech Laboratories, Inc.).
- Lentiviral vector. The pGC-RFP plasmid together with two packaging plasmids pHelper 1.0 (encoding human immunodeficiency virus [HIV] gag, pol, and rev) and the plasmid pHelper 2.0 (encoding for VSV-G envelope) (Genechem Company, Shanghai, China).
- Growth medium (normal and selective), appropriate for cell culture (Gibco BRL, China).
- LipofectAMINE 2000 (Invitrogen, Carlsbad, CA).
- HEPES buffer 20 mM, pH 7.2 (Sigma).
- G418 neomycin (Gibco BRL, China).
- Trypan blue (Sigma).
- 1% Pentobarbital sodium (Xitang, Germany).

2.2. Equipment

- Lightools Fluorescence Imaging System (Lightools Research).
- Fluorescence microscope (Nikon, Japan).
- Humidified incubator with a 5% CO_2 atmosphere (Revco).
- Culture dishes, including 6-well and 96-well dishes (Costar).
- 1-ml 29-gauge, latex-free syringe (Weigao, China).
- Image Pro Plus 5.0 software (Media Cybernetics).
- D540/40 excitation filter (Chroma Technology).
- D580 emission filter (Chroma Technology).
- FACS Calibur System (BD).

2.3. Reagent Setup

Immunodeficient mice. Caution: The animals are maintained in a specific pathogen-free environment. The animals are fed with an autoclaved laboratory rodent diet. The mice are maintained on a daily 12-h light to 12-h dark cycle. All animal experiments involve ethical and humane treatment under a license from the Jiangsu Provincial Bureau of Science.

Lentiviral pGC-RFP vector. Caution: The National Institutes of Health and Center for Disease Control have designated lentiviruses as Level 2 organisms. This requires the maintenance of a Biosafety Level 2 facility, including performing the work in a limited-access area, posting biohazard warning signs, minimizing the production of aerosols, decontaminating potentially infectious waste before disposal, and taking precautions with sharps.

2.4. Equipment Setup

Whole-body imaging equipment. A Lightools Research imaging system equipped with two 150-W mercury lamp power supplies is used. Selective excitation of RFP was produced through a D540/40-nm band-pass filter. Emitted fluorescence is collected through a long-pass filter (D580 nm, Chroma Technology) on a cooled color charge-coupled device camera. Images are processed for contrast and brightness and analyzed by IMAGE PRO PLUS 5.0 software (Media Cybernetics). High-resolution images of 1,392 × 1,040 pixels are captured, directly displayed on a monitor and digitally stored. Whole-body images of each mouse are obtained. At each imaging time point, the real-time determination of tumor burden is performed by quantifying the fluorescent surface area, as previously described (8).

Internal imaging, analysis of primary tumor, and metastasis. Mice are sacrificed and explored when they appeared premorbid. Postmortem fluorescence images are obtained. Each mouse undergoes laparotomy and median sternotomy. RFP expression is visualized in the whole-body image system described above, facilitating identification of primary and metastatic pancreatic tumor. After obtaining open images, all of the solid organs are removed and are thoroughly examined for any evidence of metastasis using both fluorescent imaging and histopathological analysis.

3. Methods

3.1. RFP Lentivirus Production

1. For an RFP-expression vector, use the pGC-RFP plasmid (encoding RFP and the neomycin-resistance gene) together with the other two packaging plasmids pHelper 1.0 (encoding human immunodeficiency virus [HIV] gag, pol, and rev) and the plasmid pHelper 2.0 (encoding the VSV-G envelope).
2. Use 293T (HEK) cells and TAT-free packaging cells to produce recombinant lentivirus. Culture 293T cells in DMEM supplemented with 10% (vol/vol) heat-inactivated FCS. It takes approximately 2–3 days for the cells to reach 70–80% confluence after seeding 3×10^5 293T cells in a 25-mm^2 flask with DMEM containing 10% FCS.
3. For lentivirus, use 293T packaging cells. One day before transfection, 1.2×10^7 cells are suspended in 20 ml growth medium without antibiotics and grown to 90–95% confluence (in 15-cm dishes). At the time of transfection, pGC-RFP (20 μg), plasmid pHelper 1.0 (15 μg), and plasmid pHelper 2.0 (10 μg) are diluted in Opti-MEM without serum and mixed gently (total volume of 2.5 ml). LipofectAMINE 2000 100 μl is diluted in Opti-MEM without serum and mixed gently (total volume is 2.5 ml) and incubated for 5 min at room temperature.
4. After 5-min incubation, the diluted DNA is combined with dilute LipofectAMINE 2000 (total volume of 5 ml), mixed gently, and incubated for 20 min at room temperature to permit the formation of the DNA–LipofectAMINE 2000 complexes. Approximately 5 ml DNA–LipofectAMINE 2000 complexes were added to each dish and mixed gently by rocking the plate back and forth. Cells are incubated in an incubator with 5% CO_2 at 37°C. After 8 h, the complexes are removed and replaced with the growth medium. Incubate at 37°C, 5% CO_2 for 48 h.
5. The virus is harvested by collecting the cell culture medium after 48 h.
6. After filtering the collected medium through 0.45-μm filters, the virus is concentrated by spinning at $4{,}000 \times g$ for 15 min followed by a second spin ($1{,}000 \times g$ for 2 min at room temperature). The concentrated virus is stored at –80°C.

3.2. RFP Gene Transduction of Cancer Cell Lines

1. For RFP gene transduction, use 30–50% confluent SW1990 human pancreatic cancer cells at the time of transduction (1×10^5 cells per well in 24-well plates).
2. For lentiviral infection, add DsRed-2 RFP-lentivirus (5 μl) at 5×10^8 TU/ml to target cells. Add polybrene to a final concentration of 8 mg/ml. Incubate cells for 24 h at 37°C.

3. Replace medium with DMEM and 10% (vol/vol) FCS after 24 h incubation. Continue to incubate cells for 48 h at 37°C.
4. Check for RFP-expressing cells by fluorescence microscopy.
5. The cells are passaged at a ratio of 1:5 into selective medium that contained 400 μg/ml G418.
6. To select brightly fluorescent cells, increase the level of G418 to 1,200 μg/ml in a stepwise manner. Culture the cells for 1–2 days in each concentration of G418, using at least four different concentrations.

Critical step: Increasing the level of G418 in a stepwise manner is very important to induce the expression of the transgene. This procedure ensures high-level production of RFP in the cells. Isolate clones expressing RFP in 96-well plates. Amplify clones in DMEM and 10% FCS in the absence of selective agent. Further select cells for brightness and stability. This step ensures that the cells will stably express RFP in the absence of antibiotic selection, which is the case in vivo.

3.3. In Vitro and In Vivo Growth Curves

SW1990 and SW1990-RFP cells are seeded at 1×10^5 in 60-mm culture dishes. At daily intervals, triplicate samples are harvested by trypsinization. An aliquot of 100 μl of cell suspension is added to 100 μl trypan blue (Sigma) and counted using a hemocytometer. The doubling time is calculated from the cell growth curve over 5 days. To determine the in vivo growth potential, 5×10^6 SW1990 or SW1990-RFP cells were injected into the flanks of nude mice, respectively. Tumor volume is measured every week over 5 weeks. Standard deviations and statistics are calculated to generate the in vitro and in vivo growth curves.

3.4. Cell Cycle Phase Distribution of Pre- and Posttransduction of RFP

SW1990 parental cells and SW1990-RFP are cultured in DMEM supplemented with 10% heat-inactivated fetal bovine serum, 100 units/ml penicillin, 100 μg/ml streptomycin, and 0.25 μg/ml amphotericin B (Gibco-BRL, Shanghai, China). Cells are harvested with 0.25% trypsinization, washed thrice with cold 1% 0.01 mol/l PBS, suspended with cold 70% ethanol, and stored at −20°C. Flow cytometry analysis is then done to determine cell-cycle distribution.

3.5. Establishment of Imageable Tumor Models

A number of options are available for establishing tumor models using fluorescent protein-expressing tumor cells, including cell injection and SOI.

1. Tail vein cell injection to establish an experimental lung metastasis model
   (a) Harvest fluorescent protein-expressing cancer cells by trypsinization using 0.25% (wt/vol) trypsin for 3 min at 37°C.

(b) Wash cells three times with cold serum-free medium using a tabletop centrifuge at 400 ×*g* for 5 min at room temperature.

(c) Resuspend the cells in approximately 0.2 ml serum-free medium.

(d) Within 30 min of harvesting, inject 6-week-old nude (nu/nu) mice with 10^6 cancer cells into the lateral tail vein in a total volume of 0.2 ml using a 1-ml 29G2 latex-free syringe.

Critical step: Cells in suspension can lose viability over time and therefore should be injected within 30 min of harvesting. Injection speed cannot be too fast, in order not to cause acute pulmonary edema.

2. Spleen cell injection to establish experimental liver metastasis model

   (a) Harvest fluorescent protein-expressing cancer cells (please see above for details).

   (b) Mice are anesthetized with 1% pentobarbital sodium (50 mg/kg) through an abdominal cavity injection.

   (c) Fixed mouse limbs with transparent gelatin after anesthesia.

   (d) The mouse abdomens are sterilized with alcohol. An incision is created through the left upper abdominal pararectal line.

   (e) Then inject 10^6 fluorescent protein-expressing cancer cells in the anus perineum of the spleen by using a 1-ml 29-gauge, latex-free syringe within 30 min of harvesting.

   (f) The abdominal wall and the skin are closed in one layer using 6-0 surgical suture.

Critical step: The puncture point should be compressed for 3 min with alcohol cotton ball to avoid bleeding.

3. Subcutaneous injection of cancer cells

   (a) Harvest fluorescent protein-expressing cancer cells (please see above for details).

   (b) Inject 6-week-old nude mice subcutaneously with 5×10^6 cancer cells that are collected and washed. This is done by inoculating cells by subcutaneous injection of the dorsal skin of the mouse in a total volume of 0.2 ml within 30 min of harvesting.

4. SOI to establish spontaneous metastasis model

   (a) Isolate fluorescent protein-expressing tumor fragments from subcutaneously-growing tumors, which are formed from injected RFP-expressing cancer cells.

(b) Tumor tissues are cut with scissors and minced into 1 mm^3 pieces in DMEM medium.

(c) Anesthesia and disinfection (please see above for details).

(d) An incision is created through the left upper abdominal pararectal line and peritoneum. The pancreas is carefully exposed and one tumor piece is transplanted onto the middle of the gland using a single 8-0 surgical suture.

(e) The pancreas is then returned into the peritoneal cavity, and the abdominal wall and the skin are closed in one layer using 6-0 surgical suture.

Critical step: Fluorescent protein-expressing cancer in the exponential growth phase, grown subcutaneously in nude mice, are resected aseptically. Necrotic tissues are cut away, and the remaining healthy tumor tissues are implanted. The above operation is carried out in a super-clean bench. The operator wears a surgical magnifying glass.

Image analysis: Measure the fluorescence intensity of RFP images of internally-growing tumors. Image size can be analyzed by using Image Pro Plus or Adobe Photo shop. The pixel area of the image is determined and converted to mm^2. It should be noted that image area and actual volume have been shown to correlate for tumors.

3.6. Steps for Quantification of Image Size

- Open the image file with Image Pro Plus. Use the manual measurement function and then mark the periphery around the image.
- Make measurements by clicking on the area icon or double-click on indicated length.
- Export the data to Microsoft Excel for statistical analysis.

3.7. Steps for Quantification of Intensity

- Open the image file with Image Pro Plus software and use the icon to outline the area of interest, and then mark the area of interest either automatically or manually.
- Open image histogram and then read the intensity of green or red channels to obtain the values for *g* as outlined above for each pixel, and multiply by the number of pixels to obtain the total intensity.
- Export the data to Microsoft Excel for statistical analysis.

It should be noted that image attenuation will occur as a function of depth. Such attenuation can be estimated by comparing the area of the whole-body image to that of the intravital image in the opened animal. Attenuation, even for a GFP tumor on the colon, is moderate (13) (see Table 1 (14)).

Analyze the obtained images (Table 1 (14)).

Troubleshooting guidance can be found in Table 2 (14).

Table 1
Comparison of whole-body and intravital open images

Tumor location	Number of pixels in image	Area of the image (mm^2)
GFP colon (whole-body image)	2,465	16.22
GFP colon (intravital image)	2,731	17.97

Table 2
Troubleshooting table

Problem	Possible reason	Solution
Autofluorescence	Use of wrong filters	It is important to minimize autofluorescence of the tissue and body fluids by using proper filters. Excitation filters should have a narrow band as close to 490 nm as possible to specifically excite GFP whose absorption peak is distinct from that of the skin, other tissues, and fluid of the animal. In addition, proper band-pass emission filters should be used with a cut-off of approximately 515 nm
Bleeding	Improper surgical procedures	Bleeding should be avoided at the surgical site as hemoglobin will absorb the incident excitation light
Dehydration	Long-term procedures on an open animal	When using open-biopsy procedures, it is essential to hydrate the animal by spraying normal saline on the open tissue
Infection	Unclean instruments and environment	When using repeat procedures such as an open biopsy or other invasive procedures, it is crucial to maintain a proper sterile operation field

4. Notes

The procedures described in this protocol enable whole-body imaging of fluorescent cells in essentially any organ. Examples include Fig. 1, which shows human pancreatic cancer line SW1990 transduced by lentivirus, resulting in high-level RFP expression. Figure 2 shows biological properties of SW1990-RFP as compared to nontransfected SW1990 cells. Figure 3 demonstrates that RFP fluorescence enabled real-time, noninvasive in vivo imaging of tumor growth of SW1990-RFP tumors after SOI. Figure 4 shows the image of the primary tumor and micrometastasis of SW1990-RFP after SOI. Figure 5 demonstrates the metastasis of the lung 30 days after tail vein injection. Figure 6 demonstrates the metastasis of the liver 45 days after spleen injection. Figure 7 shows the Lightools Research imaging system.

References

1. Holleran, J.L., Miller, C.J., and Culp, L.A. (2000) Tracking micrometastasis to multiple organs with lacZ-tagged CWR22R prostate carcinoma cells. *J Histochem Cytochem* **48**, 643–651.
2. Contag, C.H., Jenkins, D., Contag, P.R., and Negrin, R.S. (2000) Use of reporter genes for optical measurements of neoplastic disease in vivo. *Neoplasia* **2**, 41–52.
3. Prasher, D.C., Eckenrode, V.K., Ward, W.W., Prendergast, F.G., and Cormier, M.J. (1992) Primary structure of the Aequorea victoria green-fluorescent protein. *Gene* **111**, 229–233.
4. Hoffman, R.M. (2002) Green fluorescent protein imaging of tumor growth, metastasis, and angiogenesis in mouse models. *Lancet Oncol* **3**, 546–556.
5. Yang, M., Reynoso, J., Jiang, P., Li, L., Moossa, A.R., and Hoffman, R.M. (2004) Transgenic nude mouse with ubiquitous green fluorescent protein expression as a host for human tumors. *Cancer Res* **64**, 8651–8656.
6. Kishimoto, H., Kojima, T., Watanabe, Y., Kagawa, S., Fujiwara, T., Uno, F., et al. (2006) In vivo imaging of lymph node metastasis with telomerase-specific replication-selective adenovirus. *Nat Med* **12**, 1213–1219.
7. Bouvet, M., Wang, J., Nardin, S.R., Nassirpour, R., Yang, M., Baranov, E., et al. (2002) Real-time optical imaging of primary tumor growth and multiple metastatic events in a pancreatic cancer orthotopic model. *Cancer Res* **62**, 1534–1540.
8. Katz, M.H., Takimoto, S., Spivack, D., Moossa, A.R., Hoffman, R.M., and Bouvet, M. (2003) A novel red fluorescent protein orthotopic pancreatic cancer model for the preclinical evaluation of chemotherapeutics. *J Surg Res* **113**, 151–160.
9. Katz, M.H., Takimoto, S., Spivack, D., Moossa, A.R., Hoffman, R.M., and Bouvet, M. (2004) An imageable highly metastatic orthotopic red fluorescent protein model of pancreatic cancer. *Clin Exp Metastasis* **21**, 7–12.
10. Tsuji, K., Yang, M., Jiang, P., Maitra, A., Kaushal, S., Yamauchi, K., et al. (2006) Common bile duct injection as a novel method for establishing red fluorescent protein (RFP)-expressing human pancreatic cancer in nude mice. *JOP* 7, 193–199.
11. Hoffman, R.M., and Yang, M. (2006) Subcellular imaging in the live mouse. *Nat. Protocols* **1**, 775–782.
12. Shcherbo, D., Merzlyak, E.M., Chepurnykh, T.V., Fradkov, A.F., Ermakova, G.V., and Solovieva, E.A. (2007) Bright far-red fluorescent protein for whole-body imaging. *Nat Methods* **4**, 741–746.
13. Yang, M., Luiken, G., Baranov, E. and Hoffman, R.M. (2005) Facile whole-body imaging of internal fluorescent tumors in mice with an LED flashlight. *BioTechniques* **39**, 170–172.
14. Hoffman, R.M., and Yang, M. (2006) Whole-body imaging with fluorescent proteins. *Nat. Protocols* **1**, 1429–1438.

Chapter 6

Noninvasive and Real-Time Fluorescence Imaging of Peritoneal Metastasis in Nude Mice

Hayao Nakanishi, Seiji Ito, Makoto Matsui, Yuichi Ito, Kazunari Misawa, and Yasuhiro Kodera

Abstract

Peritoneal metastasis is the most important prognostic factor for gastric and ovarian cancer patients. The protocol in this chapter presents in vivo imaging procedures capable of examining the development of peritoneal metastasis from the micrometastasis stage to the advanced stage. We also describe in vivo imaging procedures for monitoring of antimetastatic agents in nude mice. In vivo imaging systems described consist of green fluorescent protein (GFP) or red fluorescent protein (DsRed) gene-tagged metastatic cancer cell lines and a handy detection device for GFP (or DsRed). This system allows both external, noninvasive, and real-time monitoring of the therapeutic effects of drugs within the animal facility and internal visualization of micrometastases at the cellular level using fluorescence microscopy. Selection of micrometastasis-positive mice and timing of drug administration after injection of tumor cells is critical for accurate evaluation of anti-metastatic efficacy. The present real-time fluorescence imaging system using GFP- and DsRed-tagged metastasis models makes it possible to overcome these problems and therefore is an indispensable tool for preclinical metastasis research and drug discovery.

Key words: Peritoneal metastasis, Micrometastasis, Chemosensitivity, Anticancer drugs, GFP, DsRed, Preclinical study, Gastric cancer, Ovarian cancer

1. Introduction

Although the survival of patients with gastric cancer has improved due to the development of new diagnostic tools, it remains one of the leading causes of cancer death in East Asian countries as well as in some western countries. Peritoneal metastasis accounts for more than 50% of the recurrence after curative surgery and is therefore the most life-threatening lesion in gastric cancer patients (1).

Robert M. Hoffman (ed.), *In Vivo Cellular Imaging Using Fluorescent Proteins: Methods and Protocols*, Methods in Molecular Biology, vol. 872, DOI 10.1007/978-1-61779-797-2_6, © Springer Science+Business Media, LLC 2012

Peritoneal metastasis can cause intestinal obstruction and malignant ascites formation which remarkably compromise the quality of life (QOL) of the affected patients.

Despite advances in therapeutic modalities for peritoneal metastases such as combination chemotherapy (2), conventional chemotherapy has limited antimetastatic efficacy because advanced deposits in the peritoneal cavity are refractory to various chemotherapeutic agents (3). Molecular-targeting therapy using small-molecular weight inhibitors and therapeutic monoclonal antibodies are potential alternatives for conventional chemotherapy. There are only few reported preclinical studies and clinical trials against peritoneal metastasis in gastric cancers (4, 5).

Conventional peritoneal metastasis models pose many difficulties for preclinical studies in that antimetastatic efficacy of the tested drugs cannot be monitored during the experiment until autopsy and micrometastasis cannot be identified without histological examination after sacrifice of the mice, due to the inability to distinguish small metastatic lesions from normal tissue (6). It is also difficult to select micrometastasis-positive mice at the start of the study in a conventional model. Since the successful tumor take rate after i.p. injection of cancer cells does not always reach 100%, and usually remains around 80–90%, such a selection of metastasis-bearing mice is an essential process in performing accurate experiments for peritoneal metastasis. It is therefore of paramount importance to develop peritoneal micrometastasis models which permit one to monitor antimetastatic efficacy of various anticancer agents throughout an experiment in real-time (7).

The use of cancer cell lines transfected with marker genes such as green fluorescent protein (GFP) and red fluorescent protein (RFP) from *Discosoma* sp. (DsRed) (8) enables visualization of metastasis development with no need for other substrates or cofactors in living animals. Hoffman and his colleagues first demonstrated that GFP-tagged metastasis models have great advantages for simple, specific, and sensitive detection of micrometastasis and for monitoring therapeutic effects of anticancer agents on metastasis in living mice (9–12). To date, using such fluorescent protein gene-tagged metastasis models, the antitumor and antimetastatic efficacy of a variety of drugs including molecular-targeting agents on the lung, breast, and gastrointestinal tract-derived cancer xenografts in nude mice have been evaluated (13–16).

Recently, an in vivo imaging system based on bioluminescence has been developed, using cancer cells tagged with luciferase, a light-emitting enzyme of the firefly. Although noninvasive quantitative monitoring of tumor growth can be achieved, this light emitting imaging system is not convenient because of the need for anesthesia, administration of the substrate luciferin, and several minutes to obtain images (17). Based on the capability of convenient and accurate real-time monitoring of metastasis, a GFP-tagged peritoneal metastasis model would be the most useful model for preclinical studies (18).

In this chapter, we describe the detailed protocol based on the authors' in vivo experiments to image metastasis especially in the peritoneal cavity.

2. Materials

2.1. Animals and Drugs

1. 6–8 weeks old male nu/nu nude mice (KSN strain, purchased from Shizuoka Laboratory Animal Center, Hamamatsu, Japan).
2. S-1 (Taiho Pharmaceutical Inc., Tokyo, Japan).
3. Paclitaxel (Bristol-Myers Squibb Japan, Tokyo, Japan).
4. Gefitinib (AstraZeneca, Macclesfield, UK).
5. Trastuzumab (Chugai Pharmaceutical Inc. Tokyo, Japan).
6. Syringe (1 ml), 27-gauge needles (Terumo, Tokyo, Japan).
7. Gastric tube (Natsume Seisakusho Co., Ltd. Tokyo, Japan).

2.2. Cell Lines

1. Metastatic cancer cell lines.
   (a) GCIY (human gastric adenocarcinoma cell line, RIKEN cell bank, Tsukuba, Japan).
   (b) MKN-28 (human gastric adenocarcinoma cell, RIKEN cell bank).
   (c) GLM-1 (human gastric adenocarcinoma cell line, HER2 gene-amplified, established in our laboratory) (4).
   (d) COLM-2, COLM-5 (human colonic adenocarcinoma cell lines, established in our laboratory) (19).
2. GFP-tagged metastatic tumor cell lines.
   (a) GFP-tagged human gastric cancer cell lines (GCIY-EGFP, MKN28-EGFP, and GLM1-EGFP) (20, 21).
   (b) GFP-tagged human colon cancer cell lines (COLM5-EGFP and COLM2-EGFP).
   (c) DsRed-tagged human colon cancer cell lines (COLM5-DsRed) (22).
3. 0.125% Trypsin/1 mM EDTA solution (Gibco, Grand Island, NY).
4. Hank's Balanced Salt Solution (HBSS) (Gibco, Grand Island, NY).

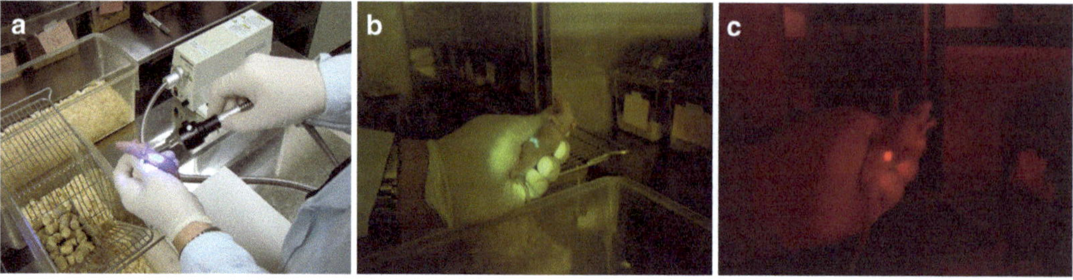

Fig. 1. Photographs of the real-time fluorescence imaging systems used for monitoring peritoneal metastases in live mice. (**a**) A handy GFP detection device, consisting of a halogen lamp with a flexible guide equipped with a band-pass filter and an eyeglass type long-pass cut filter the operator uses in the animal facility. (**b**, **c**) Mice are held in the operator's hand and the metastatic foci in the peritoneum are visualized by external illumination with the *blue* light (*green* fluorescence in **b**) or *green* light excitation (*red* fluorescence emission in **c**). Imaging is performed without anesthesia.

5. Dulbecco's Modified Eagle's Medium (DMEM) (Nissui Pharmaceutical Co., Tokyo, Japan).
6. Fetal bovine serum (Gibco, Grand Island, NY).
7. pEGFP-C1 plasmid or pDsRed2-N1 plasmid (Clontech Laboratories, Inc., Palo Alto, CA).
8. FuGENE6 transfection reagent (Roche Diagnostics, Basel, Switzerland).
9. Geneticin (G418, Wako, Osaka, Japan).

2.3. Detection and Monitoring Device

2.3.1. Device for Real-Time Monitoring of Peritoneal Metastasis in the Animal Facility

A handy GFP detection device, consisting of a 100-W halogen lamp (LG-PS2, Olympus, Tokyo, Japan) with a flexible guide equipped with a band pass filter (420–480 nm for GFP, 545–580 nm for DsRed) at the top and a eyeglass type long-pass cut filter (530 nm for GFP, 640 nm for DsRed, Kodak, Rochester, NY), was used for monitoring the anti-metastatic efficacy of drugs in tumor-bearing mice in the animal facility (see Fig. 1a). Living mice were held in the operator's hand without anesthesia, and the metastatic foci generating GFP or DsRed fluorescence could be easily visualized non-invasively by external illumination of blue light (excitation: 450–490 nm) for GFP (see Fig. 1b) or green light (excitation: 545–580 nm) for DsRed (see Fig. 1c), generated by the high-brightness halogen source. Finger manipulation of the abdominal wall by pushing or rotation improves visualization of micrometastasis in the omentum of the peritoneal cavity.

2.3.2. Device for Observation of Peritoneal Micrometastasis (See Note 1)

A macroimaging system consisting of Macro lens (J6x11, Macro, Tokyo, Japan) with a TV zoom lens (Canon, Tokyo, Japan) and a 150-W high-brightness halogen Lamp (PICL-NEX, NPI, Tokyo, Japan) with dual flexible guides was used to observe peritoneal metastases internally with laparotomy at up to 6.7× magnification (see Fig. 2a, b). Under illumination of blue light produced through a band-pass filter (for excitation: 420–480 nm) from the halogen

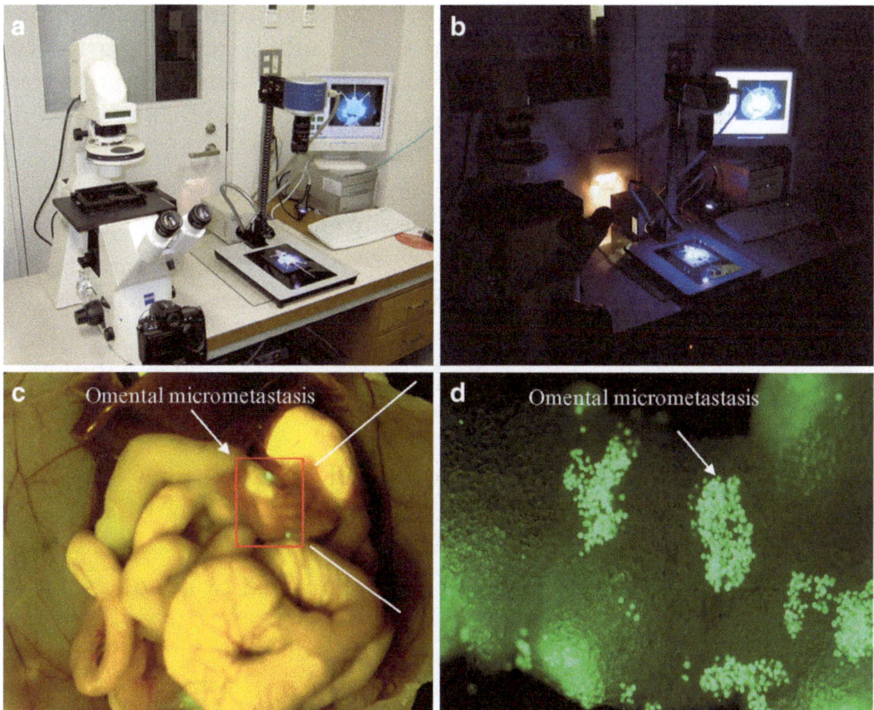

Fig. 2. Photographs of convenient fluorescence imaging devices for observation of peritoneal micrometastasis. (**a**, **b**) A macroimaging system (*right side*) consisting of a TV zoom lens and a 150-W high-brightness halogen lamp with dual flexible guides to observe peritoneal metastases internally with laparotomy at magnifications up to 6.7× under bright light (**a**) and under fluorescence (**b**). (**c**) Under *blue* light illumination, micrometastatic foci emitting GFP fluorescence in the omentum are visualized through a long-pass cut filter. (**d**) An inverted fluorescence microscope (*left side* in (**a**)) is used for observation of micrometastasis at the cellular level in the isolated omentum at high magnification ranging 6.7× to 40×.

source with a flexible guide, micrometastatic foci emitting GFP fluorescence in the peritoneal cavity were visualized through a long-pass cut filter (for emission: 530 nm<) (see Fig. 2c) and recorded on a cooled CCD camera (Penguin 600CL, Pixera, CA, USA). Images of 2,000 × 1,312 pixels were captured on a Windows PC with InStudio software.

For detecting micrometastasis at the cellular level, an inverted fluorescence microscope, Axiovert 200 (Zeiss, Germany) for GFP (DsRed) observation, was also used for observation of the isolated omentum of the mouse at high magnification ranging from 6.7× to 40× (see Fig. 2d).

3. Methods

3.1. Establishment of GFP-Tagged Metastasis Models (See Note 2)

1. GFP (DsRed)-tagged cancer cells are isolated after transfection of cancer cells with the pEGFP-C1 or pDsRed2-N1 plasmids using the FuGENE6 transfection reagent. The

pEGFP-C1or pDsRed2-N1 expression vector consists of the enhanced mutant GFP (DsRed) gene under the control of the cytomegalovirus immediate-early-gene promoter (CMV) and the neomycin-resistance gene (*neo*^{R}), as described previously (18).

2. Transfectants are first isolated in selection medium supplemented with 1.0 mg/ml geneticin.
3. G418-resistant colonies are further screened based on the intensity of their GFP fluorescence using inverted fluorescence microscopy, followed by ring cloning.
4. Resultant cell lines with bright, stable GFP fluorescence (positivity rate more than 80–90%) are used in this study.

3.2. Peritoneal Metastasis Assay

1. Exponentially-growing GFP-tagged cancer cells are harvested by treatment with 0.125% trypsin/1 mM EDTA solution for 5 min at 37°C.
2. The cells are washed with HBSS twice and are resuspended in HBSS at $3–4 \times 10^6$ per 0.4 ml. The cells are then intraperitoneally injected (i.p.) through the abdominal wall into the peritoneal cavity in nude mice using a 27-G syringe.
3. Micrometastasis formation and subsequent metastasis development are monitored in real time from day 1 post-injection throughout the experiment (usually 1–2 months) using a GFP monitoring device as described above (see Fig. 3a, b).

3.3. Preclinical Assessment of Antimetastatic Agents

1. GFP-tagged cancer cells ($3–4 \times 10^6$/0.4 ml HBSS), with peritoneal-metastatic potential, are intraperitoneally injected in mice.
2. One day post-injection, all mice are externally examined noninvasively with the handy GFP detection device to confirm micrometastasis formation on the omentum and mesentery in the peritoneal cavity. Micrometastasis-positive mice are selected. Metastasis-negative mice need to be eliminated from experimental groups, because the tumor take rate after i.p. injection of cancer cells does not always reach 100% (usually 70–90%). Such a selection of implanted mice is an essential process to perform accurate experiments for peritoneal metastasis (see Note 2). The metastasis-bearing mice are then subdivided into groups and administered drugs by either an early or late regimen. The timing of the start of administration profoundly affects the antitumor efficacy of the drugs (21).
3. In the early treatment groups, mice ($n = 5–10$) are administered the drugs from day 1 post-injection and in the late treatment groups, mice ($n = 5–10$) are administered the drug from day 8 at the same dose and schedule. Duration of drug admin-

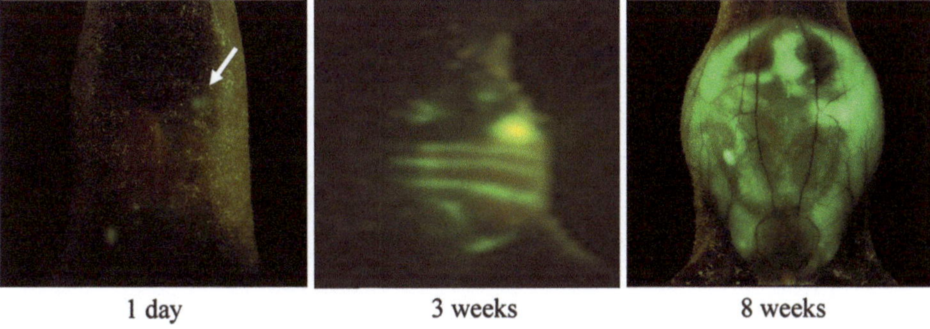

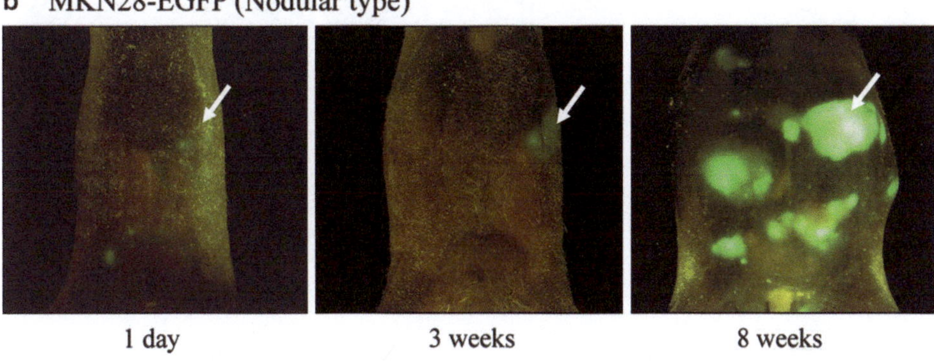

Fig. 3. Visualization of peritoneal metastasis development after intra-peritoneal injection of cancer cells in nude mice. (**a**, **b**) External visualization of peritoneal metastasis 1 day, 3 weeks, and 8 weeks after injection of GCIY-EGFP cells (**a**) and MKN28-EGFP cells (**b**) in live mice. Omental metastasis with GFP fluorescence (*arrows*) is monitored 1 day and 3 weeks post-injection. Production of ascites is demonstrated in a stripe-shaped pattern 3 weeks post-injection and in a diffuse pattern 8 weeks post-injection in (**a**). With a gastric cancer cell line of the nodular type (MKN28-EGFP), no substantial accumulation of ascites is observed throughout the experiment.

istration is usually at least for 4 weeks, but it depends on the cell lines and the aim of the study (see Fig. 4a).

4. Orally-administrative drugs such as S-1, an oral 5-FU derivative and Gefitinib, an EGFR-specific tyrosine kinase inhibitor, are administered with a gastric tube usually five times per week. For intraperitoneally-or intravenously-administered drugs, such as Paclitaxel and Cetuximab (Trastuzumab), a therapeutic monoclonal antibody for EGFR and HER2, respectively, the drugs are diluted in 0.2 ml saline and injected i.p. (i.v.) with a 27-G (30-G) syringe. In the control groups, mice are administered vehicle in the same manner.

5. Mice are monitored externally once a week, without anesthesia, for evaluation of inhibitory efficacy of the drugs on peritoneal metastases, in terms of nodular size and accumulation of ascites fluid using the handy GFP detection device (see Fig. 1a, b).

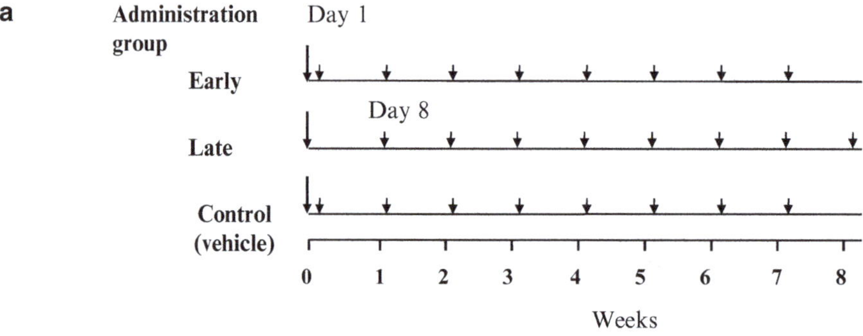

b GCIY-EGFP cells

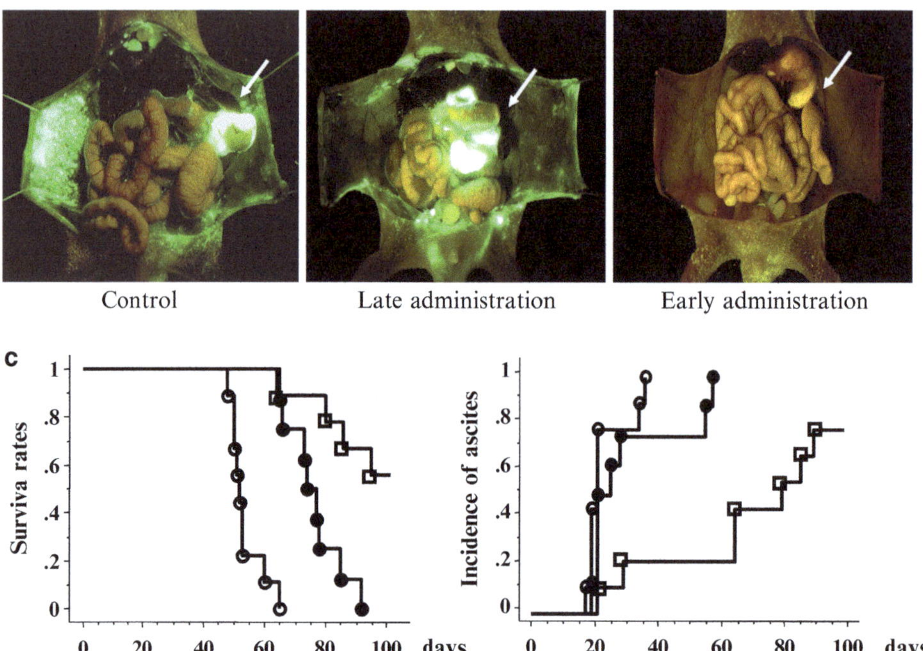

Fig. 4. Efficacy of intra-peritoneal paclitaxel treatment on survival and ascites formation in mice-bearing peritoneal micrometastases and macroscopic metastasis. (**a**) Schematic representation of experimental protocols for assessment of chemosensitivity of peritoneal micrometastasis and advanced metastasis. Mice were injected intra-peritoneally with 3×10^6 GCIY-EGFP cancer cells (*long arrow*) and divided into an early-administration group (start from day 1 post-injection) and a late-administration group (start from day 8 post-injection). Paclitaxel was administered intra-peritoneally to mice at a dose of 25 mg/kg/day or the vehicle once a week (*small arrows*). (**b**) Representative photographs of peritoneal metastasis in mice after early and late intraperitoneal paclitaxel administration. In the control and late-treatment group, accumulation of massive ascites fluid and diffuse metastasis to the entire abdominal cavity are apparent 2 months post-injection (*left* and *center*, respectively). In contrast, no ascites was accumulated and no metastatic nodules remained in the peritoneal cavities of mice treated with early paclitaxel administration (*right*). Photographs show the peritoneal cavity after washing with saline. *Arrows* indicate omentum. (**c**) Overall survival (*left*) and cumulative incidences of ascites accumulation (*right*) in mice with control, early, and late administration. Kaplan–Meier analysis and the log-rank test were used for statistical analysis. Non-treated control group (*open circle*), early-administration group (*open square*), late-administration group (*filled circle*). $P < 0.05$ (early vs. late, survival), $P < 0.05$ (early vs. late, incidence of ascites).

6. Survival of mice is also monitored (see Fig. 4c). In the GCIY-EGFP model, untreated control mice died from massive ascites accumulation and subsequent intra-abdominal hemorrhage within 10 weeks postinjection. Mice treated with the late-administrative regimen survived longer than control, but all died within 12 weeks. In contrast, some mice treated with the early-administrative regimen showed in part a complete regression without residual tumors or tumor dormancy with a prolonged survival (see Fig. 4b) (20).
7. In a separate experiment, mice are sacrificed 10 weeks after injection, and metastatic nodules with GFP fluorescence in the peritoneal cavity are removed. The tumor weights are compared between groups for quantitative assessment of the antitumor efficacy of the drugs (see Fig. 4c).

4. Notes

1. Peritoneal metastasis can be externally visualized mostly in the omentum, mesentery, and peri-gonadal fat from day 1 postinjection of cancer cells using the handy GFP detection device. Omental micrometastasis is less than 1 mm in diameter and cannot be easily recognized, even after laparotomy, without the aid of fluorescence imaging. Isolated cancer cells can be imaged using an inverted fluorescence microscope at the single cell level. However, isolated cancer cells are rarely observed in the omentum, and thus, there is no actual need for routine observation of peritoneal micrometastasis using an inverted fluorescence microscope with high magnification. It takes approximately 7 days after injection for an omental micrometastasis to grow into a metastasis, more than 2 mm in diameter and then visible under bright light.
2. For in vivo studies using a GFP-tagged metastasis model, it is very important to isolate stable transfectants with bright GFP fluorescence, strong enough to allow external monitoring without affecting biological properties. GFP expression of transfected cancer cells should be stable for more than 2 months in vivo. For this purpose, more than 80–90% of transfected cancer cells need to emit bright GFP fluorescence. All steps, including transfection with pEGFP plasmid, G418 selection and further enrichment of GFP-expressing cells affect the efficiency of successful isolation of bright GFP transfectants. Lentivirus-mediated transfection and sorting of cells with strong GFP fluorescence is also recommended. In this study, isolation of bright GFP-positive colonies using a stainless steel ring on a clean bench is used as a more convenient method for enrichment of bright cancer cells.

References

1. Moriguchi, S., Maehara, Y., Korenaga, D., Sugimachi, K., and Nose, Y. (1992) Risk factors which predict pattern of recurrence after curative surgery for patients with advanced gastric cancer. *Surg. Oncol.*, **1**, 341–346.
2. Yonemura, Y., Fujimura, T., Nishimura, G., Falla, R., Sawa, T., Katayama, K., Tsugawa, K., Fushida, S., Miyazaki, I., Tanaka, M., Endou, Y., and Sasaki, T. (1996) Effects of intraoperative chemohyperthermia in patients with gastric cancer with peritoneal dissemination. *Surgery*, **119**, 437–44.
3. Sautner, T., Hofbauer, F., Depisch, D., Schiessel, R., and Jakesz, R. (1994) Adjuvant intraperitoneal cisplatin chemotherapy does not improve long-term survival after surgery for advanced gastric cancer. *J. Clin. Oncol.*, **12**, 970–974.
4. Yokoyama, H., Ikehara, Y., Kodera, Y., Ikehara, S., Yatabe, Y., Mochizuki, Y., Koike, M., Fujiwara, M., Nakao, A., Tatematsu, M., Nakanishi, H. (2006) Molecular basis for sensitivity and acquired resistance to gefitinib in HER2 overexpressing human gastric cancer cell lines derived from liver metastasis. *Br J Cancer.* **95**, 1504–1513.
5. Hara, M., Nakanishi, H., Tsujimura, K., Matsui, M., Yatabe, Y., Manabe, T., and Tatematsu, M. (2008) Interleukin-2 potentiation of cetuximab anti-tumor activity for EGFR overexpressing gastric cancer xenografts via antibody-dependent cellular cytotoxicity. *Cancer Sci.* **99(7)**, 1471–8.
6. Yanagihara,K,. Takigahira, M,. Tanaka,H,. Komatsu, T,. Fukumoto, H,. Koizumi,F,. Nishio, K,. Ochiya, T,. Ino, Y,. Hirohashi,S. (2005) Development and biological analysis of peritoneal metastasis mouse models for human scirrhous stomach cancer. *Cancer Sci.* **96(6)**, 323–32.
7. Nakanishi, H., Ito, S., Mochizuki, Y., Tatematsu, M. (2005) Evaluation of chemosensitivity of micrometastases with green fluorescent protein-tagged tumor models in mice. In Chemosensitivity for Methods in Molecular Medicine [Blumenthal, R. D.,ed.].Humana Press, Totowa, NJ. Vol **111**, 351–362.
8. Misteli, T., and Spector, D.L. (1997) Applications of the green fluorescent protein in cell biology and biotechnology. *Nat. Biotechnol.* **15**, 961–964.
9. Chishima, T., Miyagi, Y., Wang, X., Yamaoka, H., Shimada, H., Moossa, A. R., Hoffman, R. M. (1997) Cancer invasion and micrometastasis visualized in live tissue by green fluorescent protein expression. *Cancer Res.* **57**(10), 2042–2047.
10. Yang, M., Baranov, E., Jiang, P., Sun, F. X., Li, X. M., Li, L., Hasegawa, S., Bouvet, M., Al-Tuwaijri, M., Chishima, T., Shimada, H., Moossa, A. R., Penman, S., Hoffman, R. M. (2000) Whole-body optical imaging of green fluorescent protein-expressing tumors and metastases. *Proc Natl Acad Sci USA.* **97 (3)**, 1206–1211.
11. Hoffman, R. M. (1999) Orthotopic transplant mouse models with green fluorescent protein-expressing cancer cells to visualize metastasis and angiogenesis. *Cancer Metastasis Rev.* **17**, 271–277.
12. Hoffman, R. M. (2002) Green fluorescent protein imaging of tumour growth, metastasis, and angiogenesis in mouse models. *Lancet Oncol.* **3(9)**, 546–56.
13. Hoffman, R.M. (2005) In vivo cell biology of cancer cells visualized with fluorescent proteins. *Curr Top Dev Biol.* **70**, 121–44
14. Zhou, J., Yu, Z., Zhao, S., Hu, L., Zheng, J., Yang, D., Bouvet, M., Hoffman, R.M. (2009) Lentivirus-based DsRed-2-transfected pancreatic cancer cells for deep in vivo imaging of metastatic disease. *J Surg Res.* **157(1)**, 63–70.
15. Yang, M., Jiang, P., Hoffman, R.M. (2007) Whole-body subcellular multicolor imaging of tumor-host interaction and drug response in real time. *Cancer Res.* **67(11)**, 5195–200.
16. Yokoyama, H., Nakanishi, H., Kodera,Y., Ikehara, Y., Ohashi, N., Ito, Y., Koike, M., Fujiwara, M., Tatematsu, M., Nakao, A. (2006) Biological significance of isolated tumor cells and micrometastasis in lymph nodes evaluated using a green fluorescent protein (GFP)-tagged human gastric cancer cell line. Clin. Cancer Res. **12(2)**; 361–68.
17. Buchhorn, H.M., Seidl, C., Beck, R., Saur, D., Apostolidis, C., Morgenstern, A., Schwaiger, M., Senekowitsch-Schmidtke, R. (2007) Non-invasive visualisation of the development of peritoneal carcinomatosis and tumour regression after 213Bi-radioimmunotherapy using bioluminescence imaging. *Eur J Nucl Med Mol Imaging.* **34(6)**, 841–9.
18. Nakanishi, H., Mochizuki, Y., Kodera, Y., Ito S., Yamamura, Y., Ito, K., Akiyama, S., Nakao, A., Tatematsu, M. (2003) Chemosensitivity of peritoneal micrometastases as evaluated using a green fluorescence protein (GFP)-tagged human gastric cancer cell line. *Cancer Sci.* **94(1)**, 112–118.
19. Ito, Y., Nakanishi, H., Kodera, Y., Hirai, T., Nakao, A., Kato, T. (2010) Characterization of a novel lymph node metastasis model from human colonic cancer and its preclinical use for comparison of anti-metastatic efficacy between oral S-1 and UFT/LV. *Cancer Sci.* **101(8)**, 1853–1860.

20. Nakanishi, H., Ito, S., Mochizuki,Y., Ohashi, N., Ikehara, Y., Yamamura, Y., Kodera,Y. (2007) Peritoneal carcinomatosis; A new diagnostic and therapeutic strategy. In Research Focus on Gastric Cancer. Edited by Cardinni DC., Nova Scientific Publishers. pp. 55–73.

21. Ohashi, N., Kodera, Y., Nakanishi, H., Yokoyama, H., Fujiwara, M., Koike, M., Hibi, K., Nakao, A., Tatematsu, M. (2005) Efficacy of intraperitoneal chemotherapy with paclitaxel targeting peritoneal micrometastasis as revealed by GFP-tagged human gastric cancer cell lines in nude mice. *Int. J. Oncol.* **27(3)**, 637–44.

22. Nakanishi, H., Hara, H., Ikehara, Y., Tatematsu, M., (2007) Non-invasive and real-time monitoring of molecular targeting therapy for lymph node and peritoneal metastasis in nude mice bearing xenografts of human colorectal cancer cells tagged with GFP and DsRed, *Proceeding of SPIE*, **6449**; 644910–1-9.

Chapter 7

Three-Dimensional In Vivo Imaging of Tumors Expressing Red Fluorescent Proteins

Alexander P. Savitsky, Irina G. Meerovich, Victoria V. Zherdeva, Lyaysan R. Arslanbaeva, Olga S. Burova, Darina V. Sokolova, Elena M. Treshchalina, Anatoly Yu Baryshnikov, Ilya I. Fiks, Anna G. Orlova, Michael S. Kleshnin, Ilya V. Turchin, and Alexander M. Sergeev

Abstract

3D imaging of genetically-engineered fluorescent tumors enables quantitative monitoring of tumor growth/regression, metastatic processes, including during anticancer therapy in real-time.

Fluorescent tumor models for 3D imaging require stable expression of genetically encoded fluorescent proteins and maintenance of the properties of tumor cell line including growth rate, morphology, and immunophenotype.

In this chapter, the protocol for 3D imaging of tumors expressing red fluorescent protein are described in detail.

Key words: Lipofection, Stable transfection, Immunophenotyping, Red fluorescent proteins, Solid tumor model, Photodynamic therapy

1. Introduction

The use of fluorescent proteins enables continuous visual monitoring of multiple processes in live intact organisms at the molecular and single cell level. The genes for fluorescent proteins can be introduced to any type of organism.

Experimental oncology is the most promising field for the application of molecular sensors based on fluorescent proteins. After introduction of corresponding genes into cancer cells, the cells can stably synthesize fluorescent proteins continuously. The method permits carrying out real-time, continuous visual monitoring

Robert M. Hoffman (ed.), *In Vivo Cellular Imaging Using Fluorescent Proteins: Methods and Protocols*, Methods in Molecular Biology, vol. 872, DOI 10.1007/978-1-61779-797-2_7, © Springer Science+Business Media, LLC 2012

of growth and expansion of tumors, angiogenesis, and drug response, even without using anesthesia or immobilization of animals. The high sensitivity of detection that is achieved is due to the brightness of fluorescent proteins in live organisms. Good spatial resolution of images enables observation of early responses of tumors to drug therapy when the tumor size and even the existence of the tumor cannot be detected by usual methods. Imaging with fluorescent proteins allows one to not only obtain images of a whole body of a laboratory animal but also to investigate carcinogenesis processes at the single-cell level due to the spatial resolution of instruments (1). Cell lines expressing fluorescent proteins maintain their original properties (2–4), in particular, cell proliferation rate and metastatic behavior. Cancer cells with stable expression of fluorescent proteins can be noninvasively visualized in multiple internal organs, including liver, lungs, brain, skeletal system, and lymph nodes (5, 6).

For in vivo studies of cancer, it is necessary to control key stages such as effective gene delivery, obtaining a cell line with stable expression of fluorescent proteins, preparation of tumor models, and verification of immunophenoptype of the original cell line.

1.1. Preparation of Cancer Cells with Stable Expression of Fluorescent Proteins

Preparation of cancer cells with stable expression of fluorescent protein consists of the following steps: delivery of gene encoding the fluorescence protein, selection of transduced cells, preparation and production of clones.

First, it is necessary to choose an effective gene delivery method such that the transfected gene is stably expressed during cell division in the absence of selective medium. The level of protein expression should be high enough for detection in vivo. The transfected cell line should maintain the properties of the initial cell line in vitro (cell morphology, division rate) and in vivo (metastatic behavior, invasive ability). Effective gene delivery and appropriate conditions for high gene expression are crucial conditions for successful imaging in vivo.

Lipofection is an effective chemical method of transfection. Lipofection is based on the ability of cationic lipids to have electrostatic interaction with negatively-charged surfaces of cell membranes and macromolecules and on the hydrophobic properties of lipids. Due to these properties, lipids can easily penetrate into tissues and cells (7, 8). During lipofection, DNA spontaneously binds to the external surface of cationic phosphor lipid vesicles, or liposomes, neutralizing inherent negative charge, and the resulting complexes (genosomes) interact with the cell membrane (9). An important step for DNA delivery is its transport from the cytoplasm to the nucleus. DNA in the cytoplasm can be destroyed by nucleases. The vector penetrates into a nucleus either during mitosis or through nuclear pores (10). Only approximately 1% of the introduced gene vectors penetrated into the cell, gets to the nucleus, and begins to express (11). The method can be used for the transduction of more than 90% of cell lines. As compared with viral transfection, this method is less specific (12).

The optimization of liposomal transfection procedure is necessary for each cell line. Transfection efficiency depends on such parameters as liposome morphology, composition and ratio of different liposome-forming lipids (12), and time and conditions of incubation with cells (13). Depending on these conditions, the transfection efficiency may vary within three orders of magnitude (14). Cationic liposomes are the most effective nonviral means of delivery of genes to eukaryotic cells and this is the reason for their wide application. The efficiency of transfection with liposomes varies from less than 1% to 40% depending on the plasmid, lipoplex composition, and cell line. Expression in vitro can remain stable for at least 10–20 passages (15).

1.2. Tumor Models

In vivo tumor models include subcutaneous, orthotopic, and genetically-engineered mouse models. Fluorescent models allows imaging of molecular process in real time with cellular and subcellular resolution (16, 17).

Human tumors can be successfully transplanted to mice and rats with the *nu/nu* (nude) mutation. The absence of a thymus results in the development of immunodeficiency in nude mice and rats and, therefore, in the successful grafting of heterologous tumors such as human cancer cell lines or patient tissues.

To obtain an adequate model of fluorescent tumors in laboratory animals, it is important to confirm that the expression of the fluorescent protein has no influence on the basic properties of the cancer cell line, particularly, on the growth rate, morphology, including immunophenotype of the cells.

If the fluorescent tumor model is intended to be used in the studies of antitumor action of drugs that are delivered to the tumor by antibody-mediated target transport (conjugates of drugs with antibodies, immunoliposomes, etc.), information about antigens presented on the surface of the cells of the fluorescent cell line is necessary.

1.3. Whole-Body Imaging of Small Animals

There are several methods of whole-body imaging of small animals with tumors labeled with fluorescent proteins including (1) fluorescence reflectance imaging (or epi-illumination imaging), (2) projection imaging, and (3) fluorescence-diffuse tomography (FDT). Each of the methods possesses a number of advantages which make them optimal for a wide range of biological applications (utilization of different fluorophores, accuracy of tumor size measurement, and duration of the experiment).

Fluorescence reflectance imaging allows detection and investigation of tumors by illumination with a broad homogenized light beam exciting fluorescence (epi-illumination mode). Tumor fluorescence is detected by a CCD camera with filters blocking the excitation light. Fluorescence reflectance imaging allows one to estimate lateral dimensions of tumors in a short time (1–5 s). This technique is widely used due to its simplicity and is currently used in a number of commercially available systems. The main disadvantage of fluorescence reflectance imaging is that the accuracy of tumor size estimation, in the case of deep-seated tumors (e.g., orthotopically implanted tumors), decreases due to strong scattering of the light propagating through biological tissue (18). The use of the transillumination, as well as FDT, allows one to overcome this limitation (19).

In the transillumination method, the investigated object is placed between the sources and the detector. For deep-seated tumors, the quality of images acquired by this method is higher than the one achieved using the reflectance technique; however, the technique is more complicated and the duration of image acquisition may take several minutes. In the case of transillumination, as compared with fluorescence reflectance imaging, the dependence of resolution on the depth of tumor localization is less since the emission and excitation light propagate through the animal body.

The FDT method (20–24) allows 3D reconstruction of the fluorescent tumor using information of the emitted light in various projections, achieved by changing the relative position of the source exciting the fluorescence and the detector registering fluorophore emission. 3D imaging is more complicated due to requirements for both the light source and the detector, such as the power of the source, sensitivity, and dynamic range. There is also a need for a scanning system and a 3D reconstruction program. Since photon trajectories are random in the scattering medium (they are not straight as in X-ray tomography), the problem of reconstruction of the fluorophore distribution is complicated. The inverting matrix is sparse and ill-conditioned and requires the development of a specific algorithm taking into account the particularities of light propagation. FDT provides the most accurate information about fluorescing regions in tissues; in particular, it allows the quantitation of dynamic changes in tumor size.

Excitation of currently known and available fluorescent proteins with bright emission takes place in the wavelength range of 400–600 nm (25). Strong light attenuation and high autofluorescence in this wavelength range considerably complicates the task of deep fluorescing imaging (21). Therefore, in such systems, high levels of power density of laser radiation, as well as highly sensitive detecting systems are used. Plane geometry is preferable, because it allows minimizing the thickness of the biotissue. Thus, the decay of the light propagated through the mouse is less than that of a cylindrical structure. Time necessary for scanning highly depends on fluorescence intensity, number of channels used, and number of projections necessary for reconstruction.

2. Materials

2.1. Preparation of Cancer Cell Lines with Stable Expression of Fluorescent Protein

2.1.1. Transfection

1. Human melanoma cell line melKor (Department of Experimental Diagnostics and Therapy of Tumors, N.N. Blokhin Russian Oncological Center, Russian Cell Culture Collection (Vertebra)).
2. The pTurboRFP-C plasmid (26) (Evrogen, Russia).
3. Endotoxin Free-Maxi Kit (Qiagen).
4. Lipofectamine 2000 (Invitrogen).
5. Opti-MEM I medium (Invitrogen, USA).
6. DMEM.
7. Fetal bovine serum (Hy Clone, USA).
8. Dexamethasone.

2.1.2. Selection and Cloning

9. Geneticin antibiotic (G-418) (Invitrogen).
10. Ham's F12 medium.
11. DMEM.
12. Fetal bovine serum (Hy Clone, USA).
13. Trypsin–EDTA solution 0.25%.
14. Cloning cylinders (Nunc).
15. Petri dishes.

2.1.3. PCR

1. PCR primers (sense 5′ AAGCTGTACATGGAGGGCACCG 3′ primer and antisense 5′ CCGGGCATCTTGAGGTTCTTA 3′ primer for TurboRFP).
2. 10× buffer with Mg^{2+}, ProofStart DNA Polymerase, dNTP mix (Qiagen).
3. Thermocycler.

2.1.4. Growth of Fluorescent Cancer Cell Lines

1. 24-Well plate.
2. Trypsin–EDTA solution 0.25%.
3. DMEM.
4. Fetal bovine serum (Hy Clone, USA).

2.1.5. Use of mel Kor-TurboRFP Fluorescent Cells for In Vitro Studies of Anti-tumor Action of Drugs

1. Culture mel Kor-TurboRFP cells to log-phase growth.
2. 96-well tissue culture plates (Corning Costar).
3. 15-mL centrifuge tubes (Corning Costar).
4. Trypsin–EDTA solution 0.25%.
5. DMEM.
6. Complete growth medium (DMEM supplemented with 10% fetal bovine serum (Bioclot), 100 U/mL penicillin, 100 μg/mL streptomycin).
7. Dulbecco's phosphate-buffered saline (DPBS).
8. Lisomustine—antitumor drug based on mixture of isomers of nitrosourea derivatives, stock solution in 0.9% NaCl, sterilized by filtration with a 0.22-μm filter.
9. Cyclophosphamide, stock solution in 0.9% NaCl, sterilized by filtration with a 0.22-μm filter.
10. Viability stain (0.5% trypan blue).
11. Fluorescein diacetate, 1,000× stock solution in DMSO.
12. Fluoroscan II fluorimeter (Labsystems, Finland).

2.2. Preparation and Characterization of Tumor Model

2.2.1. Preparation of Fluorescent Cancer Cells for Xeno-transplantation

1. Balb/c nu/nu mice, females, 7–10 weeks old (Shemyakin and Ovchinnikov, Institute of Bioorganic Chemistry of the Russian Academy of Sciences, Russia).
2. Animals used in these experiments were kept in a barrier facility under HEPA filtration. Mice were fed with autoclaved laboratory rodent diet (Shemyakin and Ovchinnikov, Institute of Bioorganic Chemistry of the Russian Academy of Science). Animal studies were conducted in accordance with "Guidelines for experimental (preclinical) studies of new pharmacological substances," issued by the Ministry of Health of the Russian Federation, Department of Quality Control, Effectiveness and Safety of Remedies, Scientific Center of Examination and the State Control of Medical Products, Pharmacological State Committee of the Russian Federation (Moscow, 2000).
3. Culture mel Kor-TurboRFP cells to log-phase growth.
4. Sterile 1.5-mL tubes (Eppendorf).
5. 15-mL centrifuge tubes (Corning Costar).
6. Trypsin–EDTA solution 0.25% (Paneco).
7. Cold Dulbecco's phosphate buffered saline (DPBS).
8. Viability stain (0.5% trypan blue).

2.2.2. Transplant of Fluorescent Tumors Using Matrigel

1. Balb/c nude mice, heterozygotes, females, 8 weeks old.
2. Culture mel Kor-TurboRFP cells into log-phase growth.
3. Cooled aliquots of Matrigel (BD).

4. Cooled sterile 1.5-mL tubes (Eppendorf).
5. 15-mL centrifuge tubes (Corning Costar).
6. Trypsin–EDTA solution 0.25%.
7. Cold DPBS.
8. Viability stain (0.5% trypan blue).
9. Cooled pipettes and tips.
10. Cooled 1-mL sterile syringes (BD).
11. Ice.

2.2.3. Immuno-phenotyping of the Fluorescent Tumor Model

Preparation of Cell Suspension of Cultured mel Kor-TurboRFP Cells

1. Culture mel Kor-TurboRFP cells to log-phase of growth.
2. 15-mL centrifuge tubes (Corning Costar).
3. Trypsin–EDTA solution 0.25%.
4. PBS.
5. Viability stain (0.5% trypan blue).

Preparation of Cell Suspension of mel Kor-TurboRFP Cells from the Fluorescent Tumor

1. nu/nu mouse with subcutaneous fluorescent tumor.
2. "Zoletil 100" (Virbac Sante Animale), mixture of 250 mg tiletamin hydrochloride and 250 mg zolazepam hydrochloride, stock solution in distilled water, sterile.
3. 15-mL centrifuge tubes (Corning Costar).
4. Sieves (1, 100, 20 μm).
5. Disposable plastic syringes.
6. PBS.
7. Viability stain (0.5% trypan blue).

Analysis of Immuno-phenotype of mel Kor-TurboRFP Cell Suspensions by FACS

1. Suspension of mel Kor-TurboRFP cells.
2. PBS, pH 7.4.
3. FITC-labeled sheep antiserum against murine immunoglobulins ("MedbioSpektr," Russia).
4. Anti-CD63 FITC-labeled antibodies (Serotec), anti-HLA-DR APC-labeled antibodies (Serotec), anti-HLA-ABC antibodies.
5. 1% Formalin in PBS.
6. FACSCalibur flow cytometer (BD).

2.3. 3D Imaging

1. nu/nu mice with fluorescent tumors.
2. Fluorescent imaging setup (diffuse fluorescent tomograph).
3. Anesthetics Zoletil and 2% Rometar.

2.3.1. Description of the Fluorescence Imaging Setup

In Fig. 1, the setup for small-animal fluorescence imaging which combines reflectance technique with FDT, for imaging superficial tumors as well as deep-seated tumors, is shown. For excitation of

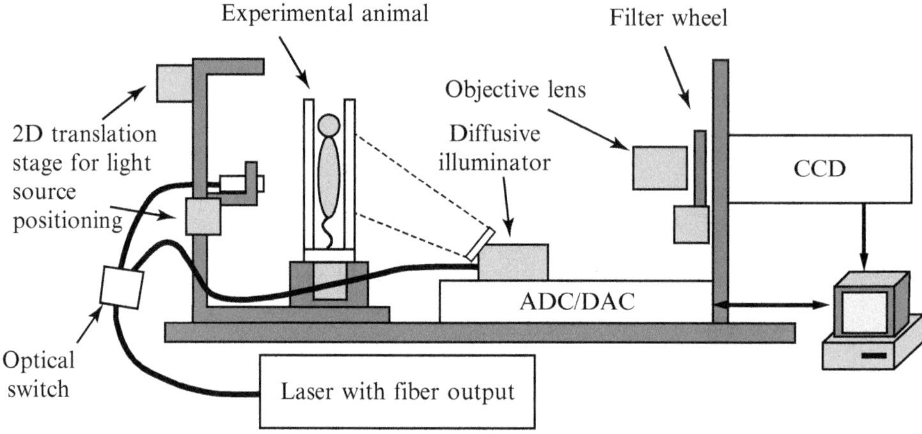

Fig. 1. Fluorescence imaging setup.

red fluorescent proteins, such as DsRed2 and TurboRFP, a Nd:YAG laser with second harmonic generation (the wavelength of 532 nm) is used. A cooled (up to −65°C) Hamamatsu CCD detector is used. An optical switch allows one to use the excitation light in reflectance configuration (for reflectance imaging) as well as in transillumination one (for tomography).

2.3.2. 3D Reconstruction

1. Software based on algorithm employing the combination of hybrid model of light propagation, Monte Carlo simulations, and algebraic reconstruction.

3. Methods

3.1. Preparation of Cancer Cell Line with Stable Expression of Fluorescent Protein

3.1.1. Transfection

1. Produce 5–10 μg of plasmid in *E. coli* cells, DH5α strain, and purify using Endotoxin Free-Maxi Kit (Qiagen) (see Note 1).
2. Seed 200×10^3 cells in log-phase on each well of a 24-well plate. Use complete growth medium (DMEM containing 10% fetal serum (see Note 2)).
3. Dilute plasmid (from 0.4 to 3.2 μg) in 150 μL Opti-MEM I.
4. Dilute Lipofectamine 2000 (from 1 to 4 μL) in 150 μL Opti-MEM I.
5. Mix 150 μL diluted DNA and 150 μL diluted Lipofectamine to obtain different ratios of DNA:Lipofectamine concentrations and incubate for 20 min at room temperature.
6. Add 100 μL Lipofectamine 2000–DNA complex and 0.5-mL culture medium without antibiotics to cells in each of three wells in 24-well plates.

7. Incubate the cells at 37°C in CO_2-incubator overnight, then change medium to complete growth medium.
8. Image the expression of fluorescent proteins using fluorescence microscopy, choose wells with viable fluorescent cells.

3.1.2. Selection and Cloning (See Note 3)

1. Add geneticin antibiotic (G-418) at 800 μg/mL, change to complete medium with G-418 every 4 days for 2 weeks (see Note 4).
2. Trypsinize and seed viable cells in complete medium at low density, approximately 100–200 cells per Petri dish in enriched Ham's F12 medium (see Note 5).
3. Use cloning cylinders for trypsinization of fluorescent clones.
4. Maintain selected clones for at least 10–15 passages to ensure stable fluorescence (see Note 6).

3.1.3. Determination of Plasmid Copy Number in Transfected Cancer Cells

1. Isolate DNA by boiling in water for 5 min (1,000 cells in 1 μL).
2. Centrifuge for 1 min at 12,000 × *g*.
3. Prepare the calibrators: add solution of pTurboRFP-C plasmid to the non-transfected cells to obtain samples containing 1, 5, 25, and 125 plasmids per cell. Use nontransfected cells as negative control.
4. Prepare the PCR mixtures containing calibrators described above and samples with unknown quantity of plasmid copies, sense and antisense primers, ProofStart DNA Polymerase, dNTPs with final concentration given in Table 1.
5. Perform the assay under temperature conditions given in Table 2.
6. Analyze the PCR products by electrophoresis in 1% agarose gel (see Fig. 2).

Table 1
Composition of PCR mixtures

Component	Final concentration
Sterile water	
10× buffer with Mg^{2+}	1×
Direct primer	1 μM
Reverse primer	1 μM
dNTPs	300 μM
Matrix	1,000 cells
ProofStart DNA Polymerase	2.5 Units

Table 2
Temperature conditions for PCR

Step	Time	Temperature (°C)
Prolonged denaturation	5 min	95
3× stage cycle		
Denaturation	30 s	94
Annealing	15 s	60
Elongation	45 s	72

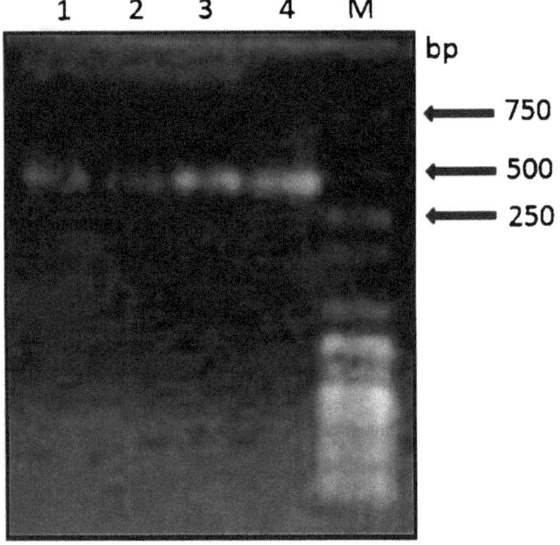

Fig. 2. Agarose gel electrophoresis of PCR products: (*1*) individual clone of mel Kor-TurboRFP cell line after selection with geneticin, (*2*) individual clone of cell line mel Kor-TurboRFP after selection with geneticin and after 20 passages of cells in non-selective medium, (*3*) mel Kor cells mixed with plasmid pTurboRFP-C plasmid in equimolar quantities, (*4*) mel Kor cells mixed with pTurboRFP-C plasmid at a ratio of 1:5 M, DNA markers.

3.1.4. Estimation of Growth Rate of Fluorescent Cells

1. Seed 200×10^3 cells in log-phase in each well of a 24-well plate phase in complete growth medium (DMEM containing 10% fetal serum).
2. Trypsinize and count cells every 8–14 h.
3. Plot data as cell number per time.
4. Compare to growth curve of non-fluorescent cell lines. The growth curves should be similar.

3.1.5. Use of mel Kor-TurboRFP Fluorescent Cells for In Vitro Studies of Anti-tumor Action of Drugs

1. Seed cells in log-phase growth in complete growth medium (DMEM containing 10% calf fetal serum, 100 U/mL penicillin, 100 μg/mL streptomycin) in 96-well culture plates (except the last vertical and horizontal rows) to obtain 80% confluence at the time of drug testing.

2. Replace the growth medium with growth medium containing different concentrations of an antitumor agent (each concentration has six replicas).
3. Incubate the cells overnight in a CO_2-incubator, replace medium with complete growth medium, and incubate the cells for another 24 h.
4. Wash wells with DPBS three times, add 100 μL DPBS to each well of the plate.
5. Measure fluorescence using a fluorescence plate reader with an interference filter with maximum spectral wavelength at 544 nm for fluorescence excitation and a filter with maximum wavelength at 607 nm for emission.
6. Test cells for viability using a fluorescein diacetate derivative (see Note 7).
7. Determine the number of surviving cells using a calibration curve, which was constructed by correlating the fluorescence signal to the amount of seeded cells.

3.2. Preparation and Characterization of Tumor Model

3.2.1. Preparation of Fluorescent Tumors In Vivo. Xenotransplantation. Monitoring of Tumor Growth

1. Prepare a cell suspension of fluorescent cells by trypsinization of cells with 3 mL Trypsin–EDTA followed by the addition of cold DPBS, centrifugation at $100 \times g$ for 5 min, and resuspension in cold DPBS.
2. Stain an aliquot of cells with an equal volume of trypan blue, load the mixture to the counting chamber of a hematocytometer, count the number of viable cells, and calculate the cell concentration in the suspension.
3. Wash the cell suspension twice with cold DPBS followed by centrifugation at $100 \times g$ for 5 min.
4. Resuspend cells in 100–200 μL of cold DPBS (per one animal). Keep suspension on ice before inoculation.
5. Mix cooled cell suspension with Matrigel (see Note 8) at a 1:2 ratio upon cooling on ice and using cooled pipettes, pipette tips, etc.
6. Inject the mixture subcutaneously in the dorsal flank of the mice, $1–3 \times 10^6$ cells per mouse.
7. Monitor the growth rate of nonfluorescent and fluorescent tumors. Measure tumor sizes (using calipers). Calculate the volume by (1):

$$V = \pi abc / 6, \qquad (1)$$

where a, b, c are tumor dimensions (see Fig. 3).

3.2.2. Immuno-phenotyping of the Fluorescent Tumor

Expression of red fluorescent proteins does not change their immuno-phenotype both in vitro and in vivo. Data on the expression

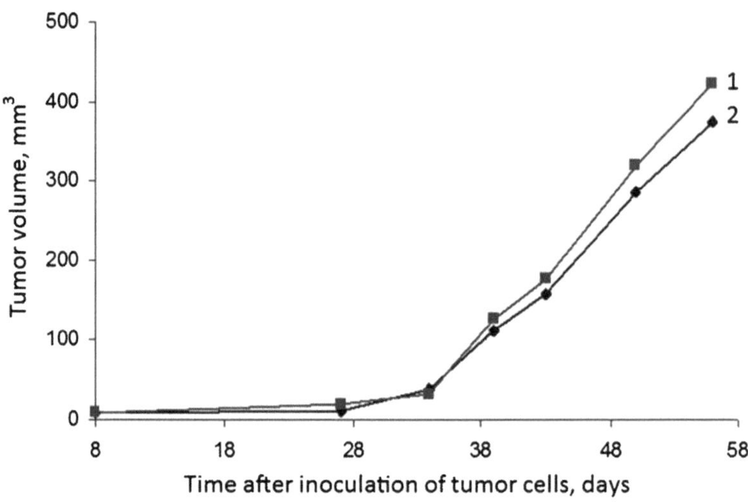

Fig. 3. Dynamics of tumor growth after implantation of three million mel Kor cells (*1*) and mel Kor-TurboRFP cells (*2*) in nu/nu mice.

Table 3
Expression level of antigens in cells of mel Kor-TurboRFP cell line (see Note 9)

Antibodies against antigen/label	Secondary antibodies/label	Expression of antigens by mel Kor line	Expression level of antigens in mel Kor-TurboRFP cell line (measured by fluorescence flow cytometry) (%)	
			In vitro cultured cells	Cells isolated from fluorescent tumor
CD63/FITC		+	96.0	97.7
Anti-HLA-DR/APC		–	4.8	5.9
Anti-HLA-ABC	$F(ab)_2$/FITC	–	1.5	6.7

level of antigens in the mel Kor-TurboRFP cell line, both cultured and isolated from tumors, are given in Table 3. mel Kor-TurboRFP cells express the CD63 melanoma cell marker that demonstrate the specificity of this cell line. Another characteristic of mel Kor-TurboRFP cells is the absence of histocompatibility antigens of first and second class. This property is rare in melanoma cells but this characterizes the mel Kor cell line (27).

Preparation of Cell Suspension of Cultured mel Kor-TurboRFP Cells

1. Prepare a cell suspension of mel Kor-TurboRFP cells by trypsinization of cells with 3 mL trypsin–EDTA followed by the addition of DMEM, centrifugation at 100 × *g* for 5 min, and resuspension in complete growth medium.
2. Stain an aliquot of the cells with an equal volume of trypan blue, load the mixture to the counting chamber of a hemato-

cytometer, count the number of viable cells, and calculate the cell concentration in the suspension.

Preparation of Suspension of mel Kor-TurboRFP Cells from the Fluorescent Tumor

1. Sacrifice the mouse by intraperitoneal injection of “Zoletil” at a lethal dose.
2. Cut the tumors from the dead mice and wash with PBS.
3. Homogenize tumor to obtain a suspension of cells using mechanical disaggregation and sieving.
4. Stain an aliquot of cells with an equal volume of trypan blue, load the mixture to the counting chamber of a hematocytometer, count the number of viable cells, and calculate the cell concentration in the suspension.

Indirect-Surface Immunofluorescence Reaction

1. Wash mel Kor-TurboRFP cells (5×10^5) with PBS.
2. Incubate cell suspension with 20 μL anti-HLA-ABC monoclonal antibodies for 30 min at room temperature.
3. Wash the cells with PBS, using centrifugation at 100 × *g* for 7 min.
4. Incubate the cells with 20 μL fluorescently-labeled sheep antiserum against murine immunoglobulins for 30 min at 4°C.
5. Wash the cells twice with PBS.
6. Resuspend the cells in PBS containing 1% formalin.
7. Use unlabeled cells fixed with 1% formalin and cells after incubation with F(ab)2-FITC and fixation with 1% formalin as controls.
8. Analyze the cell suspensions with a flow cytometer.
9. Calculate the expression level of antigens using CellQuest software.

Direct-Surface Immunofluorescence Reaction

1. Wash mel Kor-TurboRFP cells (5×10^5) with PBS.
2. Incubate cell suspension with 20 μL monoclonal antibodies (CD63/FITC or Anti HLA-DR/APC) for 30 min at room temperature.
3. Wash the cells twice with PBS.
4. Resuspend the cells in PBS containing 1% formalin.
5. Use unlabeled cells fixed with 1% formalin as controls.
6. Analyze the cell suspensions with a flow cytometer.
7. Calculate the expression level of antigens using CellQuest software.

3.3. 3D Imaging

1. Place the animal in a container consisting of a supporting plate and a covering glass plate that is slightly pressed to fix the animal. The distance between the plates is about 1–1.5 cm.
2. Make measurements using DFT-3.
3. Obtain 3D reconstruction data by appropriate software.

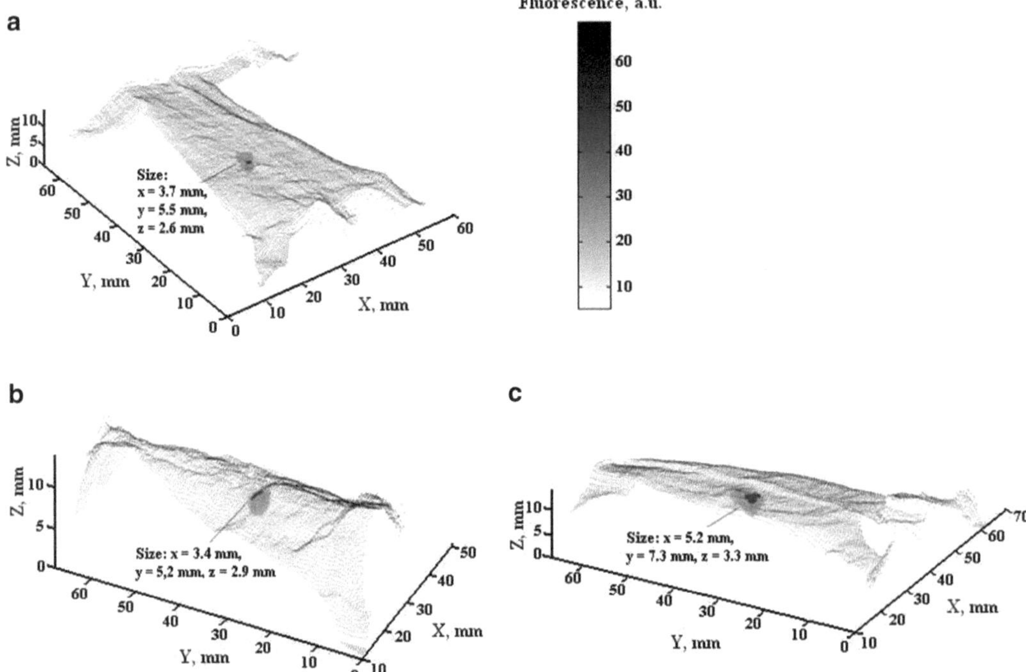

Fig. 4. 3D reconstruction of images of fluorescent tumors in mice. mel Kor-TurboRFP tumors were established by subcutaneous inoculation of a fluorescent cancer cell suspension into the dorsal flank. (**a**) Tumor before PDT, (**b**) tumor at 17 days after PDT, and (**c**) control tumor at the same time point as (**b**).

We studied the antitumor action of photodynamic therapy (PDT) in mice with mel Kor-TurboRFP tumors. The photosensitizer "Photosence" was injected intravenously in the tail vein of nu/nu mice with mel Kor-TurboRFP tumors at a dose of 1 mg/kg bodyweight. After 22 h, mice were anesthetized by injection of a mixture of "Zoletil," (50 mg/kg), and 10 μL of 2% Rometar solution. 15–20 min after anesthesia of the animals, laser irradiation (671 nm, power capacity 150 mW/cm^2) was carried out.

Photodynamic action was estimated by the change in tumor size, which was determined by mechanical measurement of the tumor with calipers, and by analysis of fluorescent images of the tumor. The results demonstrated that growth of tumor ceases 1 week after photodynamic therapy. The fluorescence field of the mel Kor-TurboRFP tumor did not increase.

After 17 days, the tumor treated with PDT was substantially smaller than tumors in mice not treated with PDT (see Fig. 4). Thus, diffusion fluorescence tomography allows noninvasive monitoring of tumor growth and analysis of tumor response to antitumor therapy.

In order to obtain data for 3D reconstruction of the fluorophore concentration, the investigated object is scanned in transillumination

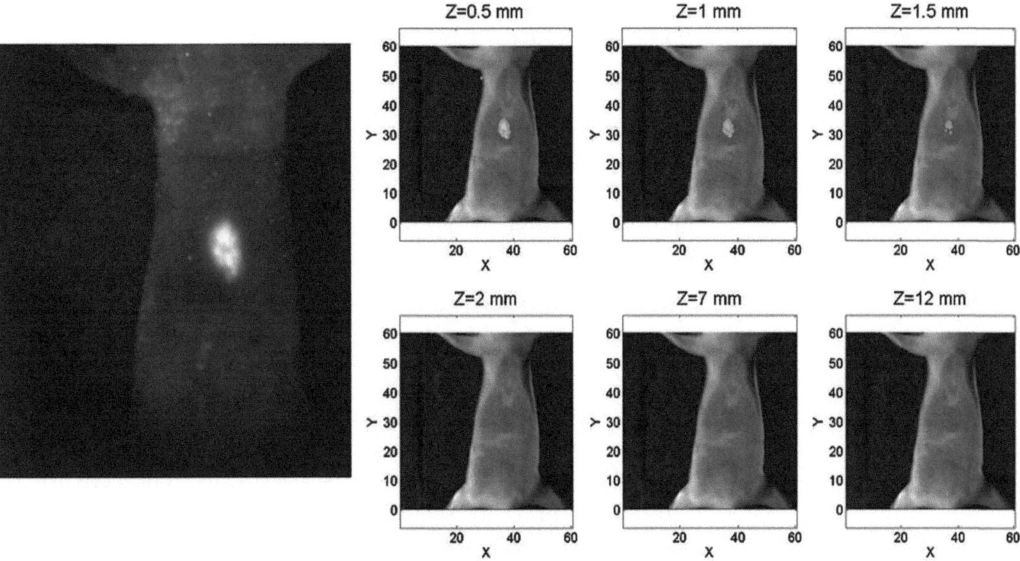

Fig. 5. Slices of the reconstruction of subcutaneous tumor at different depths.

configuration by moving the optical fiber with a collimator using stepping motors. Step sizes and the scanning area are choosen optimally by analyzing reflectance fluorescence images (see Fig. 5 left). Moreover, these images can be used in the reconstruction algorithm for improving its precision and speed. For the scanning procedure, an animal is placed so that the tumor is near the excitation source, because the decay of excitation light is less than for light emission. For reconstruction of the fluorophore distribution, an algorithm, described in (25), is used. This algorithm employs the combination of a hybrid model of light propagation, Monte Carlo simulations, and algebraic reconstruction. The results of 3D reconstruction of TurboRFP-expressing tumors in nude mice are shown in Fig. 5 right (see Note 10).

4. Notes

1. Another E. coli strain for plasmid production could also be used. The purification step is necessary to reduce the toxicity of the transfection.
2. It is recommended to seed such an amount of cells to obtain a mono-layer at 80–90% confluence at the time of transfection.
3. In some cases, the initial heterogeneity and biological properties of a cell line are maintained without cloning.
4. The pTurboRFP-C plasmid contains the Neo^r gene encoding neomicinphosphotransferase that enables selection of stably of

transfected cells resistant to neomicin. Resulting cells stably express TurboRFP.

5. Enriched Ham's F12 medium may include DMEM, Ham's F12, and conditioned medium in a 1:1:1 ratio with 10% fetal bovine serum. For some cell lines, including melanoma, dexamethasone at a concentration of 10^{-7} M may be added to the medium to increase clonal growth, if removed 5 days after plating.
6. Probability of insertion of genes in the genome of the cell by lipofection method is less than 1%. Clones are isolated that are stably transfected.
7. In both tests the number of surviving cells is determined using a calibration curve, which is plotted as a dependence of fluorescence signal on the amount of seeded cells. The cells for the calibration curve are seeded into culture microplates simultaneously with the above described cells. However, the second row of the plate used for the calibration contains the amount of cells equal to that used for the determination of anti-tumor activity. The 3rd to 11th rows contain serial two-fold dilutions of the cells. The 12th row in all plates is used as a control (without cells).
8. In order to decrease the inoculation dose and to increase the effectiveness of the implantation of tumor cells in mice, BD Matrigel™ Matrix (a substance obtained from the basal membrane of mouse Engelbreth-Holm-Swarm [EHS] sarcoma) is recommended. Tumors were obtained in a group of animals after injection of 10^6 cells per animal that had an average size of 300 mm^3 on day 36.
9. If the fluorescent tumor model to be used for determination of antitumor efficacy of drugs delivered to the tumor by antibody-mediated targeting (conjugates of drugs with antibodies, immunoliposomes, etc.), information about antigens presentation on the cell surface of the fluorescent cell line is necessary.
10. Reconstruction accuracy depends on localization, dimensions, and optical density of the tumor, fluorophore brightness, an signal-to-noise ratio in the raw transillumination images. In Fig. 6, the results of a model experiment demonstrating the accuracy of reconstruction are shown. As seen in Fig. 6, the accuracy of the reconstruction of the isolated object is quite good: the average reconstruction accuracy of the object shape is within 0.3 mm, and accuracy of the object localization is 1 mm. However, two spherical objects seated close to each other (less than 3 mm) cannot be reconstructed separately.

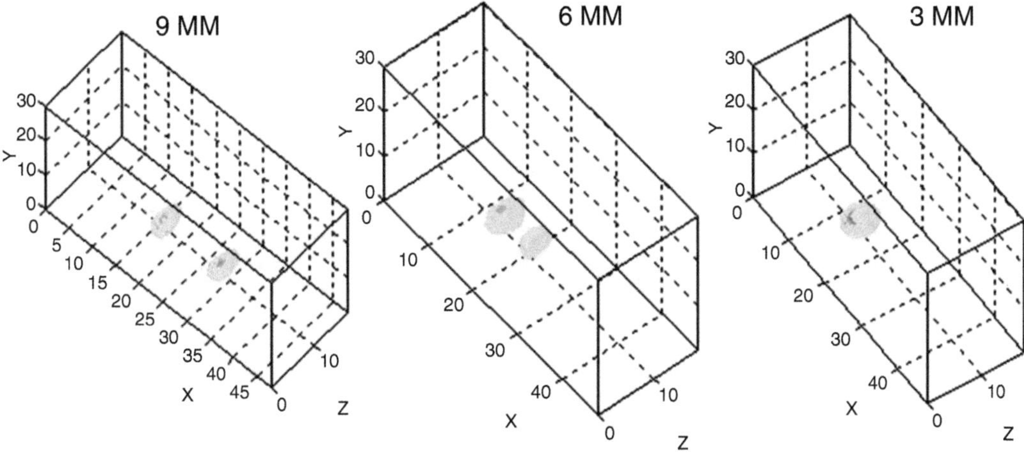

Fig. 6. The results of reconstruction of the fluorophore distribution in tissue phantom with two fluorescent objects, which simulate tumors labeled with fluorescent proteins. The distances between tumors are 9, 6, and 3 mm (from *left* to *right*).

References

1. Hoffman, R.M. (2005) Advantages of multicolor fluorescent proteins for whole-body and in vivo cellular imaging *J Biomed Optics* **10**, 41202.
2. Farina, K.L., J.B. Wyckoff, J. Rivera, H. Lee, J.E. Segall, J.S. Condeelis, J.G. Jones (1998) Cell motility of tumor cells visualized in living intact primary tumors using green fluorescent protein *Cancer Res* **58**, 2528–2532.
3. Sturm, J.W., Keese, M.A., Petruch, B., Bonninghoff, R.G., Zhang, H., Gretz, N., Hafner, M., Post, S., McCuskey, R.S. (2003) Enhanced green fluorescent protein transfection of murine colon carcinoma cells: key for early tumor detection and quantification *Clin Exp Metastasis* **20**, 395–405.
4. Lu, J., Chen, H., Chu, Y., Lin, T.E., Hsu, P., Huang, M., Tseng, C., Hsiao M., Lu, J. (2003) Establishment of red fluorescent protein-tagged HeLa tumor metastasis models: determination of DsRed2 insertion effects and comparison of metastasis patterns after subcutaneous, intraperitoneal, or intravenous injection *Clin Exp Metastasis* **20**, 121–133.
5. Chishima, T., Miyagi, Y., Li, L. (1997) The use of histoculture and green fluorescent protein to visualize tumor cell host interaction *In Vitro Cell Dev Biol* **33**, 745–747.
6. Chishima, T., Miyagi, Y., Wang, X. (1997) Metastatic patterns of lung cancer visualized live and in process by green fluorescent protein expression *Clin Exp Metastasis* **15**, 547–552.
7. Lasic, D.D. (1997) Colloid chemistry. Liposomes within liposomes *Nature* **387**, 26–27.
8. Zhdanov, R.I., Podobed, O.V., Vlassov, V.V. (2002) Cationic lipid-DNA complexes-lipoplexes-for gene transfer and therapy *Bioelectrochemistry* **58**, 53–64.
9. Felgner, J.H., Kumar, R., Sridhar, C.N., Wheeler, C.J., Tasi, Y.J., Border, R., Ramsey, P., Martin, M., Felgner, P.L. (1994) Enhanced gene delivery system and mechanism studies with a novel series of cationic lipid formulations *J Biol Chem* **269**, 2550–2561.
10. Dowty, M.E. (1995) Plasmid DNA entry into postmitotic nuclei of primary rat myotubes *Proc Nat Acad Sci USA* **92**, 4572–4576.
11. Tseng, W.C. (1997) Transfection by cationic liposomes using simultaneous single cell measurements of plasmid delivery and transgene expression *J Biol Chem* **272**, 25641–25647.
12. Lesage, D., Cao, A., Briane, D., Lievre, N., Coudert, R., Raphael, M., Salzmann, J.L., Taillandier, E. (2002) Evaluation and optimization of DNA delivery into gliosarcoma 9 L cells by a cholesterol-based cationic liposome *Biochim et Biophys Acta* **1564**, 393– 402.
13. Kim, A., Lee, E.H., Choi, S.-H., Kim, C.-K. (2004) In vitro and in vivo transfection efficiency of a novel ultradeformable cationic liposome *Biomaterials* **25**, 305–313.
14. Fraley, R., Papahadjopoulos, D. (1982) Liposomes: the development of a new carrier system for introducing nucleic acids into plant and animal cells *Cur. Top in Microbio. and Immunol* **96**, 171–192.
15. Nakanishi H., Mochizuki, Y. (2003) Chemosensitivity of peritoneal micrometastases

as evaluated using a GFP-tagged gastric cancer cell line *Cancer Sci* **94**, 112–118.

16. Hayashi, K., Yamauchi, K., Yamamoto, N., Tsuchiya, H., Tomita, K., Bouvet, M., Wessels, J., and Hoffman, R.M. (2009) A color-coded orthotopic nude-mouse treatment model of brain-metastatic paralyzing spinal cord cancer that induce angiogenesis and neurogenesis *Cell Proliferation* **42**, 75–82.
17. Amoh, Y., Katsuoka, K., and Hoffman, R.M. (2008) Color-coded fluorescent protein imaging of angiogenesis: the AngioMouse models *Curr Pharm Des* **14**, 3810–3819.
18. Ntziachristos, V., Ripoll, J., Wang, L.V. and Weissleder. R. (2005) Looking and listening to light: the evolution of whole-body photonic imaging *Nature Biotechnology* **23**, 313–320.
19. Ntziachristos, V., Turner, G., Dunham, J., Windsor, S., Soubret, A., Ripoll, J., Shih, H. A. (2005) Planar fluorescence imaging using normalized data *J Biomed Opt* **10**, 064007.
20. Deliolanis, N.C., Kasmieh, R., Würdinger, T., Tannous, B.A., Shah, K. and Ntziachristos, V. (2008) Performance of the Red-shifted Fluorescent Proteins in deep tissue molecular imaging applications *J Biomed Opt* **13**, 044008.
21. Zacharakis, G., Ripoll, J., Weissleder, R., Ntziachristos, V. (2005) Fluorescent protein tomography scanner for small animal imaging *IEEE Transactions on Medical Imaging* **24**, 878– 885.
22. Patwardhan, S.V., Bloch, S.R., Achilefu, S., Culver, J.P. (2005) Time-dependent whole-body fluorescence tomography of probe biodistributions in mice *Optics Express* **13**, 2564–2577.
23. Ntziachristos, V., Tung, C.-H. Bremer, C., and Weissleder, R. (2002) Fluorescence molecular tomography resolves protease activity in vivo *Nature Medicine* **8**, 757–761.
24. Turchin, I.V., Kamensky, V.A., Plehanov, V.I., Orlova, A.G., Kleshnin, M.S., Fiks, I.I., Shirmanova, M.V., Meerovich, I.G., Arslanbaeva, L.R., Jerdeva, V.V., Savitsky, A.P. (2008) Fluorescence diffuse tomography for detection of red fluorescent protein expressed tumors in small animals *Journal of Biomedical Optics* **13**, 041310.
25. Shaner, N.C., Steinbach, P.A., and Tsien, R.Y. (2005) A guide to choosing fluorescent proteins *Nature Methods* **2**, 905–909.
26. pTurboRFP-C vector. Description list, “Evrogen”.
27. Mikhajlova, I.N., Baryshnikov, A.Ju., Morozova, L.F., Burova, O.S., Demidov, L.V., Kozlov, A.M., Larin, S.S., Georgiev, G.P., Vorozhtsov, G.N., Gnuchev, N.V. (2006) Human melanoma cellular line mel Kor used in preparing antitumor vaccine. *Patent RU N2287578 (C1).*

Chapter 8

Real-Time Visualization and Characterization of Tumor Angiogenesis and Vascular Response to Anticancer Therapies

H. Rosie Xing and Qingbei Zhang

Abstract

In vivo angiogenesis assays provide more physiologically relevant information about tumor vascularization than in vitro studies because they take the complex interactions among cancer cells, endothelial cells, mural cells, and tumor stroma into consideration. Traditional microscopic assessment of vascular density conducted by immunostaining of tissue sections or by lectin angiogram visualization of tumor vessels is invasive and requires the sacrifice of tumor-bearing animals. Therefore, it prohibits longitudinal time-course observation in a single animal and requires a large number of animals at each time point to derive statistically-meaningful observations. Additionally, heterogenous behavior among different tumors will inevitably introduce individual biological variance that may obscure reliable interpretation of the results. While various artificial in vivo angiogenesis assays, such as the Matrigel implant assay, chick chorioallatoic membrane assay, and dorsal skin fold chamber assay have been developed and employed to more directly observe the progression of physiological angiogenesis, they can not appropriately assess tumor angiogenic progression or tumor vascular regression in response to therapeutic intervention. Here, we describe a noninvasive method and a detailed protocol that we have developed and optimized using the Olympus OV-100 in vivo imaging system for real-time high-resolution visualization and assessment of tumor angiogenesis and vascular response to anticancer therapies in live animals. We show that using this approach, tumor vessels can be monitored longitudinally through the whole vasculogenesis and angiogenesis process in the same mouse. Further, morphologic changes of the same vessel prior to and after drug treatments can be captured with microscopic high resolution. Moreover, the multichannel co-imaging capability of the OV-100 allows us to analyze and compare tumor vessel permeability before and after antiangiogenesis therapy by employing a near-infrared blood pool reagent, or by visualizing improved cytotoxic drug delivery upon tumor vessel normalization by using a fluorophore tagged drug. This noninvasive method can be readily applied to orthotopically transplanted breast cancer models as well as to subcutaneously-transplanted tumor models.

Key words: In vivo imaging, Angiogenesis, Antiangiogenesis, MDA-MB-435-GFP, Mammary fat pad injection, Vascular response, Vessel normalization

Robert M. Hoffman (ed.), *In Vivo Cellular Imaging Using Fluorescent Proteins: Methods and Protocols*, Methods in Molecular Biology, vol. 872, DOI 10.1007/978-1-61779-797-2_8, © Springer Science+Business Media, LLC 2012

1. Introduction

Angiogenesis is a hallmark of human cancers (1), and tumor progression depends on the angiogenic switch (2). Tumor angiogenesis differs significantly from normal physiological angiogenesis, and is characterized by aberrant organization and structure, compromised blood flow, increased permeability, poor perfusion, altered endothelial cell–pericyte interactions, and delayed maturation (3, 4). Vascular leakiness and the lack of functional lymphatic vessels inside tumors lead to elevated intratumor pressure (5), which hinders the efficiency of oxygenation of the tumor tissue and effectiveness of anticancer drug delivery. While the goal of the current anti-angiogenic regimens are optimized to cause vessel damage and destruction, the efficacy of antiangiogenic therapies may be also achieved through correcting the defects of tumor microvascular functions, thus achieving normalization (6).

Elucidating the mechanisms of tumor angiogenesis and characterization of the vascular response to anticancer or antiangiogenic drugs will rely on technically straightforward, reliable, and most importantly, physiologically-relevant assays (7). Among current methods studying angiogenesis, noninvasive real-time in vivo imaging has become increasingly important because it affords continuous observation of this biological event in a single living organism. Such longitudinal studies of disease progression and therapeutic response in individual subjects will significantly reduce the number of small animals required, allow assessment of individual variability, and reduce the interference of biological variance on data presentation and interpretation. We have explored, optimized, and demonstrated the feasibility of employing the Olympus OV-100 Small Animal in vivo imaging system to (1) characterize the kinetics of tumor angiogenesis, (2) to assess vascular response to therapeutic intervention, and (3) to design optimal dosing schedules of chemotherapy and anti-angiogenic combination therapies to achieve co-targeting of cancer cells and tumor vascular networks.

Among several pro-angiogenic factors, VEGF is the predominant regulator of tumor-associated angiogenesis (8, 9). Anti-VEGF therapy has demonstrated considerable therapeutic efficacy in preventing tumor growth and metastasis of solid tumors (10, 11). Avastin is a humanized anti-VEGF antibody that received U.S. FDA approval as the first antiangiogenesis treatment for cancer.

Rapamycin, an mTOR inhibitor, has demonstrated impressive growth inhibitory efficacy against a broad range of human cancers in both preclinical and clinical evaluation by inhibiting mTOR signaling in cancer cells (12, 13). The anti-tumor activity of Rapamycin and its analogues occurs partially through their inhibitory effect on tumor angiogenesis (14).

We compared tumor vasculature response to anti-VEGF antibody and rapamycin using noninvasive real-time in vivo imaging. Continuous and direct visualization of tumor vessel changes in real time allowed the demonstration of both vascular destruction and functional normalization (alleviation of tumor-vessel hyper-permeability) by anti-angiogenic therapy. The ability of rapamycin to co-target cancer cells and tumor stroma and the enhancement of the therapeutic efficacy of a cytotoxic drug (paclitaxel) by improving intra-tumor delivery after vessel normalization was also observed (15).

2. Materials

2.1. Cell Culture

1. MDA-MB-435-GFP cells are maintained in DMEM high glucose supplemented with 10% FBS + 200 μg/ml G418 (Gibco).
2. DMEM (high glucose) (Hyclone, Logan, UT) is supplemented with 10% fetal bovine serum (FBS) (Hyclone, Logan, UT), and 1% penicillin/streptomycin (Hyclone, Logan, UT).
3. Trypsin solution (0.25% trypsin/2.21 mM EDTA) (Cellgro, Manassas, VA).
4. 1× PBS (Cellgro, Manassas, VA).
5. Syringe filter (0.22 μm, Millopore).

2.2. Animals

All animal studies are carried out according to protocols approved by IACUC Committee at the University of Chicago. Female athymic Ncr/nude mice (Fredrick), 5–6 weeks old, 18–20 g are used.

2.3. Mammary Fat Pad Tumor Transplantation and Administration of Antiangiogenesis Therapy

1. 1-ml insulin syringe with a 27-gauge needle (BD).
2. Alcohol pad (Fisherbrand).
3. Ketamine/xylazine cocktail (10 mg/ml ketamine, 1 mg/ml xylazine) is prepared by mixing 1 ml ketamine (100 mg/ml, Phonenix Scientific, Inc.) and 0.5 ml xylazine (20 mg/ml, LLOYD laboratories) with 8.5 ml ddH_2O.
4. Digital caliper (Preisser, Gammertingen, Germany).
5. Rapamycin (LC laboratories, Woburn, MA) is prepared by dissolving the powder in 5.2% Tween 80 (Sigma) and 5.2% polyethylene glycol (PEG) (Sigma) and administrated i.p. at 5 mg/kg body weight daily for 14 consecutive days.
6. Anti-VEGF neutralizing antibody (R&D Systems) working solution is diluted from a stock solution with sterile 1× PBS and administrated i.p. at 5 μg/kg three times a week for 2 weeks.

2.4. OV-100 In Vivo Imaging

1. Anesthesia induction chamber (VetEquip).
2. Nose cone (VetEquip).

3. Isofluorane (VetEquip).
4. Surgical tape.
5. Warm water pad (VetEquip).
6. AngioSense^{750} (VisEN) is injected via the tail vein at 2 nmol/150 µl/mouse, 4 h prior to imaging.
7. Paclitaxel BODIPY$^{564/570}$ (Invitrogen) is first dissolved in DMSO (Sigma) to yield a 1-µM stock solution. The working solution is diluted freshly with 1× PBS and injected i.v. at 20 nmol/100 µl/mouse, 4 h prior to imaging.

2.5. Equipment for In Vivo Imaging

The Olympus OV100 allows for observation of fluorescence markers in live animals from the subcellular level to the entire animal. It also allows for time-lapse observation of a variety of physiological and pathological events. This imaging system contains the following features:

1. Four individually optimized objectives (three macrolens and one zoom lens) that are parcentered and parfocal providing a total 105-fold magnification.
2. High-sensitivity Hamamatsu Electro Multiplying (EM) CCD camera.
3. High numerical aperture (NA) optical system with rapid scanning mode that shortens exposure times and minimizes specimen damage.
4. Multiple observation channels that afford multi-channel co-imaging: bright field, GFP, RFP, 680, 750.
5. Software: scope control/image acquisition of still images and movies/image processing and analysis.

3. Methods

3.1. Preparation of MDA-MB-435-GFP Cells

1. MDA-MB-435-GFP cells are thawed from liquid nitrogen.
2. Cells are washed with 1× PBS, and cultured in DMEM (high glucose) media supplemented with 10% FBS and 200 µg/ml G418.
3. Cells are subcultured for at least three passages before harvesting at logarithimic growth phase (approximately 70–80% confluent) for mammary fat pad injection.

3.2. Mammary Fat Pad Tumor Transplantation

1. The mice are anesthetized via intraperitoneal injection of 200 µl ketamine/xylazine cocktail, and are placed abdomen-side up on a sterilized surgical board.
2. Disinfect injection area using an alcohol pad.

3. MDA-MB-435-GFP cells are trypsinized, washed, collected, and suspended in sterile PBS to a 2×10^7 stock solution. 2×10^6 MDA-MB-435-GFP cancer cells, suspended in 0.1 ml of PBS, are injected into the left inguinal mammary fat pad (m.f.p.) (see Note 1).
4. Place the mice in a dorsal recumbence position prior to awakening to make sure the injected cells accumulate around the mammary fat pad (see Note 2).
5. When tumors become palpable, approximately 2 weeks after cancer cell transplantation, cancer cell growth will be measured and recorded twice weekly by calipers and tumor volume calculated as described (16).

3.3. Anti-angiogenesis Treatment and Combination Therapy with Cytotoxic Drugs

1. When tumor volume reaches approximately 200–250 mm^3, treatment with anti-angiogenic drugs (such as anti-VEGF IgG) or anti-cancer drugs (such as rapamycin or paclitaxel) are initiated (see Note 3).
2. To visualize tumor vascular response to anti-mouse VEGF IgG and anti-cancer agents such as rapamycin, a targeted regimen, mice are randomly divided into three treatment groups: (1) vehicle treatment; (2) daily rapamycin treatment (5 mg/kg body weight) administrated i.p. for 14 consecutive days; and (3) anti-VEGF IgG treatment (5 μg/kg) administrated i.p. three times weekly for 2 weeks. Five mice from each treatment group are imaged on day 0, day 7, and day 14 (see Fig. 2).
3. To visualize the efficacy of anti-angiogenic therapy, such as anti-VEGF IgG treatment on intratumor delivery of a cytotoxic chemotherapeutic agent, such as paclitaxel, mice are treated either with vehicle (PBS), or with anti-VEGF IgG (5 μg/kg body weight) three times weekly for 1 week, followed by injection of 20 nmol of paclitaxel BODIPY$^{564/570}$ 4 h before the imaging session. Five mice from each treatment group are prepared for imaging (see Fig. 4).

3.4. In Vivo Image Acquisition (See Note 3)

1. Place the animal in the anesthesia induction chamber, turn on the Isoflurane Vaporizer Dial, and adjust the level gauge to 3%. Wait until the mouse reaches the desired depth of anesthesia by the toe-pinch test. Transfer the anesthetized mouse from the induction chamber to the OV100 imager, and immediately place the animal's muzzle into the nose cone connected to the imaging platform. Turn the isoflurane dial back to 1%. When the mouse's respiration becomes smooth and stable, imaging session can be initiated.
2. The mouse is placed on the imaging platform with abdomen-side up so that the mammary fat tumor will be imaged from the top of the tumor. Open the OV-100 image acquisition

software and set up imaging capture parameters according to manufacturer's instructions. The observation is started with 0.14×, the lowest magnification level in the GFP-LP channel. Adjust the shift control on the control module to place the tumor in the center of the view window on the computer screen. Note, from this moment on, adjustments of imaging parameters, changing the position of the imaging platform, selection of the filter sets, changing the focus, and image acquisition will all be made through the image acquisition software. Adjust the position of the platform up or down, so that the tumor will be in focus. At this level of magnification, large blood vessels innervating the surface of the tumor can be visualized and images will be acquired for selected view fields of interest.

3. To visualize at a higher resolution, increase the magnification level to 0.56×. Since the four subzoom lenses of OV-100 are par-focal, the images will largely remain focused when the magnification levels are altered at three subzoom levels (from 0.14× to 0.89×). Therefore, only very minor adjustment is required to refocus the desired imaging field. Thereafter, the light intensity and exposure time need to be adjusted so that the intensity of the histogram remains between 90% and 100% to avoid over-exposure or under-exposure. At 0.56×, both the major and small vessels on the tumor surface can be visualized clearly. The nonluminous angiogenic blood vessels appear as sharply defined dark networks against the bright tumor GFP background. In fact, no blood vessel-enhancing contrast agents are required to obtain microscopic quality high-resolution images of the tumor or the developing or resolving tumor vascular networks, from this magnification level and up.

4. Increase the magnification level to 0.89× to allow visualization of finer details of the tumor vascular network. As the magnification level increases, the intensity and exposure time need to be adjusted to avoid over- or under-exposure. At this level of magnification, the degree of extensive angiogenesis can be readily appreciated, as can the presence of a vasculature network that exhibits the classic characteristics of tumor vessels (tortuous, abrupt ends, distended, uneven in size, etc., see Fig. 1). The image depth depends largely on the abundance of blood vessels on the surface and within the tumor. We are capable of imaging the tumor vascular network as deep as 4 mm from the tip of the tumor surface. At 0.89× subzoom, the entire tumor surface fits the view screen. Thus, this level of magnification is optimal for taking high-resolution Z-stack images for 3D construction using Image-J software (see Note 4). The procedure for setting up Z-stack imaging acquisition is outlined in the software manual and can be followed easily.

5. After observations with the macrolens at subzoom magnifications, random fields of vessels or areas of interest can be chosen for observation with the zoom lens for finer details and for obtaining high-resolution planner images (see Fig. 1e, f). At the zoom level, digital magnification ranges from 1.6× to 16×. Since characteristic tumor vessels can be easily and reliably identified from one imaging session to the other, sophisticated image co-registry is not required to compare gross differences of angiogenic progression or therapeutic responses among images acquired from different sessions. Therefore, this method allows serial visual monitoring of spatial–temporal relationships between cancer cells and angiogenic sprouting over the course of angiogenic progression, and vasculature response to antiangiogenic drugs through the whole treatment course (see Figs. 1 and 2; Note 5).
6. Tumor vascular permeability and leakiness can be visualized and characterized using VisEn's near infrared (NIR) blood pool probe Angiosense^{750} with dual channel OV-100 imaging (480 nm GFP and 680 nm near infrared, see Note 6). Two nanomoles of Angiosense^{750} are injected i.v. and OV-100 imaging sessions are conducted at 4 and 24 h post-Angiosense^{750} injection for visual inspection of probe retention (4 h) and hyper-permeability of the tumor vascular network (24 h). For example, while the tumor vascular network is clearly visualized within GFP-MDA-MB-435 tumors in the GFP channel, high permeability of the tumor microvasculature is evident from the lack of retention of the probe and its accumulation near the vessels when observing 24 h after Angiosense^{750} administration (see Fig. 3a, d). In contrast, Rapamycin or anti-VEGF antibody treatment normalized tumor vascular function by reducing vascular permeability leading to increased vessel retention of Angiosense^{750} probe (see Fig. 3b, c, e, f).
7. The effect of vessel normalization on the efficiency of drug delivery can be visualized directly through OV-100 dual channel imaging (480 nm GFP and 570 nm RFP). For example, to assess the efficacy of a low-dose and short-term treatment with anti-VEGF IgG on the tumor vascular network (vessel damage or normalization) and on the efficiency of drug delivery, we treated MDA-MB-435-GFP tumors with a low dose of anti-VEGF IgG for 7 days. This time period was identified as a potential window of vessel normalization prior to the onset of more extensive vessel destruction with prolonged treatment in this model. We injected a 20 nM nontherapeutic dose of BODYPI-labeled paclitaxel (564/570) as an optical imaging tracer and conducted OV-100 imaging. At 24-h postprobe injection, while paclitaxel leaked out and accumulated as patches

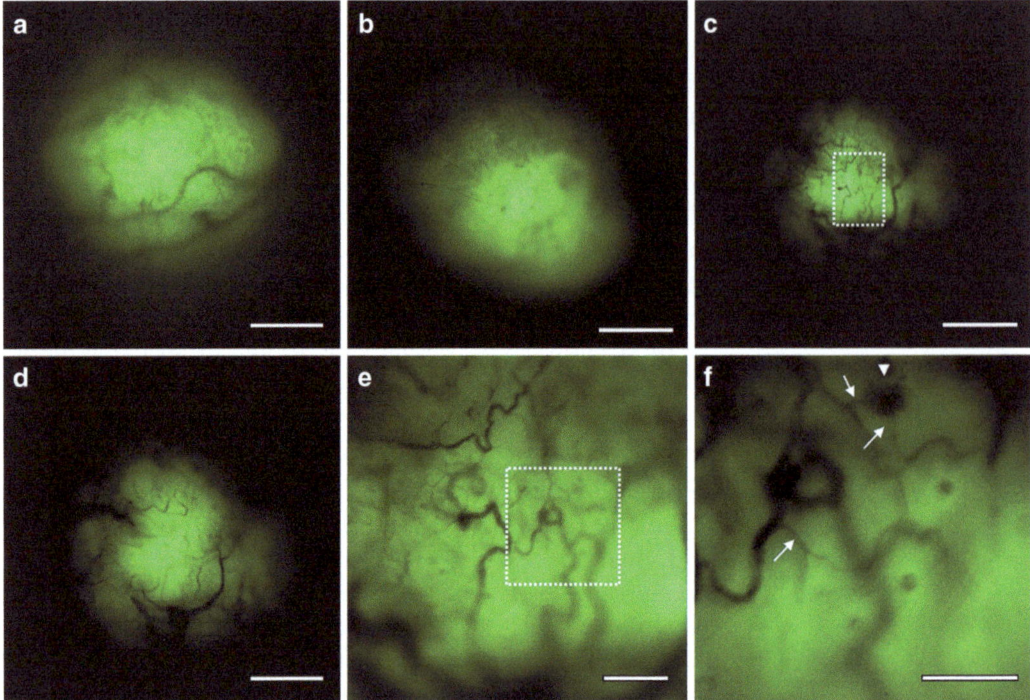

Fig. 1. Continuous and noninvasive real-time visualization of vasculogenesis and angiogenesis in a growing MDA-MB-435-GFP tumor growing in the mammary fat pad of nude mice. Four imaging sessions were conducted on the MDA-MB-435-GFP tumor to monitor the development of the tumor vascular network during the course of tumor growth. (**a–d**) Imaging of continuous angiogenesis progression using the 0.89× macro lens of the OV-100. (**a**) On day 27 posttumor inoculation, tumor volume reached 112 mm^3. Large blood vessels, likely recruited from adjacent skin were found on the surface of the tumor and microvessel sprouting from these large vessels was observed. The imaging depth was approximately 200–400 nm from the tumor surface and thus was quite shallow. (**b**) On day 34 post-tumor inoculation, tumor volume reached 202 mm^3. Straight vessels were observed to grow inwards and the imaging depth was significantly increased (approximately 400–600 nm). (**c**) On day 48, tumor volume reached 290 mm^3 and a tumor vascular network exhibiting characteristics of tumor vessels could be readily imaged and visualized. Imaging depth was increased to 2–4 mm from the tumor surface and stack images could be obtained for 3-D reconstruction. (**d**) On day 55, the tumor continued to grow and tumor volume reached 326 mm^3. Tumor vascular networks were similar to that on day 48. (**e–f**) High resolution images of fields of interest using the microlens and digital zoom of the OV-100 imager. (**e**) Magnified view of the frame in (**c**) of the tumor vascular network on day 48 post-tumor transplantation. (**f**) Further magnification of the frame in (**e**) showing the sprouting capillaries (*arrows*) from larger vessels and blood islands (*arrowhead*). Bars: (**a–d**) 2,000 μm; (**e**) 500 μm; (**f**) 200 μm. Reproducible observations were obtained from tumors of the same experiments and from one experiment to the other (15).

near the vessels in non-normalized tumors (see Fig. 4a–c), decreased vessel permeability after anti-VEGF IgG treatment led to more efficient delivery of the drug to the tumor (see Fig. 4d–f). Increased drug delivery efficiency was further confirmed by confocal microscopy of frozen tumor sections (see Note 7). Therefore, anti-angiogenic therapy-induced vessel normalization, which in turn enhanced the delivery of chemotherapeutic agents and consequently increased their antitumor efficacy. Note, this imaging procedure can be readily optimized

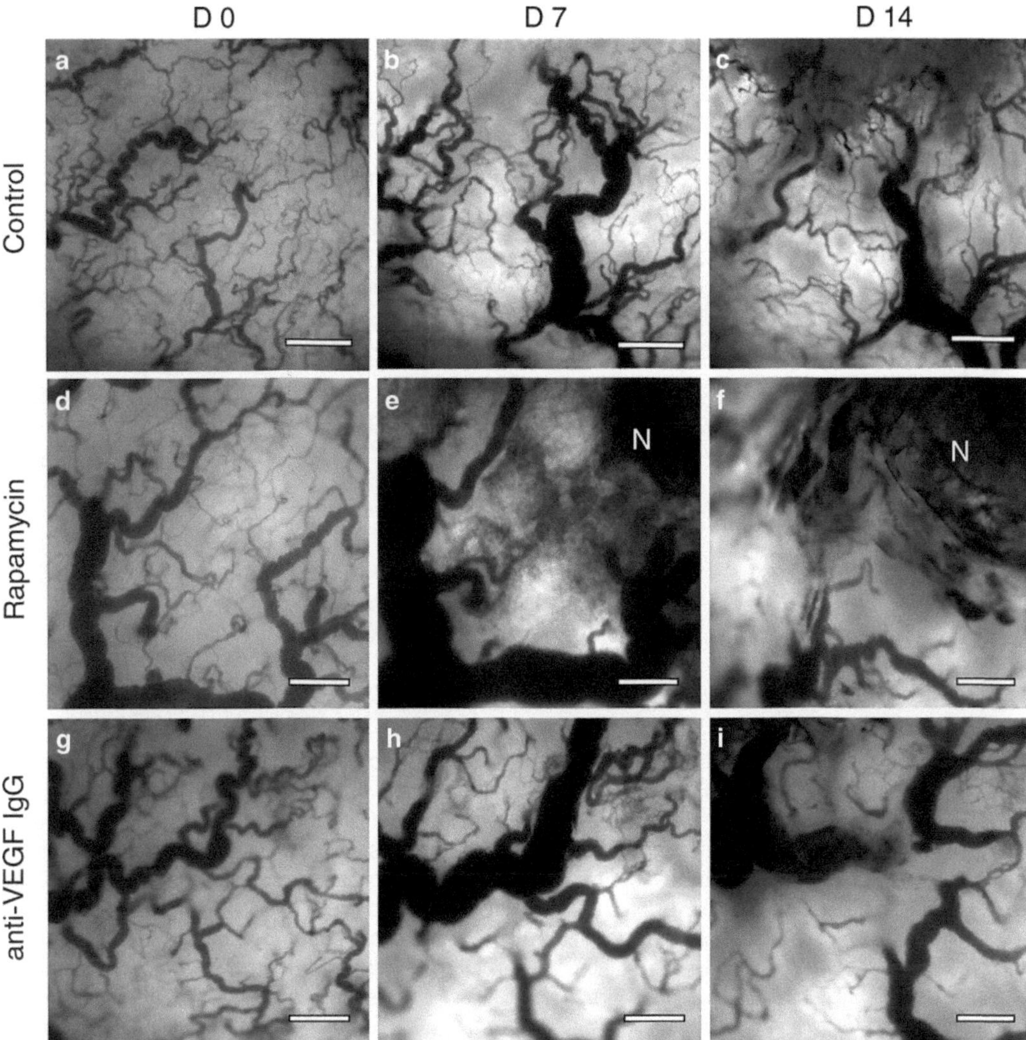

Fig. 2. Serial real-time imaging and characterization of tumor vascular response to anti-angiogenic and anti-cancer therapy. OV-100 imaging sessions were conducted at three time points on the MDA-MB-435-GFP tumor: one day prior to the initiation of treatment (day 0); on day 7 of the treatment; and on day 14, the last day of the treatment. Imaging fields of interest were identified on day 0 and morphometric changes of tumor vessels within the same field were imaged and characterized (see Note 5). (**a–c**) Control tumor received vehicle treatment. During the course of the treatment, control tumors continued to grow. Inside the tumor, large vessels became distended, while small vessels and capillaries remained abundant and intact. (**d–f**) Visualization of the anti-angiogenic efficacy of rapamycin in an MDA-MB-435-GFP tumor that received the 14-day rapamycin treatment. On day 7 (**e**), most of the small vessels were disrupted by rapamycin treatment. Large vessels became very distended. Tumor necrosis appeared (N, *right upper corner*). On day 14 (**f**), both large vessels and capillaries were damaged, and the necrotic area (N) increased in size. (**g–i**) Visualization of the anti-angiogenic efficacy of rapamycin in an MDA-MB-435-GFP tumor that received low-dose anti-VEGF IgG anti-angiogenic therapy. On day 7 (**h**), most of the vessels were intact except large vessels became distended with increased tumor volume. We identified this time point where vessel normalization occurred which increased the efficiency of chemotherapeutic drug delivery. On day 14 (**i**), tumor vessel architecture was significantly disrupted upon continued anti-angiogenic treatment. Bars: (**a–c**) 1,000 μm; (**d–i**) 500 μm. Reproducible observations were obtained from tumors of the same experiments and from one experiment to the other. Further, histology sampling and characterization were thoroughly conducted to confirm the in vivo imaging observations. CD31 staining and lectin staining both confirmed the decreased vessel density after rapamycin and anti-VEGF antibody treatment (see Note 7) (15).

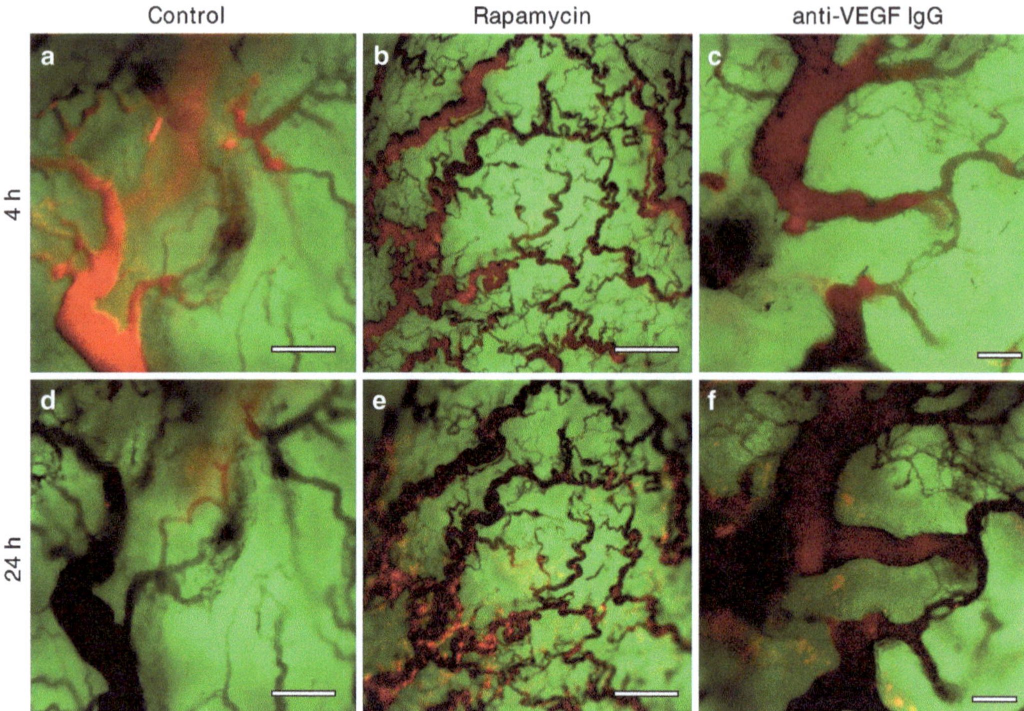

Fig. 3. Real-time dual-channel imaging of restoration of tumor vessel functional integrity upon administration of anti-angiogenic and anti-cancer therapy. Vascular permeability and leakage were visualized by retention of the blood pool agent Angiosense^{750} NIR probe. Angiosense^{750} was injected and dual channel OV-100 imaging sessions (480 nm GFP and 680 nm near infrared, see Note 6) were conducted at 4 and 24 h after Angiosense injection. (**a**–**c**) Blood vessel retention of Angiosense at 4 h after injection was observed in tumors of all treatment groups (control, rapamycin, and anti-VEGF IgG). (**d**–**f**) At 24 h after Angiosense injection, the probe almost completely leaked out due to the hyper-permeability of tumor vessels in control tumors receiving vehicle treatment (**d**). After 2 weeks of either rapamycin or anti-VEGF IgG treatment, tumor vessel damage was observed (see Fig. 2). However, restoration of vessel integrity in the nondamaged vessels was evident as a significant portion of Angiogenesis^{750} (*red*) remained in the vessels at 24 h after injection ((**d**) and (**e**), respectively). *Green* tumor; *black* tumor vessels. Bars: (**a**, **d**) 500 μm; (**b**, **e**) 1,000 μm; (**c**, **f**) 200 μm. Reproducible observations were obtained from tumors of the same experiments and from one experiment to the other (15).

and applied to characterize other anticancer agents or combination regimens to (1) characterize the anti-angiogenic properties of an anticancer drug; (2) design the optimal dosing sequence, schedule, and doses to achieve vessel damage or vessel normalization; and (3) design the most effective combination that co-targets the cancer cells and the tumor vascular network. For treatment with anti-angiogenic or anti-cancer drugs for consecutive 14 days, see Note 8 for planning imaging sessions.

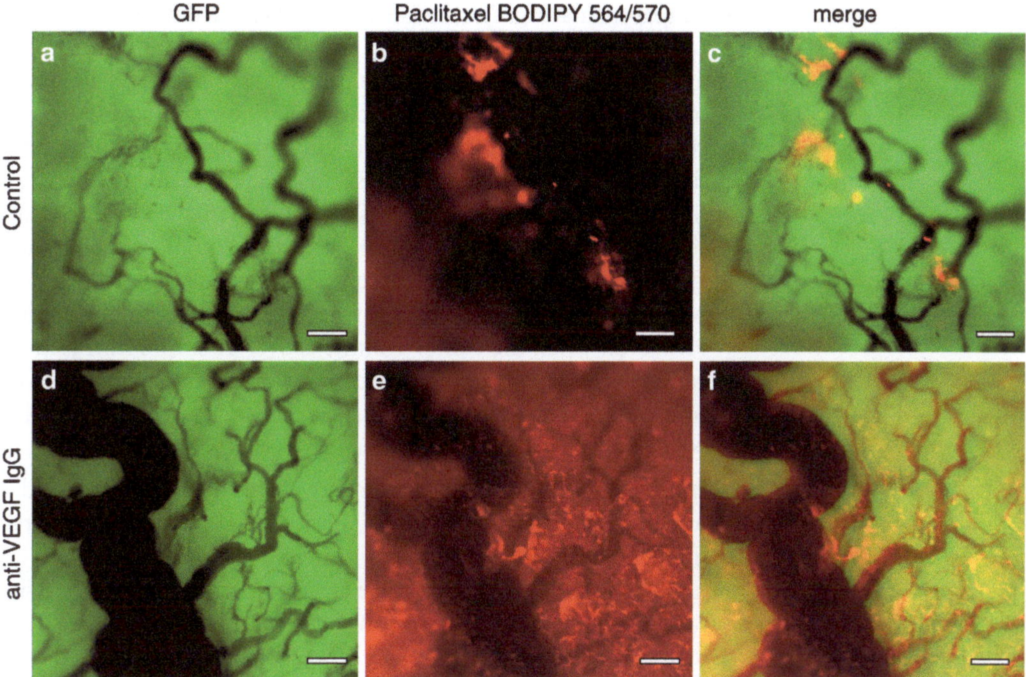

Fig. 4. Real-time dual-channel imaging of improvement of cytotoxic drug delivery upon normalization of tumor vessel function after low-dose and short-term anti-angiogenic therapy. The effect of vessel normalization on the efficiency of cytotoxic drug delivery was visualized by tumor vessel retension and distribution of paclitaxel BODIPY 564/570 in vehicle-treated control tumors (**a–c**) and in tumors treated with a low-dose anti-VEGF IgG dose schedule (three times weekly) for 1 week (**d–f**). Imaging sessions were conducted at 4 h after the injection of paclitaxel BODIPY 564/570. In a control tumor (*green*, **a–c**), paclitaxel BODIPY 564/570 (*red*) largely leaked out from tumor vessels (*black*) and accumulated in patches near the vessels. After 1 week of anti-VEGF IgG treatment, we observed minimal vessel damage (see Fig. 2h) and is thus the time point of vessel normalization. Paclitaxel BODIPY 564/570 (*red*) diffused out of vessels (*black*) and evenly distributed among tumor cells (*green*) (**d–f**). Bars: 200 μm. Reproducible observations were obtained from tumors of the same experiments and from one experiment to the other. Confocal microscopy examination of frozen tumor sections confirmed the improved delivery of Paclitaxel, consistent with OV-100 imaging observations (see Note 7).

4. Notes

1. For mammary fat pad injection, grow MDA-MB-435-GFP cells to 70–80% confluence and harvest them for injection. Over-growth of these cells in culture will lead to more invasiveness of the cell line, and metastases (lung metastasis or ascites) will develop earlier than maturation of primary tumors.
2. Suspended cancer cells are injected right under the nipple of nude mice, as close to mammary fat pad as possible. If the mammary fat pad is missed, the tumor will either not grow, or become an under vascularized subcutaneous tumor.

3. The optimal imaging window for vessel characteristics is when the tumor volume is between 100 and 300 mm^3. The preferred site for orthotopic tumor transplantation is the inguinal pair of mammary fat pads, since the skin covering the mammary fat pad is considerably thinner than that of the auxiliary pair. When tumor size reaches over 100 mm^3, the skin becomes gradually thinner, so the tumor surface and inner vessels can be visualized more clearly over a depth of 2–4 mm. From this time point, vasculogenesis and angiogenesis will take place and the vasculature network will be fully developed when the tumor volume reaches 250–300 mm^3. However, when the tumor volume is over 400–500 mm^3, significant necrosis may occur on the tumor surface, interfering with the resolution and quality of the images and thus should be avoided. Necrotic areas will appear as dark spots on the image.
4. After image data acquisition, a novel deconvolution algorithm will be applied to single and Z-stack OV-100 images to correct the "curved focal surface effect" and for enhancing axial and lateral resolution. Thereafter, use ImageJ-based image processing methods. 3D rendered morphology anaglyphs are constructed from OV-100 Z-stack images to isolate vessel features.
5. For longitudinal OV-100 imaging, special attention will be given to identifying the same imaging field of interest on consecutive imaging sessions, based upon the unique distribution and morphological pattern of tumor blood vessels that are initially mapped on day 0 and are followed in subsequent sessions.
6. The major restriction of optical imaging is the high absorption and scattering that occurs in biological tissues and, thus, limiting penetration of the light through the body. NIR (near infrared, 700–900 nm) imaging expands the capability and application of in vivo optical imaging. In our studies, we demonstrated the feasibility of conducting multi-channel imaging of tumor angiogenesis using NIR probes with the OV-100 imager, and of dual-modality co-imaging combining the OV-100 and VisEn instruments, for diffuse optical fluorescence tomography (DOT). While fluorescence planar imaging offers high-resolution visualization and anatomical details, tomographic imaging (VisEn FMT) yields 3D images and therefore 3D quantification.
7. Histology sampling and characterization are thoroughly conducted to confirm in vivo imaging observations, and to determine the mechanisms of anti-angiogenic activity of the therapeutics evaluated. For example, our preliminary investigation showed that increased pericyte coverage of the surviving tumor vessels might lead to improved vascular functions, consistent with a recent report demonstrating the role of pericytes in tumor angiogenesis (17). CD31 and lectin staining both confirmed the decreased vessel density after rapamycin

and anti-VEGF antibody treatment. Confocal microscopy examination also confirmed the improved delivery of taxol after vessel normalization.

8. Noninvasive OV-100 imaging experiments are conducted at three time points on five mice from each treatment group according to the following schedule: (1) 1 day prior to the initiation of treatment, perform OV-100 dual-channel imaging at 4 h and 24 h after the injection of blood pool contrast agent Angiosense^{750}. These studies will generate maps of OV-100 imaging fields of interest that will be used for co-registration with subsequent consecutive imaging experiments. Tumor vascular integrity prior to the treatment will also be determined; (2) on day 7, perform OV-100 imaging at 4 and 24 h after the injection of Angiosense^{750}; (3) on day 14, perform OV-100 imaging at 4 and 24 h after the injection of Angiosense^{750}.

References

1. Hanahan, D.and Weinberg R. A. (2000) The hallmarks of cancer. *Cell* **100**, 57–70.
2. Naumov, G. N., Akslen, L. A., and Folkman, J. (2006) Role of angiogenesis in human tumor dormancy: animal models of the angiogenic switch. *Cell Cycle* **5**, 1779–1787.
3. Bergers, G. and Benjamin, L. E. (2003) Tumorigenesis and the angiogenic switch. *Nat. Rev. Cancer* **3**, 401–410.
4. Burri, P. H. (2004) Intussusceptive angiogenesis: its emergence, its characteristics, and its significance. *Dev Dyn.* **231**, 474–88. doi:10.1002/dvdy.20184.
5. Fukumura, D. and Jain, R. K. (2008) Imaging angiogenesis and the microenvironment. *APMIS* **116**, 695–715.
6. Jain, R. K. (2005) Normalization of tumor vasculature: an emerging concept in antiangiogenic therapy. *Science* **307**, 58–62.
7. Staton, C. A., Stribbling, S. M., Tazzyman, S., Hughes, R., Brown, N. J., and Lewis, C. E. (2004) Current methods for assaying angiogenesis in vitro and in vivo. *Int. J. Exp. Pathol.* **85**, 233–248.
8. Leung, D. W., Cachianes, G., Kuang, W. J., Goeddel, D. V., and Ferrara, N. (1989) Vascular endothelial growth factor is a secreted angiogenic mitogen. *Science* 246, 1306–1309.
9. Grothey, A. and Galanis, E. (2009) Targeting angiogenesis: progress with anti-VEGF treatment with large molecules. *Nat Rev Clin Oncol.* **9**, 507–518.
10. Kerbel, R. and Folkman, J. (2002) Clinical translation of angiogenesis inhibitors. *Nat. Rev. Cancer* **2**, 727–739.
11. Kim, K. J., Li, B., Winer, J., Armanini, M., Gillett, N., Phillips, H. S., et al. (1993) Inhibition of vascular endothelial growth factor-induced angiogenesis suppresses tumour growth in vivo. *Nature* **362**, 841–844.
12. Rowinsky, E. K. (2004) Targeting the molecular target of rapamycin (mTOR). *Curr Opin Oncol.* **16**, 564–575
13. Noh, W. C., Mondesire, W. H., Peng, J., Jian, W., Zhang, H., Dong, J., et al. (2004) Determinants of rapamycin sensitivity in breast cancer cells. *Clin Cancer Res.* **10**, 1013–1023.
14. Phung, T. L., Ziv, K., Dabydeen, D., Eyiah-Mensah, G., Riveros, M., Perruzzi, C., et al. (2006) Pathological angiogenesis is induced by sustained Akt signaling and inhibited by rapamycin. *Cancer Cell* **10**, 159–170.
15. Zhang, Q., Bindokas, V.P., Shen, J., Fan, H., Hoffman, R.M., Xing, H.R. (2011) Time course imaging of therapeutic functional tumor vascular normalization by anti-angiogenic agents. *Mol. Cancer Ther.* **10**, 1173–1184.
16. Xing, H. R., Cordon-Cardo, C., Deng, X., Tong, W., Campodonico, L., Fuks, Z., et al. (2003) Pharmacologic inactivation of kinase suppressor of ras-1 abrogates Ras-mediated pancreatic cancer. *Nat. Med.* **9**, 1266–1268.
17. Greenberg, J. I., Shields, D. J., Barillas, S. G., Acevedo, L. M., Murphy, E., Huang, J., et al. (2008) A role for VEGF as a negative regulator of pericyte function and vessel maturation. *Nature* **456**, 809–813.

Chapter 9

In Vivo Imaging of Human Cancer with Telomerase-Specific Replication-Selective Adenovirus

Toshiyoshi Fujiwara

Abstract

Tumor-specific replication-competent viruses represent a novel approach for the treatment of neoplastic disease. These vectors can be used to directly label tumor cells in vivo, as they are designed to selectively replicate within such cells. To target cancer cells, there is a need for tissue- or cell-specific promoters that are expressed in diverse tumor types and are silent in normal cells. Telomerase activation is considered to be a critical step in carcinogenesis through the maintenance of telomeres, and its activity is closely correlated with human telomerase reverse transcriptase (hTERT) expression. We constructed a green fluorescent protein (GFP)-expressing attenuated adenovirus-5 vector, in which the hTERT promoter regulates viral replication (TelomeScan, OBP-401). We used TelomeScan to establish a new approach for visualizing metastatic or disseminated human tumors in vivo. Visualization is achieved via illumination with an excitation lamp under a three-chip color-cooled charged-coupled device camera following injection of TelomeScan into primary tumors or tumor-disseminated cavities. TelomeScan infection increases the signal-to-background ratio as a tumor-specific probe, because the fluorescent signals are only amplified in tumor cells by viral replication. This technology is adaptable to detect tumor metastasis and/or dissemination in vivo as a preclinical model of surgical navigation.

Key words: Telomerase, Adenovirus, GFP, Metastasis, Surgical navigation

1. Introduction

A variety of imaging technologies have been investigated as tools for cancer diagnosis, detection, and treatment monitoring. Improvements in external imaging methods, such as computed tomography (CT), magnetic resonance imaging (MRI), and ultrasound techniques, have increased the sensitivity for visualizing tumors and metastases in the body. A limiting factor in structural and anatomical imaging, however, is the inability to specifically

Robert M. Hoffman (ed.), *In Vivo Cellular Imaging Using Fluorescent Proteins: Methods and Protocols*,
Methods in Molecular Biology, vol. 872, DOI 10.1007/978-1-61779-797-2_9, © Springer Science+Business Media, LLC 2012

identify malignant tissues. Therefore, tumor-specific imaging is of considerable value in the treatment of human cancer because it can define the location and area of tumors without microscopic analysis. In particular, if tumors too small for direct visualization and therefore not detectable by direct inspection could be imaged in situ, surgeons could precisely excise tumors with appropriate surgical margins. This paradigm requires an appropriate "marker" that can facilitate visualization of physiological or molecular events that occur in tumor cells but not in normal cells.

Green fluorescent protein (GFP), which was originally obtained from the jellyfish *Aequorea victoria*, is an attractive molecular marker for imaging of live tissues because of its relatively non-invasive nature (1–3). We previously demonstrated the real-time fluorescence optical imaging of pleural dissemination of human non-small-cell lung cancer cells in an orthotopic mouse model using a tumor-specific replication-competent adenovirus (Telomelysin, OBP-301) (4, 5) in combination with a replication-deficient adenovirus expressing *GFP* (Ad-*GFP*) (6). We further modified OBP-301 to contain the *GFP* gene driven by the cytomegalovirus (CMV) promoter for monitoring viral replication. The resultant adenovirus, termed TelomeScan or OBP-401, efficiently and selectively replicated in human cancer cells and coordinately induced GFP expression. TelomeScan replication, however, was attenuated in normal human fibroblasts without GFP expression (7). Recently, Kishimoto et al. demonstrated that tumors labeled with TelomeScan could enable precise fluorescence-guided surgical navigation in the intraperitoneal model of disseminated cancer (8). This chapter summarizes in vivo cancer diagnostic procedures using the tumor-specific GFP-expressing TelomeScan.

2. Materials

2.1. Cell Lines and Animals

1. Human colorectal cancer cell line HT29 (American Type Culture Collection (ATCC), Manassas, VA) propagated in monolayer culture in McCoy's medium (Gibco BRL, Carlsbad, CA) supplemented with 10% heat-inactivated fetal calf serum (FCS), 100 units/mL penicillin, and 100 mg/mL streptomycin (PG/SM) (Sigma, St. Louis, MO).
2. Human tongue squamous carcinoma cell line HSC-3 (Health Science Resources Bank, Osaka, Japan) grown in Dulbecco's modified Eagle's medium containing high glucose (4.5 g/L) (D-MEM (high)) (Gibco BRL, Carlsbad, CA) with 10% FCS and PG/SM.
3. Human non-small-cell lung cancer cell line A549 (ATCC, Manassas, VA) propagated in D-MEM containing Nutrient

Mixture (Ham's F-12) and supplemented with 10% FCS and PG/SM.

4. Transformed embryonic kidney cell line 293 (Takara Bio Inc., Shiga, Japan) grown in D-MEM (high) with 10% FCS and PG/SM for adenoviral preparation.
5. Female BALB/c *nu/nu* mice purchased from Charles River Japan (Kanagawa, Japan), Clea Japan Co. (Tokyo, Japan), or Shizuoka Laboratory Animal Center (Hamamatsu, Japan).

2.2. Reagents

1. Primers for reverse transcription-PCR of an 897-bp fragment of the *E1A* gene from total cellular RNA of 293 cells: E1A-S (5′-ACA CCG GGA CTG AAA ATG AG-3′) and E1A-AS (5′-CAC AGG TTT ACA CCT TAT GGC-3′).
2. Primers for reverse transcription-PCR of a 1,822-bp fragment of the *E1B* gene from genomic DNA of 293 cells: E1B-S (5′-CTG ACC TCA TGG AGG CTT GG-3′) and E1B-AS (5′-GCC CAC ACA TTT CAG TAC CTC-3′).
3. pTA plasmid (Invitrogen, Carlsbad, CA) for subcloning amplified products.
4. pIRES vector for construction of pE1A-IRES-E1B (Clontech, Takara Bio Inc., Shiga, Japan).
5. pGL3-378 plasmid, which was kindly provided by Dr. Satoru Kyo (Kanazawa University, Kanazawa, Japan), containing a 455-bp fragment of the human telomerase reverse transcriptase (hTERT) promoter at *Mlu*I and *Bgl*II restriction sites.
6. pEGFP vector containing enhanced GFP (EGFP) cDNA (Clontech, Takara Bio Inc., Shiga, Japan).
7. Adeno-X™ Expression System for construction of TelomeScan (OBP-401) (Clontech, Takara Bio Inc., Shiga, Japan).
8. Matrigel basement membrane matrix (Becton Dickinson, Bedford, MA).
9. Pentobarbital sodium (Nembutai) (Dainippon Sumitomo Pharmaceutical Co., Ltd., Osaka, Japan) for anesthesia with intraperitoneal administration.
10. Ketamine hydrochloride (Ketalar) (Sankyo Co., Ltd., Tokyo, Japan) for anesthesia with intramuscular administration.

2.3. Equipment

1. Hamamatsu C5810 three-chip cooled color charge-coupled device (CCD) camera (Hamamatsu Photonics Systems, Hamamatsu, Japan).
2. GG475 emission filter (Chroma Technology, Brattleboro, VT).
3. Hamamatsu L8428 xenon lamp power supply (Hamamatsu Photonics Systems, Hamamatsu, Japan).

4. AQUACOSMOS software U7501 for fluorescence imaging (Hamamatsu Photonics Systems, Hamamatsu, Japan).
5. Fluorescence stereomicroscope SZX7 (Olympus, Tokyo, Japan).
6. Adobe Photoshop 4.0.1 J software for image processing (Adobe Systems Inc., Woodstock, GA).

3. Methods

Lymphatic invasion is one of the major routes for cancer metastasis. Adequate resection of locoregional lymph nodes is required for curative treatment in patients with advanced malignancies. Therefore, the use of TelomeScan for real-time imaging of tumor tissues in vivo offers a practical, safe, and cost-effective alternative to the traditional, cumbersome procedures for histopathological examination. Following intratumoral injection of TelomeScan into orthotopically-implanted human colorectal or tongue tumors in mice, para-aortic or neck lymph node metastasis could be visualized under a fluorescence imaging camera (7, 9). Histopathological analysis confirmed the presence of metastatic tumor cells in the lymph nodes by fluorescence emission, whereas GFP-negative lymph nodes contained no tumor cells. These data indicate that TelomeScan selectively replicates in neoplastic lesions, resulting in GFP expression in metastatic lymph nodes.

Progression of non-small-cell lung cancer or malignant pleural mesothelioma is characterized by the dissemination of neoplastic cells. To overcome the poor prognosis of lung cancer and mesothelioma involving the pleura, innovative methodology for early diagnosis as well as treatment is needed. We reported previously that intra-thoracic injection of replication-deficient adenovirus-expressing antisense against heparanase, which is associated with metastatic potential, inhibited pleural dissemination of human lung cancer cells in an orthotopic murine model (10). These observations suggest that intra-thoracically-administered adenovirus is capable of infecting disseminated neoplastic nodules in the thoracic cavity. Thus, we also evaluated the infection and replication of TelomeScan in human non-small-cell lung cancer cells growing intra-thoracically in athymic *nu/nu* mice.

3.1. Construction of TelomeScan (OBP-401)

1. An 897-bp fragment of the *E1A* gene and a 1,822-bp fragment of the *E1B* gene were amplified by PCR from cellular RNA and genomic DNA of 293 cells, respectively. The amplified products were subcloned into the pTA plasmid. Following confirmation by DNA nucleotide sequencing, the *E1A* (911 bp) and *E1B* (1,836 bp) genes were cloned into the pIRES vector (pE1A-IRES-E1B).

2. A 455-bp fragment of the hTERT promoter, which contains a 378-bp region upstream of the transcription start site, was ligated into the pE1A-IRES-E1B (phTERT-E1A-IRES-E1B). The 3,828-bp fragment was digested from the phTERT-E1A-IRES-E1B, and was then cloned into pShuttle after deletion of the CMV promoter.
3. A fragment containing the CMV promoter, EGFP, and simian virus 40 polyadenylation signal (CMV-EGFP-SV40pA) was also prepared and inserted into the shuttle plasmid pHM11. The *Csp45*I fragment of this plasmid was then inserted into the *Cla*I site of the pShuttle vector.
4. The resultant pShuttle vector was applied to the Adeno-X™ Expression System. Recombinant adenovirus was isolated from a single plaque and expanded in 293A cells. The resultant virus was termed TelomeScan (OBP-401) (see Fig. 1).
5. The virus was purified by ultracentrifugation in step gradients of cesium chloride. Titers were determined by plaque-forming assay using 293 cells. Storage is at –80°C (see Note 1).

3.2. Orthotopic Human Rectal Cancer Model (See Fig. 2)

1. In order to generate an orthotopic rectal cancer model, female BALB/c *nu/nu* mice were anesthetized using pentobarbital sodium or ketamine hydrochloride, and were then placed in a supine position.

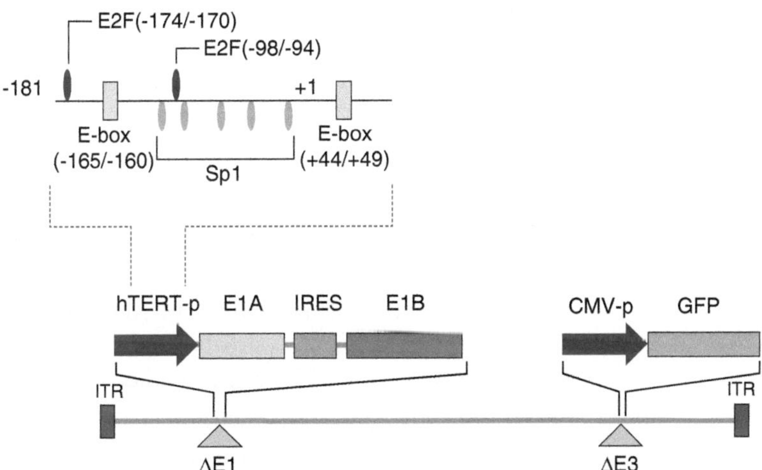

Fig. 1. Schematic DNA structure of TelomeScan (OBP-401). TelomeScan is a telomerase-specific replication-competent adenovirus variant, in which the hTERT promoter element drives expression of the *E1A* and *E1B* genes linked with an IRES. The *GFP* gene is inserted under the CMV promoter into the E3 region for monitoring viral replication. A 378-bp region upstream of the transcription start site of the hTERT promoter sequence, that contains the 181-bp core promoter region essential for transcriptional activation of hTERT, was used. Putative protein binding sites for various transcription factors are indicated.

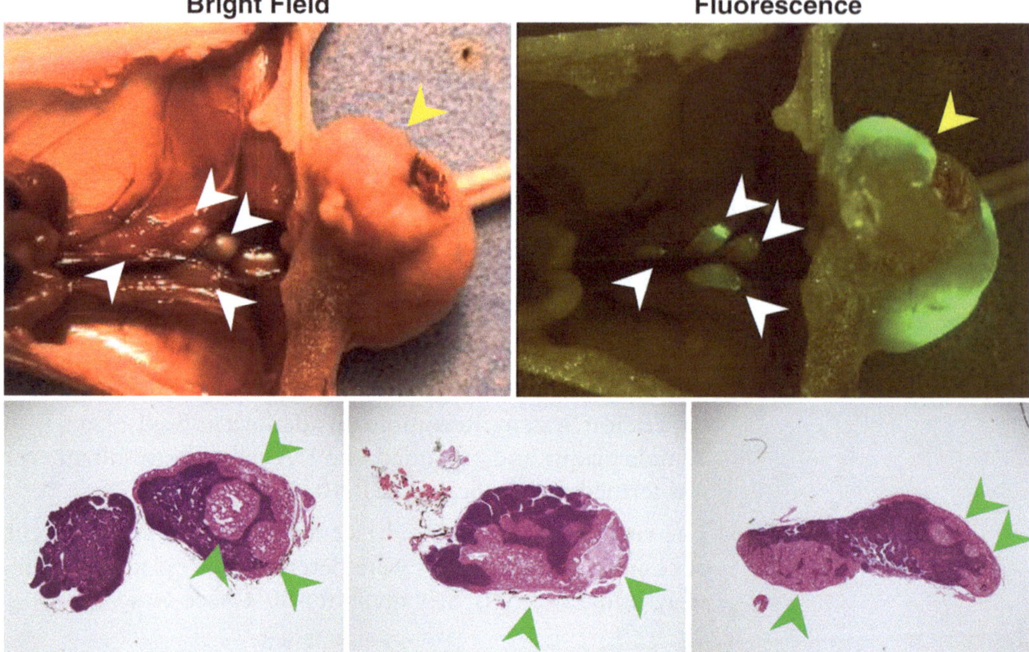

Fig. 2. Orthotopic xenografts of human colorectal cancer cells and selective visualization of lymph node metastasis. Representative internal images (*top panels*) and hematoxylin and eosin staining of lymph node sections (*bottom panels*) are shown. HT29 human colorectal cancer cells (5×10^6 cells), suspended in Matrigel, were inoculated into the rectal submucosa of athymic *nu/nu* mice. Four to six weeks later, rectal tumors were observed. Five days after injection of 1×10^8 PFU of TelomeScan into the implanted rectal tumors, the abdominal cavity was explored by laparotomy. Four lymph nodes were macroscopically identified adjacent to the aorta (bright view). Optical CCD imaging (fluorescence view) visualized these lymph nodes as light-emitting spots with GFP fluorescence. Primary tumor (*yellow arrowhead*) and metastatic lymph nodes (*white arrowheads*) are shown. Histopathological analysis confirmed the presence of metastatic adenocarcinoma cells in the lymph nodes with fluorescence emission. Metastatic foci in nodes are indicated with *green arrowheads*.

2. The ano-rectal wall was cut at a length of 7 mm to prevent colonic obstruction resulting from rectal tumor progression.
3. A cell suspension of HT29 human colorectal cancer cells, at a density of 5×10^6 cells in 100 μL of Matrigel basement membrane matrix, was injected slowly into the submucosal layer of the rectum through a 27-gauge needle.
4. Rectal tumors appeared within 7 days after injection. Histopathological examination of the excised primary tumor showed a submucosal tumor formation composed of implanted HT29 cells with a solid architecture and invasion into the muscularis propria and submucosa. Examination under high magnification showed tumor cell-filled lymphatic vessels in the muscularis propria layer.
5. Four weeks after tumor inoculation, 1×10^8 plaque-forming units (PFU)/100 μL TelomeScan were injected directly into

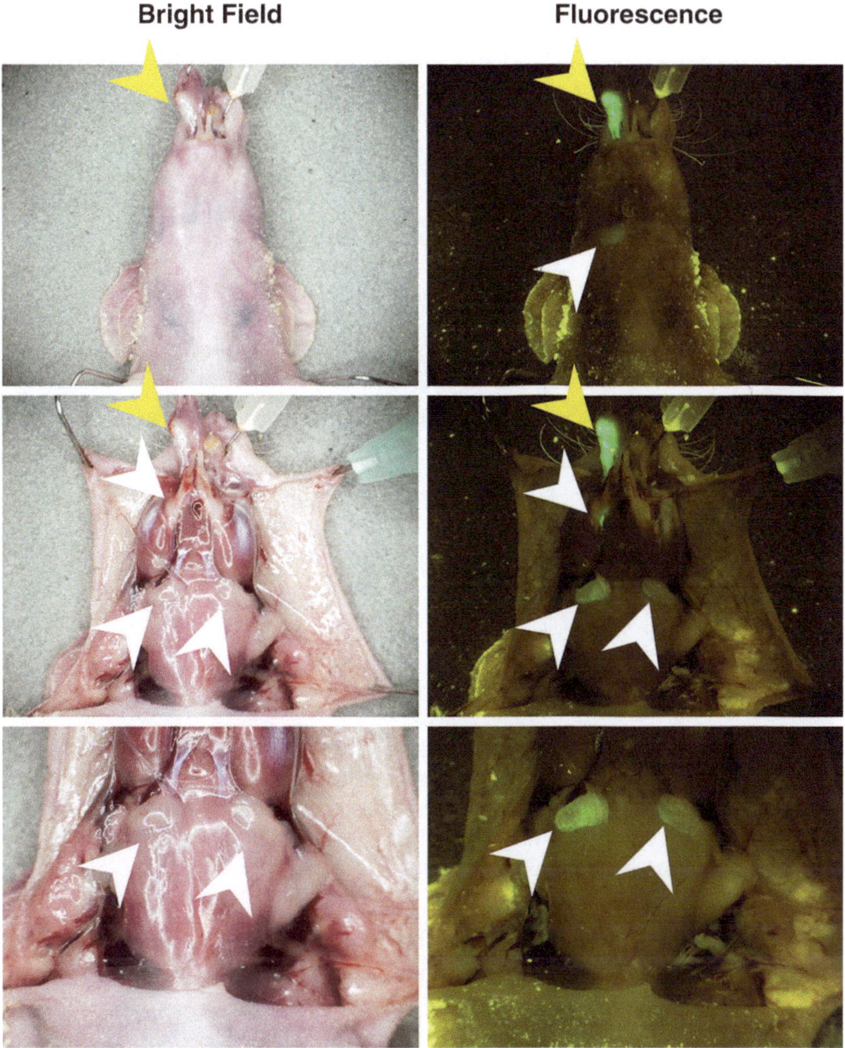

Fig. 3. Virus spread of TelomeScan, delivered into primary tumors, via lymphatics to regional lymph nodes in orthotopic human cancer xenograft models with HSC3 human oral squamous-cell carcinoma cells. The tongues of BALB/c *nu/nu* mice were inoculated with 1×10^5 cells of HSC3 human squamous-cell carcinoma. Mice received intra-tumoral injection of TelomeScan at a concentration of 1×10^8 PFU 24 days after tumor inoculation. The mice were assessed for lymph node metastasis 5 days later under a fluorescence stereomicroscope. External images (*top panels*), internal images (*middle panels*), and higher magnification of internal images (*bottom panels*) are shown. Primary tumor and metastatic lymph nodes are indicated by *yellow arrowheads* and *white arrowheads*, respectively.

the rectal tumors under anesthesia (see Notes 2 and 3). Mice were sacrificed and their abdominal spaces were examined during laparotomy 5 days after virus injection.

3.3. Orthotopic Human Head and Neck Cancer Model (See Fig. 3)

1. HSC-3 human oral squamous-cell carcinoma cells were harvested and suspended at a concentration of 5×10^6 cells/mL in medium.
2. In order to generate an orthotopic head and neck cancer model, 6-week-old female BALB/c *nu/nu* mice were anesthetized

Fig. 4. Visualization of disseminated human non-small-cell lung cancer cells in vivo by TelomeScan infection. Six weeks after intra-thoracic inoculation of 5×10^6 A549 human lung cancer cells, mice received an intra-thoracic injection of 1×10^8 PFU of TelomeScan. Internal images of pleural lung cancer dissemination visualized by intra-thoracic injection of OBP-401. GFP fluorescence was detected 6 days after virus administration using a CCD camera (*top panels*). We confirmed that excised tumor nodules were also labeled with GFP (*bottom panels*).

and injected directly with 20 μL of a cell suspension containing 10^5 cells into the right lateral border of the tongue with a 27-gauge needle.

3. Histopathological examination of excised primary tumors showed tumor formation comprised of implanted HSC-3 cells with a solid architecture.
4. When the tumor grew to 2–3 mm in diameter, approximately 5–7 days after tumor inoculation, 20 μL of a solution containing 1×10^8 PFU of TelomeScan was injected into the tumor under anesthesia.
5. Five days after virus injection, primary tongue tumors and lymph node metastases were visualized by fluorescence imaging.

3.4. Orthotopic Human Non-small-Cell Lung Cancer Pleural Dissemination Model (See Fig. 4)

1. A 27-gauge needle was used to intra-thoracically inject female BALB/c *nu/nu* mice with 100 μL of a suspension containing 5×10^6 A549 human non-small-cell lung cancer cells under anesthesia.
2. Two weeks later, disseminated tumor nodules were detected in the visceral pleura, parietal pleura, diaphragmatic pleura, and mediastinum.

3. The same technique was used for intra-thoracic injection of 1×10^8 PFU of TelomeScan under anesthesia.
4. Six days after virus injection, mice were killed and their thoracic spaces were examined by optical fluorescence imaging.

3.5. In Vivo Fluorescence Imaging

1. In vivo GFP fluorescence imaging was performed by illuminating the animal with a 150-W xenon lamp using an excitation filter (midpoint wavelength peak, 470 nm).
2. The re-emitted fluorescence was collected through a GG475 long pass filter on a Hamamatsu C5810 three-chip color CCD camera or a fluorescence stereomicroscope (SZX7).
3. High-resolution image acquisition was accomplished using an EPSON PC. Images were processed for contrast and brightness with Adobe Photoshop 4.0.1 J software.

3.6. Histological Examination

1. Tumor tissues identified with GFP fluorescence were harvested.
2. Tissues were fixed in 20% neutral-buffered formalin.
3. Paraffin-embedded sections (4 μm) were deparaffinized in xylene and rehydrated through a graded alcohol series into PBS.
4. Slides were stained with hematoxylin and eosin to detect tumor tissues.

4. Notes

1. The particle-to-PFU ratio of the adenoviral vectors should be kept low, because infectivity and replication primarily depend on virus purity. High-purity virus stock contains numerous productive and infectious virus particles. The virus particle (VP)/PFU ratio should be 100 or less, and ideally 50 or less. The VP/PFU ratio of the stock TelomeScan that we used in this study was 20.
2. Tumor cells are able to maintain telomere length predominantly due to telomerase, and its activity is detected in about 85% of malignant tumors (11). Telomerase is absent in most normal somatic tissues (12), with a few exceptions, including peripheral blood leukocytes and certain stem cell populations (13, 14). The strong association between telomerase activity and malignancy suggests that telomerase is a plausible target for the diagnosis and treatment of cancer (15). Levels of telomerase activity, however, vary among cell types. In addition, it has been reported that an alternative mechanism for telomere elongation, the so-called alternative lengthening of telomeres (ALT), is present in a certain proportion of telomerase-negative

human tumors (16). Therefore, prior to in vivo imaging experiments, the sensitivity of individual human cancer cell lines to TelomeScan should be investigated in vitro using a fluorescence microplate reader (DS Pharma Biomedical, Osaka, Japan) with excitation/emission at 485 nm/528 nm. GFP fluorescence was expressed relative to that of H1299 human non-small-cell lung cancer cells, which are sensitive to TelomeScan infection. For human cancer cells that are resistant to TelomeScan infection in vitro, higher doses of TelomeScan should be used for in vivo imaging in order to increase the multiplicity of infections (MOI).

3. Although TelomeScan showed broad sensitivity in human cancers originating from various organs, virus infectivity also depends on coxsackie-adenovirus receptor (CAR) expression, which is not expressed on the cell surface of some types of human cancer cells. Thus, tumors that lose CAR expression may be refractory to infection with TelomeScan. When CAR-negative human cancer cell lines are used for in vivo experiments, the fiber-protein of TelomeScan should be modified to contain RGD (Arg-Gly-Asp) peptide, which binds with high affinity to integrins ($\alpha v\beta 3$ and $\alpha v\beta 5$) on the cell surface, on the HI loop of the fiber protein. We previously confirmed that the fiber-protein-modified OBP-405 virus mediated not only CAR-dependent virus entry but also CAR-independent, RGD-integrin-dependent virus entry (5).

Acknowledgments

The authors would like to thank Dr. Robert M. Hoffman for his helpful discussion and encouragement. The authors also thank Tomoko Sueishi and MitsukoYokota for the excellent technical support. This work was supported in part by grants from the Ministry of Education, Culture, Sports, Science, and Technology of Japan (to T.F.) and by grants from the Ministry of Health, Labour, and Welfare of Japan (to T.F.).

References

1. Chishima, T., Miyagi, Y., Wang, X., Yamaoka, H., Shimada, H., Moossa, A.R., *et al.* (1997). Cancer invasion and micrometastasis visualized in live tissue by green fluorescent protein expression. *Cancer Res* 57(10), 2042–2047.
2. Yang, M., Baranov, E., Jiang, P., Sun, F.X., Li, X.M., Li, L., *et al.* (2000). Whole-body optical imaging of green fluorescent protein-expressing tumors and metastases. *Proc Natl Acad Sci USA* 97(3), 1206–1211.
3. Hoffman, R.M. (2005). The multiple uses of fluorescent proteins to visualize cancer in vivo. *Nat Rev Cancer* 5(10), 796–806.
4. Kawashima, T., Kagawa, S., Kobayashi, N., Shirakiya, Y., Umeoka, T., Teraishi, F., *et al.* (2004). Telomerase-specific replication-selective virotherapy for human cancer. *Clin Cancer Res* 10(1 Pt 1), 285–292.
5. Taki, M., Kagawa, S., Nishizaki, M., Mizuguchi, H., Hayakawa, T., Kyo, S., *et al.* (2005).

Enhanced oncolysis by a tropism-modified telomerase-specific replication-selective adenoviral agent OBP-405 ('Telomelysin-RGD'). *Oncogene* 24(19), 3130–3140.

6. Umeoka, T., Kawashima, T., Kagawa, S., Teraishi, F., Taki, M., Nishizaki, M., *et al.* (2004). Visualization of intrathoracically disseminated solid tumors in mice with optical imaging by telomerase-specific amplification of a transferred green fluorescent protein gene. *Cancer Res* 64(17), 6259–6265.
7. Kishimoto, H., Kojima, T., Watanabe, Y., Kagawa, S., Fujiwara, T., Uno, F., *et al.* (2006). In vivo imaging of lymph node metastasis with telomerase-specific replication-selective adenovirus. *Nat Med* 12(10), 1213–1219.
8. Kishimoto, H., Zhao, M., Hayashi, K., Urata, Y., Tanaka, N., Fujiwara, T., *et al.* (2009). In vivo internal tumor illumination by telomerase-dependent adenoviral GFP for precise surgical navigation. *Proc Natl Acad Sci USA* 106(34), 14514–14517.
9. Kurihara, Y., Watanabe, Y., Onimatsu, H., Kojima, T., Shirota, T., Hatori, M., *et al.* (2009). Telomerase-specific virotheranostics for human head and neck cancer. *Clin Cancer Res* 15(7), 2335–2343.
10. Uno, F., Fujiwara, T., Takata, Y., Ohtani, S., Katsuda, K., Takaoka, M., *et al.* (2001). Antisense-mediated suppression of human heparanase gene expression inhibits pleural dissemination of human cancer cells. *Cancer Res* 61(21), 7855–7860.
11. Shay, J.W., Bacchetti, S. (1997). A survey of telomerase activity in human cancer. *Eur J Cancer* 33(5), 787–791.
12. Dong, C.K., Masutomi, K., Hahn, W.C. (2005). Telomerase: regulation, function and transformation. *Crit Rev Oncol Hematol* 54(2), 85–93.
13. Hiyama, K., Hirai, Y., Kyoizumi, S., Akiyama, M., Hiyama, E., Piatyszek, M.A., *et al.* (1995). Activation of telomerase in human lymphocytes and hematopoietic progenitor cells. *J Immunol* 155(8), 3711–3715.
14. Tahara, H., Yasui, W., Tahara, E., Fujimoto, J., Ito, K., Tamai, K., *et al.* (1999). Immunohistochemical detection of human telomerase catalytic component, hTERT, in human colorectal tumor and non-tumor tissue sections. *Oncogene* 18(8), 1561–1567.
15. Shay, J.W., Wright, W.E. (2002). Telomerase: a target for cancer therapeutics. *Cancer Cell* 2(4), 257–265.
16. Muntoni, A., Reddel, R.R. (2005). The first molecular details of ALT in human tumor cells. *Hum Mol Genet* 14 Spec No. 2, R191–R196.

Chapter 10

Real-Time Fluorescence Imaging of Abdominal, Pleural, and Lymphatic Metastases

Susanne Carpenter and Yuman Fong

Abstract

Virally-directed fluorescence imaging has the potential to revolutionize intra-operative oncologic staging and tumor resection. Many viruses genetically engineered to specifically infect tumor cells as cancer therapy can be further modified to have a visible marker gene for cancer staging. In this chapter, we describe such a herpes simplex virus (HSV) modified to be detected by fluorescence. Other viruses so designed can be similarly used in cancer detection and staging. Replication-competent, tumor-specific HSV NV1066 expresses green fluorescent protein (GFP) in infected cancer cells. One single dose of NV1066 administered via intratumor, intracavitary, or systemic injection can spread within and across body cavities to target tumor cells while sparing normal tissue cells from infection. Tumors otherwise invisible by conventional laparoscopy appear green with the use of an endoscope equipped with a fluorescent filter. Furthermore, with GFP expression easily visualized by stereomicroscopy, microscopic, and pathologic analysis is significantly enhanced. This chapter addresses NV1066-directed visualization of peritoneal, pleural, and lymphatic metastases. This chapter also provides protocols for the production of tumor models in various body cavities in rodents.

Key words: Fluorescence imaging, Green fluorescent protein, Metastases, Herpes simplex virus, NV1066, Oncolytic virotherapy, Virally directed imaging

1. Introduction

Recent advances in radiology have ushered in a new era in cancer staging and resection. In abdominal cancers, many advanced imaging modalities assist surgeons in characterizing malignancy and metastases preoperatively. However, even with judicious and thorough pre-operative workups, 20–30% of patients undergoing attempted resection of hepatic colorectal metastases have unresectable disease determined at surgical exploration (1). Similarly, 20–30% of peritoneal and pleural malignant implants are first

Robert M. Hoffman (ed.), *In Vivo Cellular Imaging Using Fluorescent Proteins: Methods and Protocols*,
Methods in Molecular Biology, vol. 872, DOI 10.1007/978-1-61779-797-2_10, © Springer Science+Business Media, LLC 2012

diagnosed intraoperatively (2–4). Cytoreductive surgery has been shown to improve survival in patients with peritoneal metastases of many origins (5–7). Surgery is also a mainstay of treatment for malignant pleural mesothelioma (MPM), a disease highly resistant to chemotherapy and requiring high-dose radiation of the entire hemithorax (8, 9). Even with tri-modal therapy combining aggressive chemotherapy, radiation, and surgery, prolonged disease-free survival is uncommon (9–11). Even though completeness of resection or cytoreduction is the most important prognostic factor for patients with peritoneal carcinomatosis (5, 12, 13), few tools other than direct tumor visualization, which can often be incomplete, are available to the surgical oncologist intraoperatively to assist in the identification of abdominal malignancy (14). The same is true of disseminated pleural malignancy, rendering detection of thoracic micrometastases extremely difficult (15).

Although sentinel lymph node biopsy has revolutionized evaluation of axillary lymph node status, sentinel-node metastases are still missed intraoperatively, and nonsentinel-node metastases are left unidentified (16, 17). Lack of detection and aggressive metastatic disease contribute to frequent recurrences among patients with apparently complete oncologic resections. Virally-directed fluorescence endoscopic imaging and fluorescence microscopy offer simple and efficient means of significantly improving intraoperative direct and pathologic visualization, and hold promise for multi-modal intraoperative imaging and treatment.

Oncolytic viruses are engineered to infect, replicate in, and lyse cancer cells. In the last decade, oncolytic viral therapy has emerged at the forefront of novel cancer therapy (18). Specifically, oncolytic herpes simplex virus (HSV) has been successfully tested against numerous cancer types, including mesothelioma, head and neck, lung, esophageal, gastric, colorectal, gallbladder, hepatic, pancreatic, and bladder cancer (19–25).

NV1066 is an attenuated oncolytic herpesvirus that expresses enhanced green fluorescent protein (GFP) utilizing a cytomegalovirus (CMV) promoter described in detail previously (21). This promoter is expressed 2–6 h following viral entry into cells (21). Viral selectivity for tumor cells, correlation of GFP expression with NV1066 infection, and use of this expression to identify and treat cancer have been clearly demonstrated in both in vitro and in vivo models (24, 26–28). Real-time identification of NV1066-infected tumors and metastases (including micrometastases <1 mm in diameter) using endoscopic fluorescent light filters has been confirmed in 110 cell types and 16 different organs (29). This chapter focuses on real-time NV1066-directed imaging of abdominal, peritoneal, pleural, and lymphatic metastases in murine models.

2. Materials

2.1. Establishment of Abdominal Carcinomatosis in a Mouse Model (28)

1. Eight-to-ten-week-old athymic mice are housed in a temperature- and light-controlled animal facility. The mice received food and water ad libitum (see Note 1).
2. For all invasive manipulations, animals are anesthetized with inhalational isoflurane or methoxyflurane. Sacrifice is achieved by CO_2 inhalation at experiment's end or per protocol.
3. 1×10^7 cancer cells (see Note 2). We describe the use of OCUM-2MD3 gastric cancer cells (obtained as a gift from Dr. Masaka Yashiro, Osaka City University Medical School, Japan) (28).
   (a) Cell culture medium is Dulbecco's Modified Eagle's Medium (DMEM), supplemented with high glucose, 2 mM L-glutamine, 0.5 mM sodium pyruvate, 100 U/mL penicillin, 100 mg/mL streptomycin, and 10% fetal calf serum (FCS).
   (b) All cell cultures herein are placed in a humidified incubator with 5% CO_2 at 37°C.
4. 500 μL phosphate-buffered saline (PBS) for injection of 1×10^7 cancer cells into the animal peritoneal cavity.
5. Insulin syringe with 28.5-gauge hypodermic needle.

2.2. Establishment of Malignant Pleural Mesothelioma in a Mouse Model

1. Athymic mice are utilized. Standard animal care and sacrifice for this process are congruent with steps 1 and 2 of Subheading 2.1.
2. Anesthesia is induced with a mixture of isoflurane (2 L/min) and oxygen (2 L/min) in an induction chamber and is maintained by a nasal cone.
3. 1×10^7 thoracic cancer cells (MSTO-211 H is described here) are mixed with 100 μL PBS.
   (a) As above, cells are cultured in appropriate medium and placed in a humidified incubator with 5% CO_2 at 37°C.
4. Surgical materials required include:
   (a) 10% Povidone/iodine solution for preparation.
   (b) Number 11 surgical scalpel blade.
   (c) 27-gauge needle.
   (d) Insulin syringe.

2.3. Establishment of Axillary Lymphatic Metastases in a Mouse Model

1. Six-to-eight-week-old female athymic mice are utilized. Standard animal care and sacrifice is congruent with steps 1 and 2 of Subheading 2.1.
2. 1×10^7 breast cancer cells (LN435 is described here) are mixed with 100 μL PBS.
   (a) As above, cells are cultured in appropriate medium and placed in a humidified incubator with 5% CO_2 at 37°C.

3. 1% Isosulfan blue dye (50 μL) is used to confirm the normal lymphatic drainage pattern of the primary tumor site.
4. Surgical tools of choice are utilized for biopsy of established LN435 tumor and implantation into third thoracic mammary fat pad.

2.4. Viral Stock and Propagation

1. Virus (NV1066) obtained from MediGene, Inc (Martinsried, Germany). NV1066 is an attenuated, replication-competent oncolytic herpesvirus with deletions of virulence genes ICP0, γ134.5, and ICP4. Viral construction has previously been described in detail (21). The virus has a marker gene for enhanced GFP (EGFP) that is controlled by a constitutively-expressed CMV promoter (see Note 3).
2. Viral stocks are propagated on Vero cells.
   (a) African green monkey kidney cells (Vero cells) obtained from American Type Culture Collection (ATCC) (Rockville, MD).
   (b) Vero cell culture medium: minimal essential medium (MEM) containing 10% FCS supplemented with penicillin and streptomycin (see Note 4).

2.5. Viral Titer by Standard Plaque Assay (See Note 5)

1. 42×10^5 Vero cells (7×10^5 cells per well in 2 mL Vero cell medium [MEM]).
2. Six-well plates—sufficient for control, and for each desired multiplicity of infection (MOI), with experiments repeated in triplicate (e.g., at least three wells per MOI).
3. Agarose preparation:
   (a) 1 g agarose Type VII (Sigma-Aldrich, St. Louis, MO).
   (b) 50 mL distilled de-ionized water.
   (c) Microwave.
   (d) Warm 2× DMEM + 10% FCS with penicillin and streptomycin (without phenol red).
   (e) 500-mL bottle with 0.2-μm bottle-top filter.
   (f) Water bath at 37°C.
4. Neutral red solution preparation:
   (a) 1 mL neutral red solution (Sigma-Aldrich).
      - Neutral red solution must be protected from light and stored at 4°C.
   (b) 25 mL distilled de-ionized water.
   (c) 25 mL 2× DMEM + 10% FCS with penicillin and streptomycin (without phenol red).

2.6. Viral Administration

Abdominal Model:

1. A minimum of 5×10^6 plaque-forming units (PFU) NV1066 suspended in 100 μL PBS (see Note 6).
2. Insulin syringe with 28.5-gauge hypodermic needle.

Pleural Model:

1. 1×10^7 PFU NV1066 suspended in 100 μL PBS.
2. 1 mL syringe with 27-gauge needle.

Axillary Model:

1. 5×10^6 PFU NV1066 suspended in 50 μL PBS
2. Insulin syringe with 28.5-gauge hypodermic needle.

2.7. Endoscopy

1. Fluorescence thoracoscopic/laparoscopic system developed in collaboration with Olympus America, Inc. (Scientific Equipment Division, Melville, NY). A control button on the scope handle enables rapid exchange between bright-field and fluorescence light modes.
2. Light source is derived from Olympus Visera CLV-U40 model (Olympus America, Inc.) adapted with an interchangeable excitation filter set at 470 ± 20 nm to accommodate a GFP excitation peak at 475 nm. Emission filter fixed at 500 nm for GFP emission peak at 509 nm.
3. Camera processor: Olympus Visera OTV-S7V with emission filter at 510 nm (Olympus America, Inc.).
4. Number 11 surgical scalpel blade.

2.8. Open Imaging

1. Zeiss LSM 510 confocal laser scanning microscope (Carl Zeiss, Inc., Oberkochen, Germany) and MetaMorph Imaging System (Downingtown, PA) is used to visualize GFP-expressing cancer cells with both bright-field and fluorescence modes. This imaging system is used both for pathologic analysis and for open imaging following mouse laparotomy or thoracotomy.
   (a) For experiments involving axillary metastases, lymph nodes were imaged using a Leica MZFL3 Stereomicroscope (Leica Microsystems, Germany).
2. Image-capture system: Retiga EX digital CCD camera (Qimaging, BC, Canada).

2.9. Pathologic Analysis

1. For tissue sectioning and staining to confirm GFP expression in tumors:
   (a) Samples emitting EGFP are frozen in Tissue-Tek embedding medium (Sakura Finetek, Torrance, CA) and visualized under fluorescence stereomicroscopy.

(b) Cryotome for sectioning.
(c) 10% Phosphate-buffered formalin fixative.
(d) Paraffin for embedding.
(e) Hematoxylin and eosin (H&E) for confirmation of fluorescent correlation with tumor presence.

2. Immunohistochemistry (IHC) to confirm that GFP localizes to virus:
(a) Rabbit anti-HSV-1 polyclonal antibody (Ready-to-Use, Biogenex, San Ramon, CA) to detect the presence of virus in harvested tissues (see Note 7).
(b) Biotinylated secondary antibody can be added and visualized with streptavidin-labeled horseradish peroxidase and chromogen solutions (Super Sensitive Ready-to-Use Detection System, Biogenex).
(c) Harris hematoxylin for counterstaining.

2.10. Confirmation of Viral Specificity for Tumors by Real-Time Polymerase Chain Reaction

(This step is an optional addition to histologic confirmation of viral specificity for tumor).

1. Samples of infected and uninfected tissue from same organ.
2. Liquid nitrogen.
3. Wizard Genomic DNA Purification Kit (Promega, Madison, WI) is used to isolate genomic DNA from animal tissue samples.
(a) Materials required in addition to the kit (as per manufacturer's protocol):
   - 1.5-mL microcentrifuge tubes.
   - 15-mL centrifuge tubes.
   - Small homogenizer (alternatively mortar and pestle).
   - PBS.
   - 95°C water bath.
   - 37°C water bath.
   - Isopropanol, room temperature.
   - 79% Ethanol, room temperature.
4. Real-time polymerase chain reaction (RT-PCR) is performed using an ABI Prism 7900HT Sequence Detection System (Applied Biosystems, Foster City, CA):
(a) To amplify and detect the 111-bp fragment of the HSV immediate-early gene: forward (5′-ATGTTTCCCGTCTGGTCCAC-3′) and reverse (5′-CCCTGTCGCCTTACGTGAA-3′) primers and a dual-labeled fluorescent TaqMan probe (5′-FAM-CCCCGTCTCCATGTCCAGGATGG-TAMRA-3′).

(b) To amplify the 87-bp coding sequence of 18s rRNA to normalize the total amount of DNA present: forward (5′-CGCCTACCACATCCAAGGAA03′) and reverse (5′-GCTGGAATTACCGCGGCT-3′) primers and a dual-labeled fluorescent TaqMan probe (5′-VIC-TGCTGG CACCAGCTTGCCCTC-TAMRA-3′).

3. Methods

One advantage of virally-directed real-time in vivo imaging is its utility in multiple tumor models. Several authors in our group have achieved successful real-time visualization of metastases via virally directed fluorescence imaging with thoracic (see Fig. 1) and abdominal carcinomatosis (see Fig. 2), as well as in metastatic lymph node identification (see Fig. 3) (21, 24, 26, 28). While not discussed here, several authors have also demonstrated the ability of NV1066 to identify tumors via systemic, intra-tumor, and intra-cavitary injection. Furthermore, NV1066 has been noted to traverse body cavities, identifying lesions in the abdomen when injected only into the thoracic cavity (29).

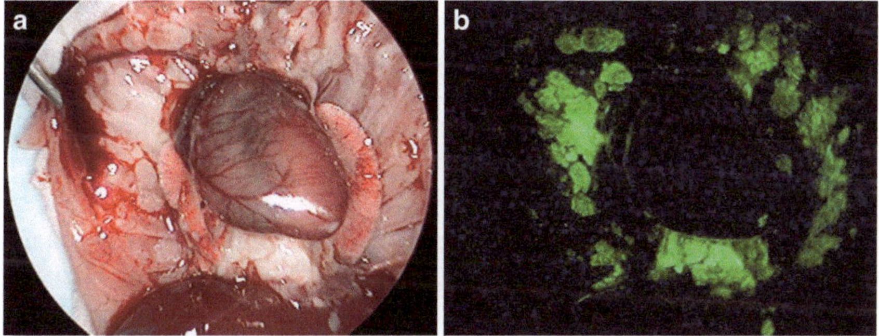

Fig. 1. (**a**, **b**) Fluorescence thoracoscopy following NV1066 treatment of pleural metastases in a murine model.

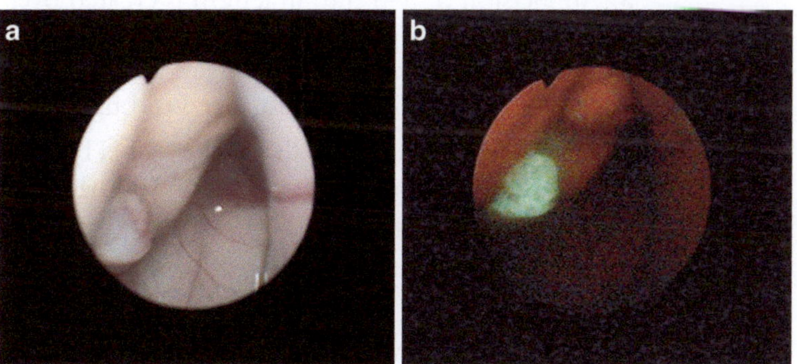

Fig. 2. (**a**, **b**) Fluorescence abdominal laparoscopy showing peritoneal macrometastasis following NV1066 treatment of induced peritoneal carcinomatosis in a murine model.

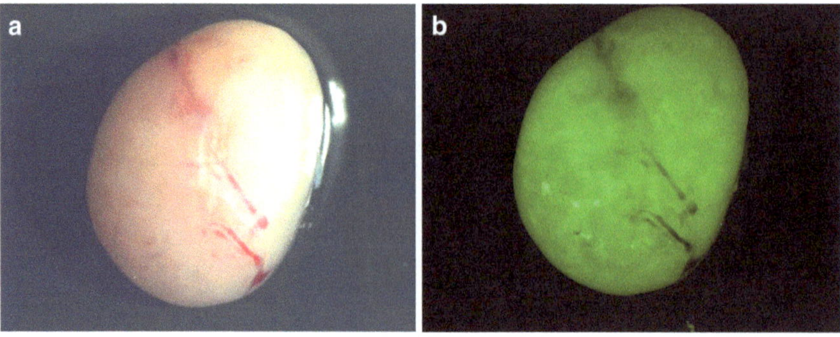

Fig. 3. (**a**, **b**) Ex vivo microscopy of murine axillary lymph node metastasis.

It is important to mention that establishing peritoneal carcinomatosis in the murine model is not a benign procedure. Mice reliably develop severe carcinomatosis including cachexia and ascites. When designing in vivo experiments one must consider this issue when delineating time points for viral treatment and criteria for animal sacrifice in order to minimize suffering.

3.1. Establishment of Abdominal Carcinomatosis in a Mouse Model

1. 1×10^7 cancer cells (OCUM-2MD3 in this example) are suspended in 500 μL PBS and then injected into the peritoneal cavity of 8–10-week-old athymic mice (29).
2. Mice are monitored for signs of discomfort and cachexia and are sacrificed if either of these signs appears or as deemed necessary by animal caretakers to maximize humane treatment of the mice.

3.2. Establishment of Malignant Pleural Mesothelioma in a Mouse Model

1. Mice are anesthetized with isoflurane and oxygen in an induction chamber and maintained by nasal cone. They are placed in the left lateral position.
2. The right chest is disinfected with 10% povidone/iodine solution.
3. An incision of 3–5 mm in length is created over the fourth to fifth intercostal space, and sharp dissection is utilized to expose, but not penetrate the parietal pleura, such that the underlying expanding lung is easily visualized through the thin pleura.
4. 1×10^7 mesothelioma cells (MSTO-211H) are slowly injected into the pleural cavities of mice in 100 μL PBS suspension as described above.
   (a) Lung puncture is easily avoided by visualizing the needle tip through the transparent parietal pleura.
5. Skin is closed with surgical staples following injection, and mouse recovery is observed for 15 min.

3.3. Establishment of Axillary Lymphatic Metastases in a Mouse Model

1. Mice are anesthetized as described in step 1, Subheading 3.2 for all invasive procedures, other than injection.
2. Confirmation of normal lymphatic drainage pattern of the future tumor site in the third thoracic mammary fat pad is achieved by injecting three mice with 50 μL of 1% isosulfan blue dye directly at the anticipated tumor site.
   (a) Two minutes following injection, animals are sacrificed and all axillary and superficial thoracic lymph node basins are exposed. Nodes are identified by uptake of blue dye.
3. Using LN435 cell suspensions, tumors are established in the primary mammary fat pad.
4. After desired tumor growth, 1 mm^3 biopsies of existing LN435 primary mammary fat pad tumors are implanted into the third thoracic mammary fat pad of the same mouse.

3.4. Viral Propagation on Vero Cells

1. Vero cells are passaged in 225-cm^2 flasks, when approaching confluence, using 0.25% trypsin with 0.2% ethylenediaminetet-raacetic acid (EDTA). Cells are maintained in a humidified incubator with 5% CO_2 at 37°C.
2. Vero cells are grown in 24-well flat-bottom plates and are infected with NV1066 at MOIs of 0.01, 0.1, and 1.0 in 100 μL PBS.
3. Culture supernatants are collected each day for 7 days after infection.
4. All samples are tested in triplicate, and experiments are repeated at least three times.

3.5. Viral Titer by Standard Viral Plaque Assay

(This process spans 5 days; we like to start on a Monday so we can finish on a Friday).

1. *Monday*: Under semi-sterile hood conditions, seed Vero cells onto six-well plates in the late afternoon (7×10^5 cells per well in 2 mL of MEM with 10% FCS + PCS).
2. Incubate six-well plates overnight.
3. *Tuesday*: Make a tenfold serial dilution of a sample of the virus for desired MOI:
   (a) MOI 0.1 = 1:10 dilution: 100 μL sample + 900 μL media.
   (b) MOI 0.01 = 1:100 dilution: 100 μL above sample + 900 μL media, etc.
4. Aspirate media from Vero cell plates.
   (a) Use extreme care, and try not to touch the bottom of the well in order to avoid disturbing the cells.

5. Add 800 μL of desired MOI of viral sample per well.
   (a) Again use extreme caution, and attempt to deliver the sample to the plate by projecting the sample only toward the side of the well rather than the bottom.
6. Incubate for 4 h in a humidified incubator with 5% CO_2 at 37°C.
7. Prepare agarose:
   (a) Add 1 g agarose to 50 mL distilled deionized water in a beaker and heat on "high" in a microwave until melted. Melting time is typically 3–5 min.
   (b) Once melted, mix the above 50 mL water/agarose mixture with 50 mL warm 2× DME with 10% FCS, penicillin, and streptomycin (without phenol red).
   (c) Transfer the mixture from the beaker to a 500-mL bottle via a 0.22-μm bottle-top filter.
   (d) Keep bottle warm in 37°C water bath.
8. After 4 h incubation, again under semisterile hood conditions, aspirate supernatant from six-well plates (again careful not to disturb the cells).
9. Wash gently with 1 mL warm Vero-cell medium (MEM with 10% FCS with penicillin and streptomycin).
10. Gently add 2 mL agarose into each well.
    (a) Use caution to avoid producing bubbles. If bubbles form, it is necessary to attempt to pop them without disturbing the contents of the well.
11. Leave the plates in the hood for 10–15 min to let the gel solidify.
12. Once the gel has solidified, incubate for 2 days at 37°C.
13. Prepare neutral red solution:
    (a) In a beaker, add 1 mL neutral red to 25 mL distilled deionized water.
    (b) Add 25 mL 2× DME with 10% FCS, penicillin, and streptomycin (without phenol red).
14. *Thursday*: After 2 days incubation, add 2 mL of neutral red solution per well, protect wells from light, and place back in the incubator.
15. Wait for countable plaques to form (this usually takes 1 day), then count.
16. *Friday*: Calculate titer: avg # of plaques/well × dilution factor × (1 mL/0.8).

3.6. Viral Administration

Abdominal Model:

1. Seven to twelve days following implantation of cancer cells (see Note 8), 5×10^6 PFU, suspended in 100 μL PBS, are injected into the peritoneal cavity. Control mice are injected with 100 μL PBS alone.
2. Following viral administration, mice are divided into two groups which undergo laparoscopy to identify peritoneal tumors at 48 h or 7 days (see Note 9).
3. Animals are sacrificed immediately following laparoscopy.

Pleural Model:

1. Intra-pleural treatment: 1×10^7 PFU of NV1066 in 100 μL PBS is instilled intra-pleurally, as described above for establishment of pleural metastases. This is performed at a desired time interval following tumor cell instillation—in this example 5 days to examine microscopic disease, and 11 days to examine the NV1066 effect on macroscopic disease.
2. Mice are regularly assessed for weight loss and tachypnea.

Axillary Model:

1. Ten weeks following tumor implantation into the tertiary mammary fat pad, animals are treated with intratumoral injections of 5×10^6 PFU NV1066 in 50 μL PBS using an insulin syringe with 28.5-gauge hypodermic needle.

3.7. Endoscopy (See Note 10)

Abdominal Model:

1. With mice anesthetized, supine, and appropriately secured, a 3-mm incision is made with a number 11 blade scalpel. The 3-mm laparoscope is inserted into the abdomen and secured with purse-string-style sutures around the incision. These sutures also serve to simulate abdominal insufflation by lifting anteriorly.
2. The abdomen can be evaluated under standard white light bright-field or fluorescence modes. To evaluate the abdomen with a fluorescence filter, a button on the laparoscope handle is utilized and the image is converted instantly without compromising image dimensions or focus. Tissues containing tumor cells or deposits will appear green on the screen (see Fig. 2) (see Note 11).

Pleural Model:

1. Mice are sacrificed 48–72 h or 1 week following viral injection.
2. The 3–5-mm incision previously created over the fourth to fifth intercostal space is extended to 8–10 mm, and the lung is

retracted using a blunt instrument of choice. The endoscope is then inserted into the fourth to fifth intercostal space and secured with purse-string-style sutures around the incision.

3. Metastases are visualized under white light and fluorescence modes, as described above in the abdominal-metastasis model.

3.8. Open Imaging

Abdominal Model:

1. If background illumination can be effectively minimized, the fluorescence laparoscope can also be employed to image mice, during or after laparotomy, by simply aiming the laparoscope at the areas of interest.
2. However, more commonly performed is open imaging using a stereomicroscope containing both excitation and emission filters to detect GFP and standard visible light.
   (a) Images are captured using the Retiga EX digital CCD camera described above.

Pleural Model:

1. Following thorascopic examination, microscopic imaging was utilized both before and after resection of various thoracic contents. The diaphragm was the area least accessible by thorascopic examination but was amenable to microscopic examination, as described above in the abdominal model.
   (a) Microscopic examination of tissues was also performed ex vivo as described below.

Axillary Model:

1. 48 h after viral intra-tumoral injection, mice are sacrificed and lymph node basins are surgically exposed.
2. Lymph nodes are imaged using the Leica MZFL3 stereomicroscope in both bright-field and fluorescence modes (see Fig. 3).
   (a) Images are captured using the Retiga EX digital CCD camera described above.

3.9. Pathologic Analysis (Histologic Confirmation of Viral Specificity)

1. Areas of tissue that are green fluorescent and adjacent non-fluorescent sections are fixed in 10% phosphate-buffered formalin and embedded in paraffin for histologic analysis.
2. Serial 8-μm sections of all tissue blocks are cut and stained with H&E to assess for tumor presence.
   (a) Sections are compared under bright-field and fluorescence microscopy to evaluate the correlation of GFP expression with H&E indication of carcinoma.

3. Immunohistochemistry is performed to further confirm the presence of virus. Additional slides are stained with rabbit anti-HSV-1 polyclonal antibody. A biotinylated secondary antibody is added and visualized with streptavidin-labeled horseradish peroxidase and chromogen solutions.
   (a) Counterstaining with Harris hematoxylin is also performed.
   (b) All slides are compared for correlation of tumor with GFP expression and HSV localization.
   (c) An institutional animal pathologist confirms all findings.

3.10. Confirmation of Viral Tumor Specificity by RT-PCR

(This step is optional and the process is dependent upon the equipment available. We use absolute quantitative analysis).

1. Using liquid nitrogen, NV1066-infected and non-infected hepatic tissues are snap-frozen.
2. Genomic DNA is isolated using standard protocols for "Isolating Genomic DNA from Tissue Culture Cells and Animal Tissue" found at http://www.promega.com/tbs/Tm050/tm050.pdf (Wizard Genomic DNA Purification Kit; Promega).
3. Construction of standard curves is performed by spiking uninfected samples with known quantities of NV1066 prior to extraction.
4. RT-PCR is performed to quantify herpes ICP0 immediate-early gene and to normalize the amount of DNA present:
   (a) PCR conditions:
      - Stage 1: 50°C for 2 min.
      - Stage 2: 95°C for 10 min.
      - Stage 3 (35 cycles): 95°C for 15 s and 60°C for 1 min.
      - Stage 4: 25°C soak.

4. Notes

1. In our lab, we have typically used 6–10-week-old athymic mice. However, other investigators studying fluorescence in a peritoneal model using ALA have used WAG/Rij rats (30). In our lab, we have demonstrated that NV1066 specificity for cancer is not species specific, and does not require an immunocompromised host. We have shown this by testing the development of spontaneous tumors from topical application of dimethyl benzanthracene (DMBA) in SENCAR mice and hamsters, and

observing NV1066 specificity for tumor even when the virus is administered at sites distant to the tumor (29).

2. We have also successfully used 2×10^7 OCUM gastric cancer cells in 250 μL PBS (26) and 1.5×10^7 OCUM cells in 1 mL PBS (27) to establish peritoneal carcinomatosis.

3. GFP absorbs blue light with a major peak at 395 nm, and emits green light with a major peak at 509 nm (31). Enhanced GFP is a red-shift variant of GFP that fluoresces 35 times brighter than its unenhanced counterpart (32).

4. Viral stock propagated on Vero cells: when transfecting cells, media without penicillin and streptomycin are used, as addition of antibiotics results in cell death in this setting.

5. Plaque assay measurements usually underestimate the actual number of PFUs of a virus. For instance, a PFU may fail to enter a cell or may stick to the plate at a point not occupied by a cell, and may be obscured by overlying cells and thus fail to form a visible plaque. However, this assay is the closest and most efficient means of viral titering with which we have had experience.

6. We have yet to establish a standard volume of PBS or concentration of cells required to facilitate peritoneal carcinomatosis. We have established peritoneal carcinomatosis in 8-week-old athymic mice with as little as 5×10^6 OCUM-2MD3 gastric cancer cells suspended in 500 μL PBS as above, and have used as much as 2×10^7 OCUM-2MD3 gastric cancer cells suspended in 250 μL PBS (26). Furthermore, timing of viral injection is subject to experimental design, with consideration of the morbidity associated with the induced disease. For instance, we have shown that NV1066 can visualize thoracic lymph node metastases 4–8 weeks after tumor implantation (33).

7. We have also had success staining with rabbit polyclonal HSV-1 antibody using a Histomouse-SP Bulk Staining Kit (Zymed Laboratories Inc., San Francisco, CA) (29).

8. Optimum time length between initiation of peritoneal carcinomatosis and viral administration has varied in our experience, based on aggression of the cell line or type selected and experimental goals. For instance, when studying our ability to identify micrometastases (<1 mm diameter), we treated athymic mice with virus just 1 day following tumor injection and then waited 4 weeks to sacrifice the animals and harvest their organs en bloc (24). In our experience with gastric cancer cell lines, peritoneal carcinomatosis established with $1–1.5 \times 10^7$ cells dependably develops within 10–14 days (26, 27), though we have waited as long as 21 days following tumor cell implantation to administer virus (28). Frequency of viral administration

in our experiments is also varied. We have experienced success in all attempts with just one dose of virus. We typically use 1×10^7 PFU suspended in 100 μL PBS per dose, though as mentioned above we have had success with as little as 5×10^6 PFU in 100 μL PBS (29).

9. We have previously established that peak GFP expression and thus peak cell infection with NV1066 occurs at 5 days postinjection of virus. Cancer cell death is optimized at 7 days postinfection (27).

10. Topographic maps of GFP expression generated from digital images obtained with fluorescence microscopic or endoscopic systems, enable quantification of in vivo fluorescence intensity and facilitate demonstration of selective infection and tumor fluorescence in different abdominal quadrants. These maps enable the system to account for autofluorescence. Thus, one must be looking through the microscope or at the laparoscope screen in order to visualize the fluorescence. Simply looking with the naked eye and the fluorescence lamp will not work.

11. Endoscopy: GFP images should be taken with minimal background illumination to illustrate the surrounding organs. Thus, if an incision is larger than the laparoscope and stay-sutures do not create a sufficiently dark environment in the abdominal cavity, it may be necessary to decrease the background light both in the field (a standard blue towel over the incision and scope will do) and in the operating room (lowering the lights as in a standard laparoscopic case). In our experience with this peritoneal carcinomatosis model, the most common sites of residual disease are located in the retroperitoneum, pelvis, peritoneum, and liver.

References

1. Jarnagin, W.R., Conlon, K., Bodniewicz, J., Dougherty, E., DeMatteo, R.P., Blumgart, L.H., et al. (2001) A clinical scoring system predicts the yield of diagnostic laparoscopy in patients with potentially resectable hepatic colorectal metastases. *Cancer* **91(6)**, 1121–1128.
2. Paraskeva, P.A., Purkayastha, S., Darzi, A. (2004) Laparoscopy for malignancy: current status. *Seminars in Laparoscopic Surgery* **11(1)**, 27–36.
3. Parra, J.L., Reddy, K.R. (2004) Diagnostic laparoscopy. Endoscopy **36(4)**, 289–293.
4. Corvera, C.U., Blumgart, L.H., Darvishian, F., Klimstra, D.S., DeMatteo, R., Fong, Y., et al. (2005) Clinical and pathologic features of proximal biliary strictures masquerading as hilar cholangiocarcinoma. *J Am Coll Surg* **201(6)**, 862–869.
5. Sugarbaker, P.H. (2004) Managing the peritoneal surface component of gastrointestinal cancer. Part 2. Perioperative intraperitoneal chemotherapy. *Oncology* (Williston Park) **18(2)**, 207–219.
6. Sugarbaker, P.H. (2001) Cytoreductive surgery and peri-operative intraperitoneal chemotherapy as a curative approach to pseudomyxoma peritonei syndrome. *European Journal Of Surgical Oncology* **27(3)**, 239–243.
7. Yonemura, Y., Bandou, E., Kinoshita, K., Kawamura, T., Takahashi, S., Endou, Y., et al. (2003) Effective therapy for peritoneal dissemination in gastric cancer. *Surgical Oncology Clinics of North America* **12(3)**, 635–648.
8. Rusch, V.W., Rosenzweig, K., Venkatraman, E., Leon, L., Raben, A., Harrison, L., et al. (2001) A phase II trial of surgical resection and

adjuvant high-dose hemithoracic radiation for malignant pleural mesothelioma. *J Thorac Cardiovasc Surg* **122(4)**, 788–795.

9. Jaklitsch, M.T., Grondin, S.C., Sugarbaker, D.J. (2001) Treatment of malignant mesothelioma. *World J Surg* **25(2)**, 210–217.
10. Rusch, V.W. (1999) Indications for pneumonectomy. Extrapleural pneumonectomy. *Chest Surg Clin N Am* **9(2)**, 327–338.
11. Rusch, V.W., Venkatraman, E.S. (1999) Important prognostic factors in patients with malignant pleural mesothelioma, managed surgically. *Ann Thorac Surg* **68(5)**, 1799–1804.
12. Sugarbaker, P.H. Managing the peritoneal surface component of gastrointestinal cancer. Part 1. Patterns of dissemination and treatment options. *Oncology* (Williston Park) **18(1)**, 51–59.
13. Verwaal, V.J., van Ruth, .S, Witkamp, A., Boot, H., van Slooten, G., Zoetmulder, F.A. (2005) Long-term survival of peritoneal carcinomatosis of colorectal origin. *Annals of Surgical Oncology* **12(1)**, 65–71.
14. D'Angelica, M., Fong, Y., Weber, S., Gonen, M., DeMatteo, R.P., Conlon, K., et al. (2003) The role of staging laparoscopy in hepatobiliary malignancy: prospective analysis of 401 cases. *Annals of Surgical Oncology* **10(2)**, 183–189.
15. Jiao, X., Krasna, M.J. (2002) Clinical significance of micrometastasis in lung and esophageal cancer: a new paradigm in thoracic oncology. *Ann Thorac Surg* **74(1)**, 278–284.
16. Veronesi, U., Galimberti, V., Mariani, L., Gatti, G., Paganelli, G., Viale, G., et al. (2005) Sentinel node biopsy in breast cancer: early results in 953 patients with negative sentinel node biopsy and no axillary dissection. *European Journal Of Cancer* **41(2)**, 231–237.
17. Kelley, M.C., Hansen, N., McMasters, K.M. (2004) Lymphatic mapping and sentinel lymphadenectomy for breast cancer. *American Journal Of Surgery* **188(1)**, 49–61.
18. Liu, T.C., Galanis, E., Kirn, D. (2007) Clinical trial results with oncolytic virotherapy: a century of promise, a decade of progress. *Nat Clin Pract Oncol* **4(2)**, 101–117.
19. Reinblatt, M., Pin, R.H., Fong, Y. (2007) Herpes viral oncolysis: a novel cancer therapy. *J Am Coll Surg* **205(4 Suppl)**, S69–S75.
20. Bennett, J.J., Delman, K.A., Burt, B.M., Mariotti, A., Malhotra, S., Zager, J., et al. (2002) Comparison of safety, delivery, and efficacy of two oncolytic herpes viruses (G207 and NV1020) for peritoneal cancer. *Cancer Gene Therapy* **9(11)**, 935–945.
21. Wong, R.J., Joe, J.K., Kim, S.H., Shah, J.P., Horsburgh, B., Fong, Y.M. (2002) Oncolytic herpesvirus effectively treats murine squamous cell carcinoma and spreads by natural lymphatics to treat sites of lymphatic Metastases. *Human Gene Therapy* **13(10)**, 1213–1223.
22. Wong, R.J., Kim, S.H., Joe, J.K., Shah, J.P., Johnson, P.A., Fong, Y. (2001) Effective treatment of head and neck squamous cell carcinoma by an oncolytic herpes simplex virus. *Journal of the American College of Surgeons* **193(1)**, 12–21.
23. McAuliffe, P.F., Jarnagin, W.R., Johnson, P., Delman, K.A., Federoff, H., Fong, Y. (2000) Effective treatment of pancreatic tumors with two multimutated herpes simplex oncolytic viruses. *Journal of Gastrointestinal Surgery* **4(6)**, 580–588.
24. Stiles, B.M., Bhargava, A., Adusumilli, P.S., Stanziale, S.F., Kim, T.H., Rusch,V.W., et al. (2003) The replication-competent oncolytic herpes simplex mutant virus NV1066 is effective in the treatment of esophageal cancer. *Surgery* **134(2)**, 357–364.
25. Stanziale, S.F., Petrowsky, H., Adusumilli, P.S., Ben Porat, L., Gonen, M., Fong, Y. (2004) Infection with oncolytic herpes simplex virus-1 induces apoptosis in neighboring human cancer cells: a potential target to increase anticancer activity. *Clinical Cancer Research* **10(9)**, 3225–3232.
26. Adusumilli, P.S., Eisenberg, D.P., Stiles, B.M., Hendershott, K.J., Stanziale, S.F., Chan, M.K., et al. (2006) Virally-directed fluorescent imaging (VFI) can facilitate endoscopic staging. *Surgical Endoscopy* **20(4)**, 628–635.
27. Stanziale, S.F., Stiles, B.M., Bhargava, A., Kerns, S.A., Kalakonda, N., Fong, Y. (2004) Oncolytic herpes simplex virus-1 mutant expressing green fluorescent protein can detect and treat peritoneal cancer. *Hum Gene Ther* **15(6)**, 609–618.
28. Adusumilli, P.S., Eisenberg, D.P., Chun, Y.S., Ryu, K.W., Ben-Porat, L., Hendershott, K.J., et al. (2005) Virally directed fluorescent imaging improves diagnostic sensitivity in the detection of minimal residual disease after potentially curative cytoreductive surgery. *Journal of Gastrointestinal Surgery* **9(8)**, 1138–1147.
29. Adusumilli, P.S., Stiles, B.M., Chan, M.K., Eisenberg, D.P., Yu, Z., Stanziale, S.F., et al. (2006) Real-time Diagnostic Imaging of

Tumors and Metastases by Use of a Replication-competent Herpes Vector to Facilitate Minimally-Invasive Oncological Surgery. *FASEB J* **20(6)**, 726–728.

30. Gahlen, J., Stern, J., Laubach, H.H., Pietschmann, M., Herfarth, C. (1999) Improving diagnostic staging laparoscopy using intraperitoneal lavage of delta-aminolevulinic acid (ALA) for laparoscopic fluorescence diagnosis. *Surgery* **126(3)**, 469–473.
31. Chalfie, M., Tu, Y., Euskirchen, G., Ward, W.W., Prasher, D.C. (1994) Green fluorescent protein as a marker for gene expression. *Science* **263(5148)**, 802–805.
32. Cormack, B.P., Valdivia, R.H., Falkow, S. (1996) FACS-optimized mutants of the green fluorescent protein (GFP). *Gene* **173(1 Spec No)**, 33–38.
33. Adusumilli, P.S., Eisenberg, D.P., Stiles, B.M., Chung, S., Chan, M.K., Rusch, V.W., et al. (2006) Intraoperative localization of lymph node metastases with a replication-competent herpes simplex virus. *J Thorac Cardiovasc Surg* **132(5)**, 1179–1188.

Chapter 11

Real-Time Imaging of Tumors Using Replication-Competent Light-Emitting Microorganisms

Yong A. Yu, Stephanie Weibel, and Aladar A. Szalay

Abstract

Early detection of cancer and metastases is pivotal to the success of subsequent treatment intervention. In recent years, the use of live microorganisms, such as viruses and bacteria, has gained substantial research and clinical interest in both detection and therapy of cancer. Many of these live microorganisms have shown remarkable tumor-specific replication following systemic delivery. With the aid of modern molecular technologies, modified live microorganisms can be engineered to carry additional diagnostic and therapeutic capabilities. We have shown that when armed with light-emitting protein genes, such as genes for luciferase and green fluorescent protein, the entry and specific amplification of systemically-delivered vaccinia virus and bacteria in tumors can be visualized in real time using a low-light imager, or using macro- and micro-fluorescence microscopes. Therefore, through optical imaging, the location of tumors and metastases could be revealed by these light-emitting microorganisms. The tumor-colonization capability has been demonstrated in both immuno-competent as well as immuno-compromised rodent models with syngeneic and allogeneic tumors. Based on their "tumor-finding" nature, bacteria and viruses could be further designed as "vehicles" to carry multiple genes for detection and therapy of cancer.

Key words: Vaccinia virus, Bacteria, Systemic delivery, Tumors, Metastases, GFP, Luciferase, Fluorescence and luminescence imaging

1. Introduction

Tumor-specific entry and replication by systemically-delivered virus and bacteria have been demonstrated in recent years (1–5). We have engineered a vaccinia virus and bacteria to express light-emitting proteins, such as green fluorescent protein (GFP), dsRed protein, and *Renilla* luciferase, so that the location of the infected tumors could be visualized noninvasively (1, 6, 7). Tumor metastases as small as 0.5 mm in diameter could be detected by the modified vaccinia virus (1). Virus-mediated fluorescence imaging also facilitates

Robert M. Hoffman (ed.), *In Vivo Cellular Imaging Using Fluorescent Proteins: Methods and Protocols*, Methods in Molecular Biology, vol. 872, DOI 10.1007/978-1-61779-797-2_11, © Springer Science+Business Media, LLC 2012

intraoperative detection of lymph node metastases from malignant tumors, demonstrating the diagnostic utility for tumor staging (8). To take advantage of the deep-tissue imaging capability of positron emission tomography (PET), we have also engineered vaccinia virus to express the human norepinephrine transporter (hNET), which allows imaging of virus-infected tumors using PET (9, 10). Through endogenous thymidine kinase and radiolabeled pyrimidine nucleoside analogues, *E. coli*-infected tumors could also be imaged by PET (11).

In this protocol, we describe materials and methods for virus- or bacteria-mediated tumor detection in mice using optical imaging (fluorescence and luminescence). These imaging models not only allow detailed examination of the mechanism of tumor colonization but also facilitate the design of microorganism-based tumor detection and therapy for future clinical applications. Furthermore, we present a method for microscopic analysis of microbial distribution within tumor tissue. This approach also enables the detailed study of the interaction of microorganisms with the tumor microenvironment.

2. Materials

2.1. Cell Culture

1. CV-1 African green monkey kidney fibroblasts, C6 rat glioma, PANC-1 human pancreatic carcinoma, MCF-7 human breast carcinoma, A549 human non-small cell lung carcinoma, and 4T1 murine mammary carcinoma cell lines were obtained from the American Type Culture Collection (ATCC, Manassas, VA). Cell line GI-101A, a highly metastatic derivative of GI-101 human breast ductal adenocarcinoma, was kindly provided by Dr. A. Aller, Rumbaugh-Goodwin Institute for Cancer Research, Inc.
2. Dulbecco's modified Eagle's medium (DMEM) (Gibco/BRL, Bethesda, MD) is supplemented with antibiotic–antimycotic solution (100 units/mL penicillin G, 250 ng/mL amphotericin B, 100 units/mL streptomycin; from Gibco/BRL) and 10% fetal bovine serum (FBS; Invitrogen Corporation).
3. RPMI-1640 medium (Mediatech, Manassas, VA) is supplemented with 10% FBS, 100 units/mL penicillin G, 250 ng/mL amphotericin B, and 100 units/mL streptomycin.
4. Trypsin solution (0.25%) is obtained from Gibco/BRL.
5. Phosphate-buffered saline (PBS) is obtained from Gibco/BRL.

2.2. Vaccinia Virus

1. Modified vaccinia virus GLV-1h68 was constructed by inserting three expression cassettes, *Renilla* luciferase–*Aequorea* green fluorescent protein, (*RUC-GFP*) fusion (12),

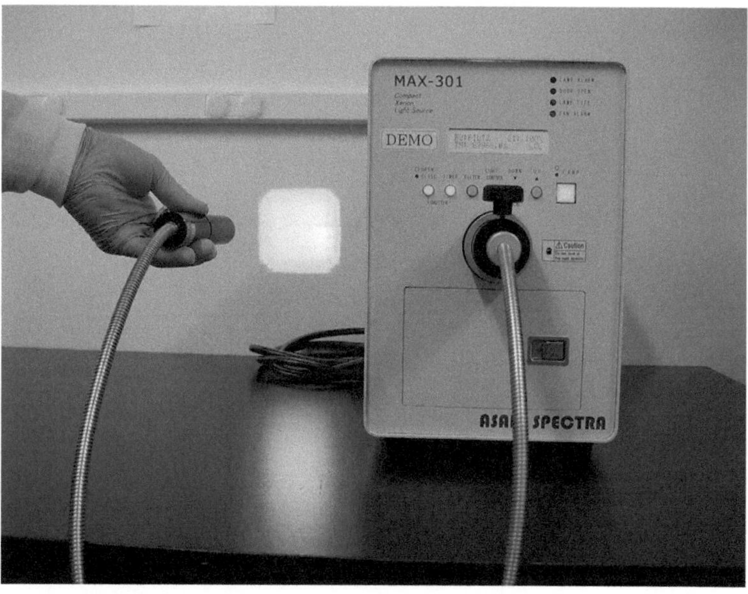

Fig. 1. MAX-302 fluorescence excitation system.

beta-galactosidase, and beta-glucuronidase into the F14.5L, J2R, and A56R loci of the viral genome, respectively (3).

2. The entire genome of GLV-1h68 was sequenced and characterized (13).

2.3. Bacteria

1. *E. coli* (DH5α, K-12; ATCC) was transformed with the plasmid pLITE201 (14), containing *luxCDABE*, to produce light-emitting bacterial strains or with the plasmid pMW211 (15) encoding the dsRed protein.
2. The bacterial luciferase operon *luxCDABE* could be easily recombined into transfer plasmids to transform Gram-negative bacteria. Other strains of bacteria for tumor targeting include, but not limited to, *Salmonella typhimurium*, *Vibrio cholerae*, *Shigella* ss., etc. (7).

2.4. In Vivo Fluorescence Imaging Instruments

1. MAX-302 (Asahi Spectra USA, Torrance, CA) (see Note 1; Fig. 1).
   - Components and features:
     - MAX-302 unit (with cool Xenon light source 300 W).
     - Quartz light guide 2 M.
     - Light guide adapter up to 2 m in length.
     - Quartz collimator lens.
     - Bandpass filter 480 nm for GFP excitation.
     - 515-nm green fluorescence emission filter.

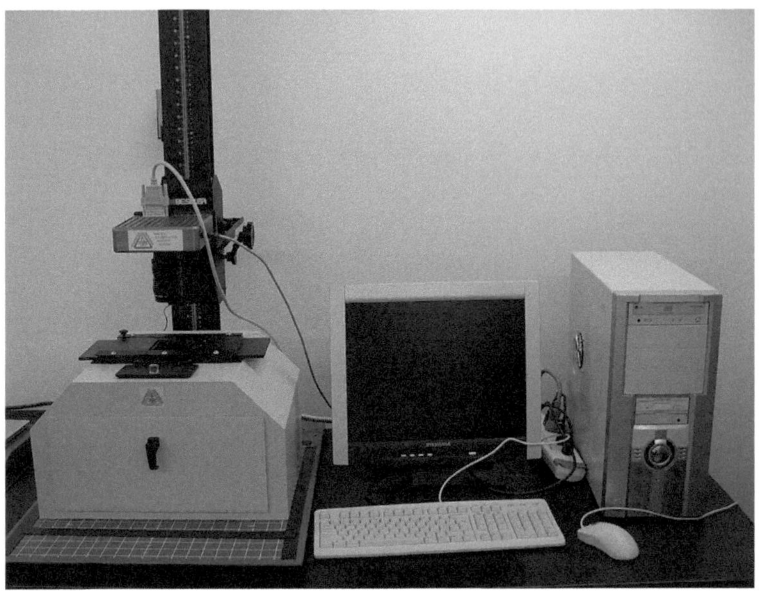

Fig. 2. Lightools fluorescence imaging system.

2. Lightools fluorescence imaging system (Lightools Research, Encinitas, CA) (see Note 1; Fig. 2).
   - Components and features:
     - Dual-excitation light path from top (two 470/40 nm excitation filters).
     - Newer version of the system allows two-color (e.g. GFP and RFP) controllable excitation.
     - 515-nm green fluorescence emission filter.
     - Low cost.
3. Maestro 2 in vivo imaging system (CRi, Woburn, MA) (see Note 1).
   - Components and features:
     - Dual optical imaging capability (fluorescence and luminescence imagings).
     - Automated filter selection (wavelength range: 500–950 nm, tunable).
     - Cooled CCD camera (0°C)—Sony ICX285.
     - Able to separate up to five overlapping fluorophores from background autofluorescence.
     - Image up to three whole mice simultaneously or zoom down to 25 μm per pixel for near-single-cell resolution.
     - Multiple view capability—view three sides of an animal simultaneously using Maestro's C3 Mirror.
     - Heated imaging chamber.

2.5. In Vivo Luminescence Imaging Instrument

1. IVIS Imaging System (Caliper Life Sciences, Hopkinton, MA) (see Note 2).
   - Components and features:
     - Dual optical imaging capability (fluorescence and luminescence imaging).
     - Twenty eight filters spanning 430–850 nm to distinguish multiple bioluminescent and fluorescent reporters.
     - Cooled CCD camera (–90°C).
     - In vivo luminescence imaging sensitivity: as few as 50 cells.
     - Quantitation of fluorescence reporters.
     - Image up to five mice simultaneously.
     - Spatial resolution: down to 20 μm with 3.9 cm field of view.
     - 3-D image option.
2. NightOWL II LB 983 low-light imager (Berthold Technologies, Bad Wildbad, Germany) (see Note 2; Fig. 3).
   - Components and features:
     - Dual optical imaging capability (fluorescence and luminescence imaging).
     - Camera spectral range: 300–1,100 nm, with quantum efficiency up to 90% in the spectral range between 500 and 660 nm, which is optimal for firefly luciferase, GFP, and its derivatives.

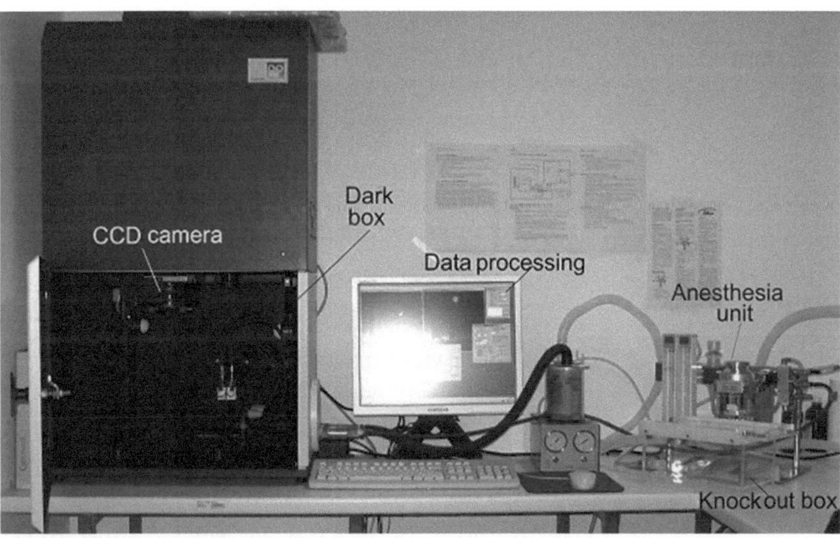

Fig. 3. NightOWL II LB 983 low-light imager.

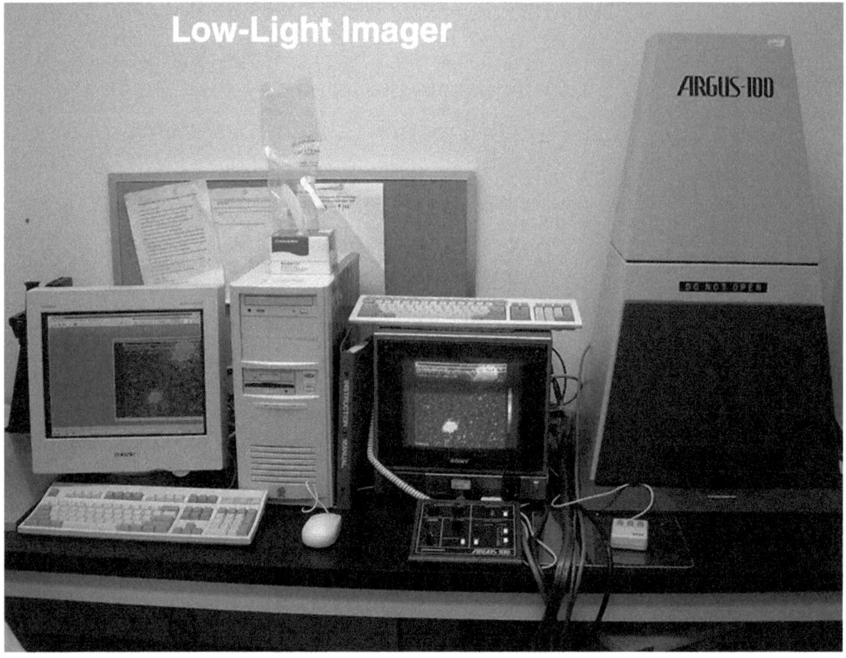

Fig. 4. ARGUS-100 low-light imager.

- Motor-driven camera inside the light-tight cabinet.
- Cooled CCD camera (–80°C to –90°C).
- Image up to five mice simultaneously.
- Spatial resolution: down to 10 μm with 1 cm field of view.
- 3-D image option.

3. ARGUS-100 low-light imager (Hamamatsu Photonics, Hamamatsu, Japan) (see Note 2; Fig. 4).
   - Components and features:
     - Single-photon-counting camera (ARGUS-100) linked to a macroscopic video camera in a light-tight box (or to a microscopic video camera C2400-20).
     - Controller.
     - Mouse image processor.
     - Color video monitor.

2.6. Luminescence Imaging Reagents

1. Coelenterazine solution: 0.5 μg/μL benzyl-coelenterazine (h-CTZ; NanoLight Technology, Pinetop, AZ) diluted in 100% ethanol.
2. Luciferase assay buffer: 0.5 M NaCl, 1 mM EDTA, 0.1 M potassium phosphate pH 7.4.

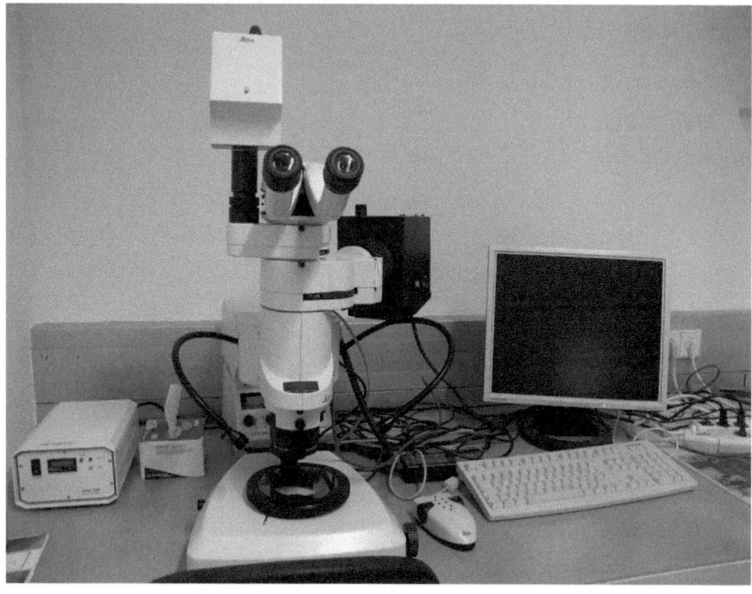

Fig. 5. MZ16 FA planapochromatic stereo fluorescence microscope (Leica, Heerbrugg, Switzerland) equipped with a digital CCD camera (DC500, Leica).

2.7. Histology and Microscopy

1. PBS: Prepare a 10× stock solution with 1.37 M NaCl, 27 mM KCl, 100 mM Na_2HPO_4, 18 mM KH_2PO_4, adjust to pH 7.4.
2. Fixation buffer: prepare a 4% (w/v) paraformaldehyde solution in 1× PBS, pH 7.4. Store solutions at –20°C.
3. Embedding: 5% (w/v) low-melt agarose (Applichem, Darmstadt, Germany) in 1× PBS, pH 7.4.
4. Wash buffer: 1× PBS, pH 7.4.
5. Permeabilization buffer: 0.3% (v/v) Triton-X 100 in 1× PBS, pH 7.4.
6. Staining solution: For actin labeling Phalloidin-TRITC and Phalloidin-FITC (Sigma Aldrich, Taufkirchen, Germany), respectively are diluted 1:200 in 1× PBS, pH 7.4. Cell nuclei are labeled using Hoechst 33342 1:400.
7. Mounting medium: 60% (v/v) and 80% (v/v) glycerol in 1× PBS, pH 7.4.
8. MZ16 FA planapochromatic Stereo-Fluorescence microscope (Leica, Heerbrugg, Switzerland) equipped with a digital CCD camera (DC500, Leica).
9. TCS SP2 AOBS confocal microscope (Leica) equipped with an argon, helium–neon and UV laser (see Fig. 5).

3. Methods

3.1. Preparation of Vaccinia Virus

1. CV-1 cells are cultured in supplemented DMEM at 37°C under 5% CO_2.
2. Viruses were propagated in CV-1 cells, and up to 7×10^9 pfu/mL GLV-1h68 could be purified from 2×10^8 infected CV-1 cells through sucrose gradients (16).
3. Purified GLV-1h68 virus is stored at –80°C freezer (see Note 3).
4. On the day of viral injection, thaw a vial of virus at room temperature. Dilute the virus using PBS.
5. One live virus is considered as 1 pfu (plaque-forming unit).

3.2. Vaccinia Virus Replication in Cancer Cell Culture

1. MCF-7 breast cancer cells (1×10^5 pfu) are seeded onto 12-well titer plates.
2. After culturing for 24 h, cells are infected with virus at a multiplicity of infection (MOI) of 0.001.
3. Incubate the cells at 37°C for 1 h with brief agitation every 10 min.
4. Remove the infection medium, and replace with fresh medium.
5. Harvest cells at 24, 48, 72, or 96 h after infection.
6. Apply three rounds of freeze and thaw cycle to release viral particles from infected cells.
7. Determine the titers (pfu/mL) in duplicate by standard plaque assay in CV-1 cell monolayers.

To determine virus-mediated luminescence activity in cell culture:

1. At 48 h after virus infection, aspirate off the supernatant.
2. Trypsinize and release the cells from the bottom of the plate.
3. Remove trypsin and resuspend the cells in 1 mL luciferase assay buffer.
4. Transfer the suspension into a clear glass vial (scintillation-counting vial works the best).
5. Count luminescence emission in a luminometer (Turner TD-20e; Turner BioSystems, Sunnyvale, CA).

3.3. Preparation of Bacteria

1. All Gram-negative bacteria are cultured in LB medium (Difco Laboratories) at 37°C, supplemented with antibiotics (100 μg/mL ampicillin, if transformed with pLITE201).
2. An aliquot (10 μL) of overnight culture is used to inoculate 100 mL LB media supplemented with antibiotics.
3. At mid-log phase (OD600 = 0.4–0.7), spin down the bacteria at 3,000 rpm (1465 × *g*) for 5 min.

4. Resuspend the pellet in PBS for i.v. injection into animals. Injection volume is between 100 and 200 μL.
5. One live bacterium is considered as 1 cfu (colony-forming unit).

3.4. In Vivo Tumor Models (See Note 4)

Subcutaneous tumor models:

1. C6 cells are cultured in supplemented RPMI 1640 media. PANC-1 and A549 cells are cultured in supplemented DMEM at 37°C under 5% CO_2. GI-101A is cultured in RPMI 1640 supplemented with 5 ng/mL h-estradiol and 5 ng/mL progesterone (Sigma, St. Louis, MO), 10 mmol/L HEPES (Mediatech, Inc.), 1 mmol/L sodium pyruvate (Mediatech, Inc.), 20% FBS, an antibiotic/antimycotic solution.
2. Five- to six-week-old male BALB/c athymic nu^-/nu^- mice (25–30 g body weight) are purchased from Harlan (Frederick, MD). Eight animals are included in each group (i.e., $n=8$).
3. The cancer cell lines are propagated by standard cell culture techniques to provide sufficient cells for implantation.
4. On the day of tumor implantation, cells are trypsinized and resuspended. A viable cell count is performed and the suspension is diluted in saline to a density of $\sim 5 \times 10^7$ cells/mL. A volume of 0.1 mL ($\sim 5 \times 10^6$ cells) is delivered subcutaneously (s.c.) to the right lateral thigh (see Note 5) of each mouse. The use of anesthesia is not necessary.
5. Tumor growth is recorded twice weekly in 3-D using digital calipers. Tumor volume is calculated as [(length×width×height)/2] and reported in mm^3.

Orthotopic breast tumor model:

1. MCF-7 cells, stably transformed with a plasmid carrying pro-IGF-II cDNA (from D. deLeon, Loma Linda University, Loma Linda, CA), are cultured in DMEM/F12 medium supplemented with 5% FBS and 560 μg/mL G418 (Life Technologies, Carlsbad, CA).
2. Four-to six-week-old female nude mice are first implanted with 17β-estradiol pellets (0.72 mg per pellet 90-day release; Innovative Research of America) in the dorsal skin to facilitate estrogen-dependent breast tumor development and metastases.
3. MCF-7/pro-IGF-II cells (1×10^6) are injected directly into the second left mammary fat pad of each mouse through an incision just below the second nipple.
4. For surgical orthotopic implantation (SOI), tumors developed from implanted cells were resected and minced into 1 mm^3 cubes for tissue transplantation into the mammary fat pad. Solid tumor metastases as large as 50 mm^3 may appear in other mammary fat pads 3–6 months after cellular implantation. Extensive lung metastases may also occur within 3–6 months after cellular implantation.

3.5. Systemic Delivery of Vaccinia Virus or Bacteria

1. Anesthetize the animals with ketamine (40 mg/kg)/xylazine (12 mg/kg)
2. 1×10^8 cfu light-emitting bacteria, suspended in 100 μL PBS, are injected with a 1-cm³ insulin syringe equipped with a 29½-gauge needle (Becton Dickinson, Franklin Lakes, NJ) through the surgically-exposed superficial femoral vein (located on the inner thigh) (see Note 6).
3. To inject vaccinia virus, 5×10^6 pfu virus are suspended in 100 μL PBS and injected using an insulin needle.
4. After each injection, the incision is re-approximated with 5-0 nylon sutures (Harvard Apparatus, Holliston, MA).

3.6. In Vivo Fluorescence Imaging

1. Seven days (see Note 7) following systemic (intravenous) delivery of vaccinia virus, mice are imaged for green fluorescence.
2. Under isoflurane anesthesia, mice are positioned under blue excitation light (from either the MAX-302 unit or from the Lightools fluorescence imaging system) in complete dark.
3. Through the yellow emission filter, surface green fluorescence should be visible with the naked eye.
4. Digital image acquisition is performed using a CCD camera (Lightools) or using a digital camera (Sony DSC-F707) (see Fig. 6).

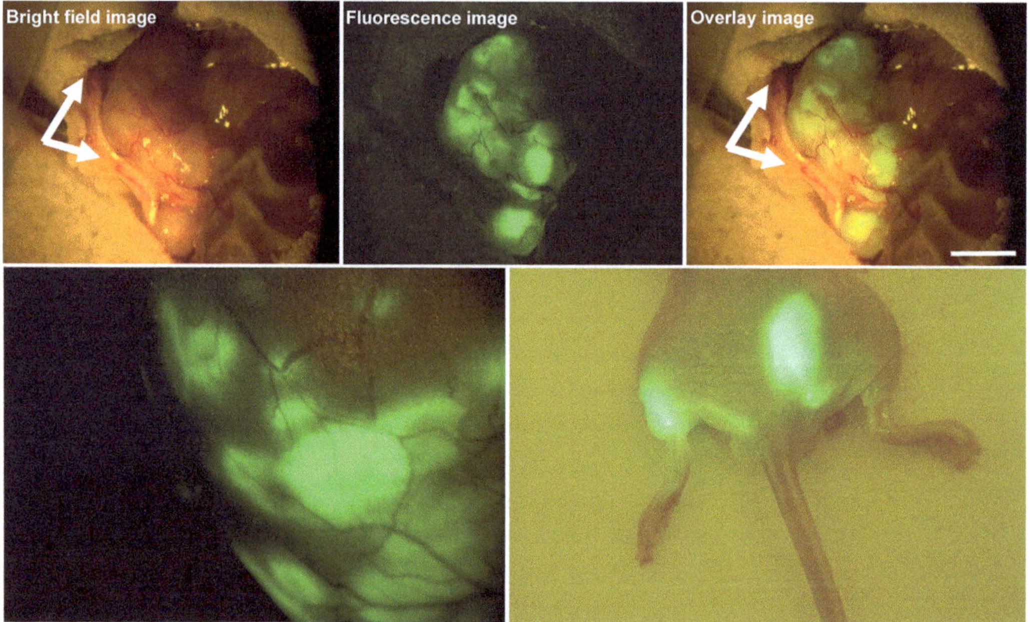

Fig. 6. Visualization of tumors using modified vaccinia virus encoding green fluorescent protein. Images of orthotopic MCF-7/pro-IGF-II breast carcinoma (*top panels*), s.c. C6 glioma (*lower left*), and PANC-1 pancreatic carcinoma (*lower right*) are shown. Bar = 5 mm.

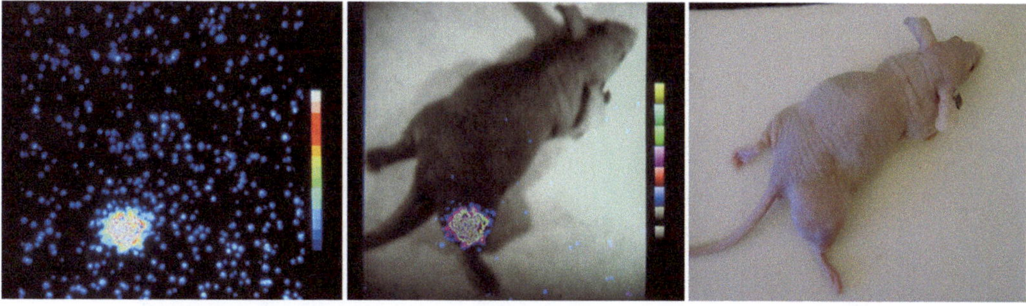

Fig. 7. Visualization of s.c. A549 tumor using i.v. injected *E. coli*/pLITE-201. Low light (*left*), low light and bright-field overlay (*middle*), and bright-field (*right*) images of the same animal are shown.

3.7. In Vivo Luminescence Imaging

1. Seven days following systemic (intravenous) delivery of vaccinia virus or bacteria, mice with tumors are imaged for luminescence.
2. For mice injected with vaccinia virus, diluted coelenterazine solution (for each animal, add 5 μL of 0.5 μg/μL coelenterazine into 95 μL of PBS) is injected into the tail vein.
3. Since pLITE-201-transformed bacteria are spontaneously luminescent, no injection of substrate in the animals is required.
4. Under ketamine/xylazine anesthesia, mice are positioned in the dark box of ARGUS-100 low-light imager.
5. Photon emission is measured for 1 min (see Note 8), and the images were recorded using Image Pro Plus 3.1 software (Media Cybernetics, Silver Spring, MD).
6. A digital photographic image is also recorded, which is then superimposed with the low-light image to identify the luminescent regions on the animals. Depending on the location of the tumors, mice could be imaged from dorsal, ventral, right lateral, and left lateral views (see Figs. 7 and 8).

3.8. Titrate Vaccinia Viral and Bacterial Titers in Tissue Samples

To analyze vaccinia viral distribution in organs and tissues in the animal host after systemic delivery, we titrated viral titers in tissue samples.

1. Typically, four mice from each group are sacrificed for analysis. Organs/tissues (e.g., liver, spleen, kidneys, lung, heart, brain, bladder, and tumors) are excised and weighed (see Note 9).
2. The excised tissues are homogenized using MagNA Lyser (Roche Diagnostics, Indianapolis, IN) at a speed of 6,500 for 30 s in ice-cold PBS supplemented with Complete Proteinase Inhibitor (Roche Molecular Biochemicals, Indianapolis, IN).
3. After three cycles of freeze and thaw, the samples are sonicated for 1 min, three times, on ice water.
4. The supernatants are collected by centrifugation at 1,000 × *g* for 5 min.

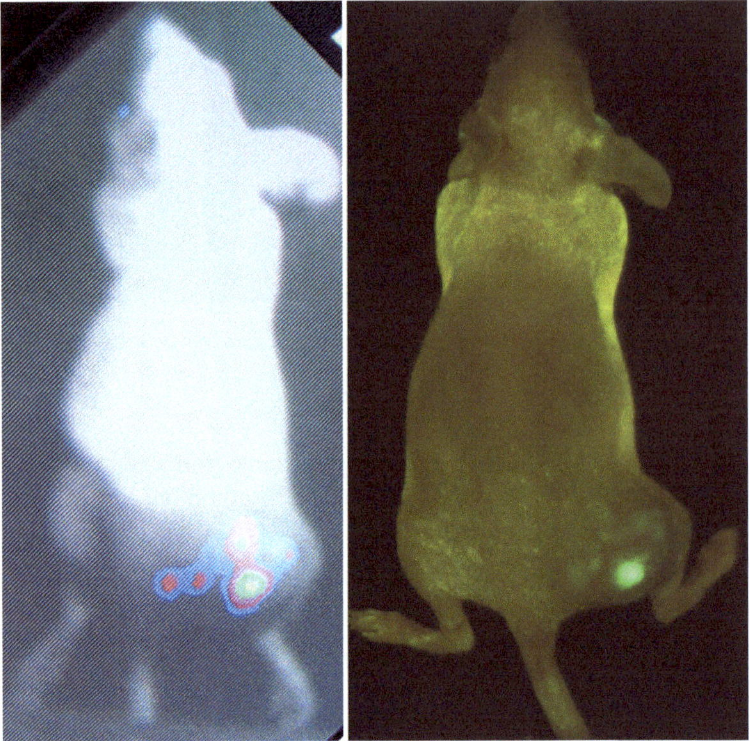

Fig. 8. Visualization of s.c. GI-101A tumor using i.v.-injected vaccinia virus encoding RUC-GFP. Virus-mediated luminescence signal (*left*) is pseudo-colored in the tumor in the presence of luciferase substrate coelenterazine. Virus-mediated GFP expression images of the virus-infected tumor (*right*).

5. The virus titers are determined in duplicate by standard plaque assays using CV-1 cells.
6. Viral titer is either reported as pfu/g of tissue or as pfu/whole organ.

To analyze bacterial distribution in organs and tissues:

1. Prepare organ/tissue samples as shown above (steps 1 and 2).
2. Each sample is serially diluted using PBS and plated on selective agar with 100 μg/mL ampicillin.
3. After overnight incubation at 37°C, colonies are counted. Bacterial titers per organ or tumor are then calculated based on dilution factor.
4. Bacterial titer is either reported as cfu/g of tissue or as cfu/whole organ.

3.9. Analysis of Microbial Distribution Within the Tumor Tissue

To study viral and bacterial distribution, as well as the interaction of these microorganisms with the host tumor microenvironment, we prepared whole-tumor cross-sections for microscopic analysis (17). Using stereo fluorescence microscopy with 10× magnification, we are able to create "overview" images of whole tumor cross-sections to visualize the microbial distribution pattern. Detailed microscopic analysis on the cellular level is then performed with a confocal microscope.

1. Tumors of infected animals are excised without bones/fibers, and immediately snap-frozen in liquid nitrogen for 5 min followed by incubation in fixation buffer (10 mL/g tumor tissue) for 12–15 h at 4°C under constant shaking.
2. Fixed samples are rinsed three times in wash buffer at room temperature for 30 min under shaking, followed by embedding in 5% low-melt agarose and incubation for 1 h at 4°C.
3. Sectioning (100 μm) of agarose blocks with embedded tumors is performed with a Vibratom VT1000S (Leica). Tissue sections are immediately transferred into 1× PBS.
4. Tumor slices are incubated in 48-well plates with 0.5 mL permeabilization buffer for 1 h at room temperature under shaking.
5. For labeling of actin and cellular nuclei, the tumor sections are incubated in 0.2 mL staining solution containing fluorescence-conjugated phalloidin (FITC in combination with red fluorescent bacteria; TRITC with green-fluorescent virally-infected tumors) and Hoechst 33342 for 12–15 h at room temperature under constant shaking.
6. The sections are rinsed (3× 10 min) in wash buffer followed by incubation in 0.5 mL 60% glycerol/PBS for 30 min.
7. Tissue slides are mounted in 80% glycerol/PBS on glass slides and sealed with nail polish.
8. Samples are viewed with the MZ16 FA Stereo-Fluorescence microscope (Leica) equipped with the DC500 digital camera and adequate filters (GFP3, dsRed; Leica). Images are captured with low (10×) magnification to represent whole tumor cross-sections.
9. To perform studies on the cellular level, 100-μm tumor sections are investigated with the Leica TCS SP2 AOBS confocal microscope up to a magnification of 630×. Fluorescence signals of double- and triple-labeled specimens are serially recorded with appropriate excitation and emission filters to avoid bleed-through.
10. Digital images (1,300 × 1,030 pixels, stereo fluorescence microscope; 1,024 × 1,024 pixels, confocal microscope) are processed with Photoshop 7.0 (Adobe Systems, Mountain View, CA) and merged to yield pseudo-colored images (see Fig. 9).

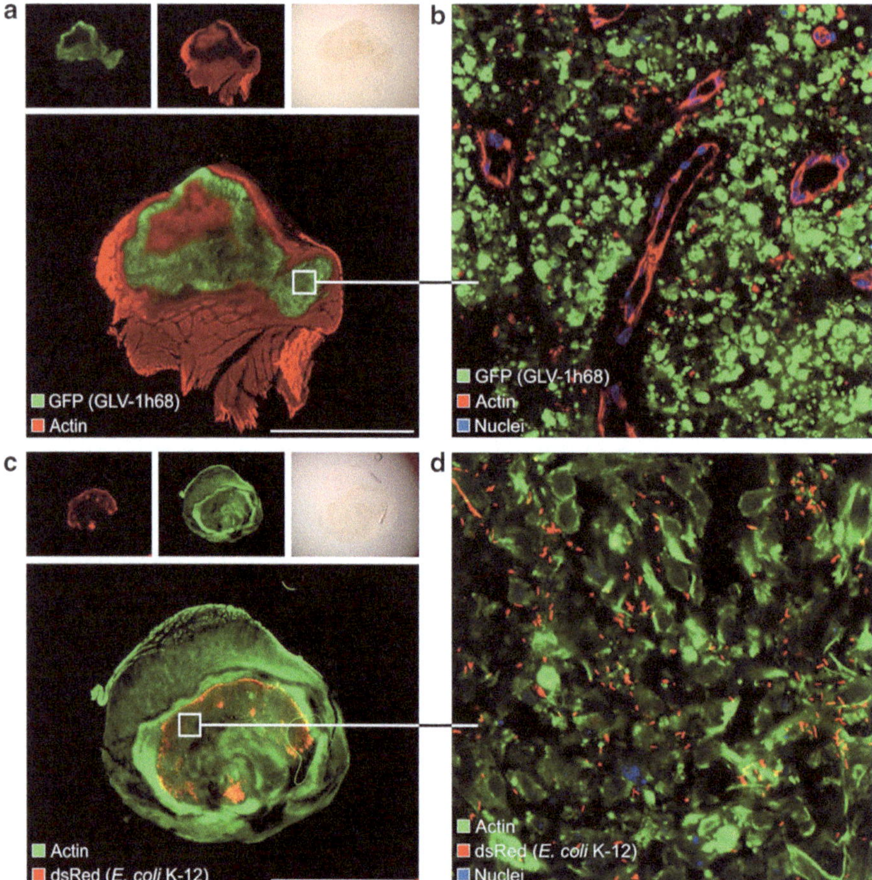

Fig. 9. Localization of microorganisms within tumor tissue. (**a**, **c**) Overview images obtained with a stereo fluorescence microscope (10×); (**b**, **d**) Confocal images. (**a**) GI-101A tumor infected with vaccinia virus (GLV-1h68) for 42 days. Phalloidin-TRITC-labeled actin cytoskeleton (*red*) indicates viable tissue. (**b**) In vaccinia virus-infected tumor areas, only the blood vessels (*red*) are morphologically intact as indicated by actin and nuclear staining. (**c**) 4T1 tumor infected with *E. coli* K-12 (*red*) for 3 days. The area of Phalloidin-FITC-labeled actin cytoskeleton (*green*) is concentrated around the bacterially-infected tumor center. (**d**) Bacteria-infected areas show signs of necrotic tissue destruction. Scale bars = 5 mm.

4. Notes

1. With in vivo fluorescence imaging gaining popularity, more choices of in vivo fluorescence imaging systems are available on the market. Anyone of the systems listed in this section could serve the function of detecting fluorescence in live animals, and potentially even on human patients.
2. There are several choices of low-light imagers on the market for imaging photon emission as well as fluorescence. Any one of the systems listed in this section could serve the function of detecting luminescence in live animals.
3. Avoid frequent freeze and thaw of the virus, which will result in a decrease in viral titer. Virus should be dispensed into small

aliquots (e.g., 1 mL/vial) before transferring to a –80°C freezer. It is recommended that the virus thawed from the same vial be used when performing viral titration in parallel to each imaging experiment.

4. In this protocol, only tumor models in immuno-compromised nude mice are presented. For details of tumor models in immunocompetent mice, please see refs. 1, 8. Please see the following references for additional examples of imaging in different tumor types (21, 22) and in orthotopic models (23).

5. We routinely use the right lateral thigh as the site for subcutaneous implantation of tumor cells. A more common place for subcutaneous implantation is the upper dorsum of mice. However, since we do not use Matrigel (BD Biosciences, San Jose, CA) to assist in tumor implantation, a subcutaneous implantation in the lateral thigh results in more consistent tumor sizes than the dorsal flank site.

   However, if subcutaneous implantation at the dorsal flank is desired, an alternative way to conduct tumor implantation is to first develop tumors in a few "tumor-production animals" using the cell-implantation method. Once tumors develop in these animals, harvest the tumors and dice them into 30–40-mg tumor fragments. Then inject these tumor fragments using a 12-gauge trocar needle, into the subcutaneous space of dorsal flank. This will result in consistent tumor development among the animals.

6. Alternative routes for intravenous delivery include injection in the tail vein or retro-orbital venous plexus. In both of these routes, anesthesia is not required, although transient anesthesia using isoflurane could help when conducting retro-orbital injection.

7. Green fluorescence (small spots) could be detected as early as the third day after virus injection. Due to the noninvasive nature of optical imaging, surface green fluorescence could be imaged everyday for the duration of the study. At day 7, virus-mediated green fluorescence appears in patches in large tumors (>1,500 mm^3 in size; see lower left in Fig. 6), indicating the sites of viral infection. Over time, when viral-infected tumor cells die and the infection spreads to neighboring cells, fluorescent-ring-like structures are observed. Green fluorescence disappears when all tumor cells are killed by the virus.

8. The luminescence signal is saturated for the first 15 min after injection of substrate. Beyond that point, light emission will gradually decrease. Therefore, there is ample time for multiple image acquisitions before low light emission fades off.

9. To avoid cross contamination of the virus from each sample, scissors and forceps should be treated with 50% bleach for 10 s and rinsed briefly with PBS in between dissection of each organ/tissue.

Acknowledgements

This work was supported by an internal research grant from Genelux Corporation. The authors would like to thank Qian Zhang, Nanhai Chen, and Terry Trevino for scientific and technical contributions, and Kim Duffy for editorial assistance.

References

1. Yu, Y. A., Shabahang, S., Timiryasova, T. M., Zhang, Q., Beltz, R., Gentschev, I., Goebel, W., Szalay, A. A. (2004) Visualization of tumors and metastases in live animals with bacteria and vaccinia virus encoding light-emitting proteins. *Nat. Biotechnol.* **22**, 313–320.
2. Woo, Y., Adusumilli, P. S., Fong, Y. (2006) Advances in oncolytic viral therapy. *Curr. Opin. Investig. Drugs.* **7**, 549–559.
3. Zhang, Q., Yu, Y. A., Wang, E., Chen, N., Danner, R. L., Munson, P. J., Marincola, F. M., and Szalay, A. A. (2007) Eradication of solid human breast tumors in nude mice with an intravenously injected light-emitting oncolytic vaccinia virus. *Cancer Res.* **67**, 10038–10046.
4. Hoffman, R. M. (2009) Tumor-targeting amino acid auxotrophic *Salmonella typhimurium*. *Amino Acids.* **37**, 509–521.
5. Yu, Y. A., Galanis, C., Woo, Y., Chen, N., Zhang, Q., Fong, Y., Szalay, A. A. (2009) Regression of human pancreatic tumor xenografts in mice after a single systemic injection of recombinant vaccinia virus GLV-1h68. *Mol. Cancer Ther.* **8**, 141–151.
6. Yu, Y. A., Timiryasova, T., Zhang, Q., Beltz, R., and Szalay, A. A. (2003) Optical imaging: bacteria, viruses, and mammalian cells encoding light-emitting proteins reveal the locations of primary tumors and metastases in animals. *Anal. Bioanal. Chem.* **377**, 964–972.
7. Yu, Y. A., Zhang, Q, and Szalay, A. A. (2008) Establishment and characterization of conditions required for tumor colonization by intravenously delivered bacteria. *Biotechnol Bioeng.* **100**, 567–578.
8. Kelly, K., Brader, P., Woo, Y., Li, S., Chen, N., Yu, Y. A., Szalay, A. A., and Fong, Y. (2009) Real-time intraoperative detection of melanoma lymph node metastases using recombinant vaccinia virus GLV-1h68 in an immunocompetent animal model. *Int. J. Cancer.* **124**, 911–918.
9. Chen, N., Zhang, Q., Yu, Y. A., Stritzker, J., Brader, P., Schirbel, A., Samnick, S., Serganova, I., Blasberg, R., Fong, Y., Szalay, A. A. (2009) A novel recombinant vaccinia virus expressing the human norepinephrine transporter retains oncolytic potential and facilitates deep-tissue imaging. *Mol. Med.* **15**, 144–151.
10. Brader, P., Kelly, K. J., Chen, N., Yu, Y. A., Zhang, Q., Zanzonico, P., Burnazi, E. M., Ghani, R. E., Serganova, I., Hricak, H., Szalay, A. A., Fong, Y., Blasberg, R. G. (2009) Imaging a genetically engineered oncolytic vaccinia virus (GLV-1h99) using a human norepinephrine transporter reporter gene. *Clin. Cancer Res.* **15**, 3791–3801.
11. Brader, P., Stritzker, J., Riedl, C. C., Zanzonico, P., Cai, S., Burnazi, E. M., Ghani, E. R., Hricak, H., Szalay, A. A., Fong, Y., Blasberg, R. (2008) *Escherichia coli* Nissle 1917 facilitates tumor detection by positron emission tomography and optical imaging. *Clin. Cancer Res.* **14**, 2295–2302.
12. Wang, Y., Yu, Y. A., Shabahang, S., Wang, G., and Szalay, A. A. (2002) *Renilla* luciferase-*Aequorea* GFP (Ruc-GFP) fusion protein, a novel dual reporter for real-time imaging of gene expression in cell cultures and in live animals. *Mol. Genet. Genomics.* **268**, 160–168.
13. Zhang, Q., Liang, C., Yu, Y. A., Chen, N., Dandekar, T., Szalay, A. A. (2009) The highly attenuated oncolytic recombinant vaccinia virus GLV-1h68: comparative genomic features and the contribution of F14.5L inactivation. *Mol. Genet. Genomics.* **282**, 417–435.
14. Voisey, C. R., and Marincs, F. (1998) Elimination of internal restriction enzyme sites from a bacterial luminescence (*luxCDABE*) operon. *Biotechniques* **24**, 56–58.
15. Sorensen, M., Lippuner, C., Kaiser, T., Misslitz, A., Aebischer, T., Bumann, D. (2003) Rapid maturing red fluorescent protein variants with strongly enhanced brightness in bacteria. *FEBS Lett.* **552**, 110–114.
16. Joklik, W. K. (1962) The purification of four strains of poxvirus. *Virol.* **18**, 9–18.
17. Weibel, S., Stritzker, J., Eck, M., Goebel, W., Szalay, A. A. (2008) Colonization of experimental murine breast tumors by *Escherichia coli* K-12 significantly alters the

tumor microenvironment. *Cell Microbiol.* **10**, 1235–1248.

18. Lin, S. F., Price, D. L., Chen, C. H., Brader, P., Li, S., Gonzalez, L., Zhang, Q., Yu, Y. A., Chen, N., Szalay, A. A., Fong, Y., Wong, R. J. (2008) Oncolytic vaccinia virotherapy of anaplastic thyroid cancer *in vivo. J Clin Endocrinol. Metab.* **93**, 4403–4407.
19. Yu, Z., Li, S., Brader, P., Chen, N., Yu, Y. A., Zhang, Q., Szalay, A. A., Fong, Y., Wong, R. J. (2009) Oncolytic vaccinia therapy of squamous cell carcinoma. *Mol. Cancer.* **8**, 45.
20. Kelly, K., Woo, Y., Brader, P., Yu, Z., Riedl, C., Lin, S. F., Chen, N., Yu, Y. A., Rusch, V. W., Szalay, A. A., and Fong, Y. (2008) Novel oncolytic agent GLV-1h68 is effective against malignant pleural mesothelioma. *Hum. Gene Ther.* **19**, 774–782.
21. Hoffman, R. M. (2005) The multiple uses of fluorescent proteins to visualize cancers in vivo. *Nat. Rev. Cancer* **5**, 794–806.
22. Hoffman, R. M. and Yang, M. (2006) Whole-body imaging with fluorescent proteins. *Nat. Protoc.* **1**, 1429–1438.
23. Hoffman, R. M. (1999) Orthotopic metastatic mouse models for anticancer drug discovery and evaluation: a bridge to the clinic. *Invest. New Drugs* **17**, 343–359.

Chapter 12

GFP-Transgenic Animals for In Vivo Imaging: Rats, Rabbits, and Pigs

Takashi Murakami and Eiji Kobayashi

Abstract

Specifically, gene-encoded biological probes serve as stable and high-performance tools to visualize cellular fate in living animals. The rat, as with the mouse, has offered important animal models for biology and medical research, and has provided a wealth of physiological and pharmacological data. The larger-body animals, in comparison to the mouse have allowed the application of various physiological and surgical manipulations that may prove to have biological significance. We have further extended the techniques of genetic engineering to rats, rabbits, and pigs, and have created corresponding GFP-transgenic animals. The GFP-positive organs of these animals provide valuable sensors in preclinical settings for cell therapy and transplantation studies. In this chapter, we highlight expression profiles in these animal resources and describe examples of preclinical applications.

Key words: GFP, Imaging, Animal resource, Rats, Rabbits, Pigs, Gene expression, Transplantation, Cell therapy

1. Introduction

In vivo imaging has become an indispensable technology in both the biological sciences and medicine. In the past decades, there has been a huge increase in the number of imaging technologies and their applications (1). In particular, fluorescence imaging has been rapidly adapted for in vitro and in vivo analysis of biological processes (1, 2). Visualization of processes occurring in the complex environment of the cell and/or tissue needs an appropriate cellular marking procedure. Fluorescent dyes have often been used for such purposes. However, since fluorescence intensity may decrease during in vivo cellular proliferation, the use of fluorescent dyes may not always be suitable for in vivo imaging (3). In contrast,

Robert M. Hoffman (ed.), *In Vivo Cellular Imaging Using Fluorescent Proteins: Methods and Protocols*,
Methods in Molecular Biology, vol. 872, DOI 10.1007/978-1-61779-797-2_12, © Springer Science+Business Media, LLC 2012

genetically-encoded biological fluorescent probes serve as stable and high-performance tools to visualize cellular fate in living animals.

The development of genetic molecular tags such as green fluorescent protein (GFP) from the jellyfish (*Aequorea victoria*) has largely accelerated the revolution occurring in this field over the past decade. In fact, GFP, as the most popular biological light source, has offered important opportunities for the investigation of a wide variety of biological processes in living cells and animals (2, 4).

Mice represent a convenient experimental animal for basic research. Genetic engineering in mice has elucidated a plethora of molecular mechanisms that explain important biological processes (5, 6).

The field of stem cell biology has undergone a remarkable evolution, and much effort has been devoted to developing ways to exploit the potential of adult stem cells for therapeutic purposes (7, 8). In this light, new translational research fields in medicine (such as regenerative medicine) demand experimental animals of a larger body size, and such animal resources allow us to test various techniques in clinical medicine (3). Based on this concept, we have developed GFP-transgenic (Tg) animals, including rats, rabbits, and pigs (3, 9, 10). In this chapter, we highlight these animal resources and describe examples of preclinical applications.

2. GFP-Transgenic (Tg) Rats

2.1. Transgenic Rat System

Recent advances in rat genetic engineering are catching up to mouse technology (3, 11–13). GFP transgenic (Tg) rats have been shown to be a major asset in preclinical regenerative medicine (3) to trace the fate of transplanted cells and tissues. Tg rats have been employed as valuable cellular and organ sources in preclinical settings for cell therapy and transplantation studies (14–16).

Kobayashi et al. have developed nine genetically-marked rats (color-engineered rats, see Table 1) since the first generation of GFP-Tg Wistar rat was established in 2001 (17). These Tg rats have been used worldwide to investigate specific biological events in a variety of biomedical research fields such as neuroscience (16, 18, 23), cardiology (15, 19–21), dermatology (14, 22, 24), transplantation (18, 25–28), and tissue engineering (19–21). Many studies have demonstrated potential applications for Tg rats such as cellular sources for cellular trafficking, trans-differentiation, and tissue repair. Since these Tg rats are capable of crossbreeding, it is possible to generate a double-marker rat, and to visualize cellular fate by both fluorescence and luminescence (29, 30).

In order to obtain transgenic animals that ubiquitously express a particular gene under a general promoter, the injected transgene must properly integrate in the genome. Furthermore, despite

Table 1
Color-engineered rat colonies

Rat Strain	Promoter	Marker protein	References
Wistar	CAGGS[a]	GFP[b]	(17)
Lewis	CAGGS	GFP	(18)
Lewis	Albumin	GFP	NP[c]
Dark Agouty	CAGGS	LacZ (β-galactosidase)	(56)
Lewis	ROSA26	LacZ (β-galactosidase)	(18)
Wistar	Albumin	DsRed2[d]	(31)
Wistar	CAGGS	DsRed2, GFP (in Cre/LoxP system)	(33)
Lewis	ROSA26	Luciferase[e]	(57)
Lewis	ROSA26	DsRed monomer	NP[c]

[a]Cytomegalovirus enhancer/chicken β-action promoter
[b]Green fluorescnt protrein (*Aequorea victoria*)
[c]Not published: the Tg animal is available from the National Bio Resource Project (NBRP) for the Rat in Japan (http://www.anim.med.kyoto-u.ac.jp/nbr/)
[d]DsRed (*Discosoma*)
[e]Luciferase from the firefly (*Photinus pyralis*)

advances in genetic engineering, it remains difficult to handle rat eggs for DNA microinjection. Therefore, it is more difficult to construct a Tg rat system. The most remarkable feature in the color-engineered rat system (see Table 1) represents the establishment of the GFP-Tg Lewis rat (18). This line expresses GFP ubiquitously and strongly in all tissues (see Table 2), and cell sources such as neural progenitor cells provide a high-performance tool for the investigation of cellular fate (16, 18, 23).

2.2. GFP Expression Profile and Experimental Applications

2.2.1. Tissue/Organ Transplantation

The first generation of the GFP-Tg rat was on a Wistar background (17). Use of Wistar rats was indeed useful and effective for the monitoring of cellular fate in certain experiments including organ transplantation (17, 25–28) and cutaneous biology (14, 22). For example, it is known that the liver is a tolerogenic organ in immunology and contains abundant hepatic lymphocytes. The study of the long-term tracking of rat donor passenger leukocytes (DPLs) (26) showed that a small number of MHC class II$^+$GFP$^+$ DPLs (from GFP-Wistar rats) were present at the transplanted graft site and in the recipient spleen of rats, but not in the bone marrow. This study also demonstrated a substantial correlation of "leukocyte parking" with spontaneous tolerance in rat liver transplantation.

Miyashita et al. (22) also demonstrated the usefulness of GFP-Wistar rats by characterizing the role of hair follicles as a source of stem cells by implantation of cultured keratinocyte sheets. This study

Table 2
Expression profile of GFP-expressing TG animals (CAGSS[a] promoter)

	Rat		Rabbit	Pig
Organ/tissue	*Wistar*[b]	*LEW*[b]	*Kbs:JW*[c]	*Jinhua*[d]
Brain	+/−	+++	+	++
Eye	++	+++	N.D.	+
Lung	+/−	+++	+/−	++
Heart	++	+++	++	+++
Thymus	+	+++	N.D.	+
Liver	+	+++	+	++
Pancreas	++	+++	+++	+++
Samll intesine	+	+++	+	+
Colon	+	+++	++	+
Kidney	++	+++	+	++
Skeletal muscle	++	+++	N.D.	+++
Skin	+	+++	++	++

GFP expression was determined under a 489-nm excitation light
"+/−" weak, "+" moderate, "++" strong, "+++" very strong, N.D. not described
[a]Cytomegalovirus enhancer/chicken beta-action promoter
[b]ref. 18
[c]ref. 9
[d]ref. 10

suggested that stem cells are present in the induced follicle and that the induced follicle consists of polyclonally derived cells.

A liver-specific transgenic Wistar reporter rat (Alb-DsRed2, see Table 1) was used to demonstrate the role played by bone marrow-derived cells during liver regeneration (31, 32). The DsRed2/GFP Wistar double-reporter rat (see Table 1) was used to investigate the control of gene expression under Cre/LoxP site-specific recognition (33, 34), and study cellular fusion.

Transgenic rats derived from a Wistar background, however, have some shortcomings: all of the tissues do not express enough marker genes in the established animal lines, even though reporter genes are driven by a ubiquitous CMV enhancer/chicken beta-actin promoter (CAG promoter) (17, 35). Furthermore, transplanted cells or tissues are occasionally rejected by immune responses due to a mismatch of the minor histocompatibility complex (mHC) derived from an outbred strain of Wistar rats (17). To address these problems, Inoue et al. (18) developed GFP-expressing Lewis (LEW) rats harboring the same genetic background (MHC haplotype: $RT1^{l}$). The new CAG/GFP-LEW line

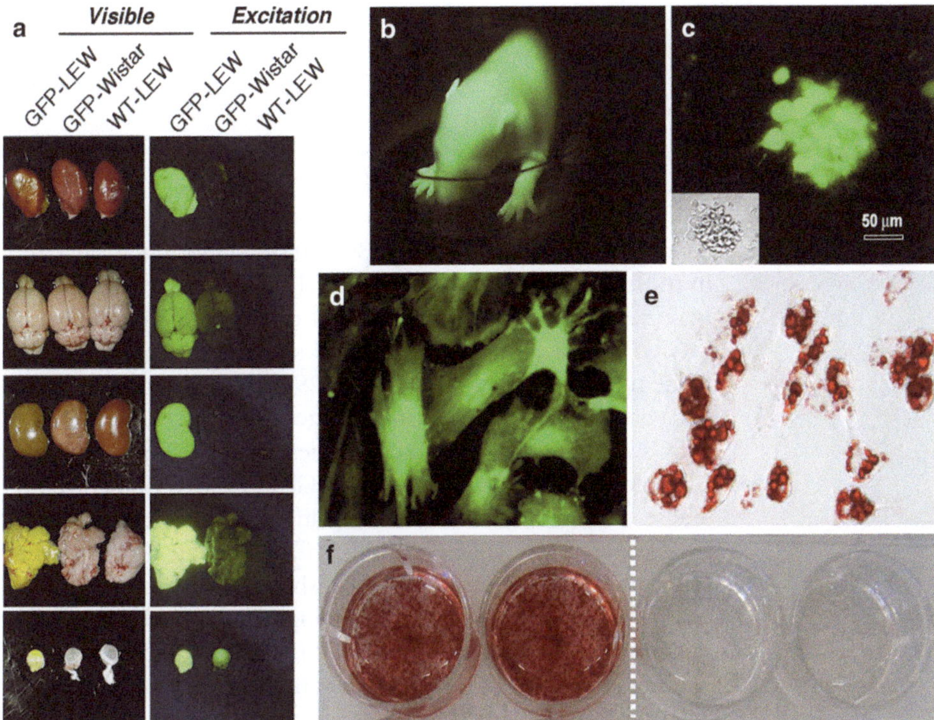

Fig. 1. GFP expression profile of GFP-transgenic rats. (**a**) Differential GFP-expression pattern between GFP-Wistar and GFP-LEW Tg rats. Representative organs (heart, brain, kidney, small intestine, eye) were removed from CAG/GFP-LEW Tg rats (18), CAG/GFP-Wistar Tg rats (17), and wild-type LEW rats, and examined under visible (*left panels*) or 489-nm excitation light (*right panels*). (**b**) Representative image of a newborn GFP-LEW Tg rat under excitation light. (**c**) Neural progenitor cells from CAG/GFP-LEW Tg rats express substantial levels of GFP under a 489-nm excitation light (*green*). (**d**) GFP expression of MSCs isolated from CAG/GFP-LEW Tg rats. These MSCs were capable of differentiating into (**e**) adipocytes (stained with oil red O for lipid droplets; original magnification 200×) and (**f**) osteocytes (*left panel*: stained with alizarin red for mineral deposition, *left panel*: controls).

expressed GFP strongly and ubiquitously in most of the organs, compared with the first generation GFP-Tg Wistar line (17, 18). As shown in Fig. 1a, representative organs such as the heart, brain, and kidney demonstrated higher levels of GFP-expression in the new GFP-LEW Tg line. The strong expression was observed both in neonate and newborn rats (see Fig. 1b).

2.2.2. Cell Transplantation

Inoue et al. (18) examined the cellular fate of neural progenitor cells from CAG/GFP-LEW Tg rats in a rat cerebral infarction model.

Neural progenitor cells can be established from E14.5 of CAG/GFP-LEW Tg rats and maintained in vitro for 20 days under appropriate culture conditions (16, 18). Neurospheres strongly expressed GFP (see Fig. 1c) and nestin, and maintained the phenotype as the neural progenitor. Spheres could be transplanted stereotactically into the cerebral ventricle space of wild-type LEW rats 5 days post-cerebral infarction. Inoue et al. (18) demonstrated that GFP-positive cells were accumulated in the cerebral infarction area and were able to survive. Similar results in the case of spinal cord injury were also observed (36). Cellular migration of stem cells in the brain is strongly associated with the sphingosine-1-phosphate/S1P receptor axis (36, 37).

Mesenchymal stem cells (MSCs) are thought to participate in tissue homoeostasis, remodeling, and repair by ensuring the replacement of mature cells lost by physiological turnover, senescence, injury, or disease (38, 39). Stem cell populations are found in most adult tissues and, in general, their differentiation potential may reflect the local cell population. MSCs migrate to various in vivo locations, including injury sites and tumors (38, 40, 41). Bone marrow-derived MSCs (24, 29) and synovium-derived MSCs (30) have been isolated from transgenic rats listed in Table 1. These MSCs were used in preclinical studies for tissue injury. For example, MSCs can be isolated from CAG/GFP-Tg LEW rats (see Fig. 1d). These cells possess the ability to differentiate into adipocytes (see Fig. 1e) and osteocytes (see Fig. 1f). Focusing on their angiogenesis potential, Inoue et al. (24) showed that transplanted bone marrow-derived MSCs were retained at wound sites during the healing process in a diabetic rat model, and promoted wound healing through angiogenesis. In an another example, Horie et al. (30) studied the differentiation potential of the local MSC population and demonstrated that synovium-MSCs injected into the massive meniscectomized knee adhered to the lesion, differentiated into meniscal cells directly, and promoted meniscal regeneration without migration to distant organs. These data suggest that transgenic rats play important roles to dynamically image the fate of transplanted cells.

3. GFP-Transgenic Rabbits

The intermediate body size of rabbits is appropriate for surgical and transplantation studies. GFP-transgenic rabbits were described previously (42–44). Al-Gubory and Houdebine (45) have reported that the use of fibered confocal fluorescence microscopy with an optical mini-probe allowed detection of fluorescent signals and the visualization of detailed tissue architecture and cell morphology of a living GFP transgenic rabbit, indicating that GFP is a powerful imaging marker in rabbits. Takahashi et al. (9) showed detailed profiling data in ubiquitous GFP-expressing transgenic rabbits (Japanese White [Kbs:JW]). Takahashi et al. created a GFP-Tg rabbit that ubiquitously expresses the GFP gene by introducing the CAG expression vector (17, 35) into the male pronucleus in the rabbit egg (9). The overall production efficiency of transgenic lines of rabbits (0.1%; 1/945) was lower than that reported for the mouse (2.1%; 4/192) and rat (3.2%; 3/95) using the same DNA preparation (9). Although 75% of the pups died before sexual maturation, a transgenic line was successfully established. An examination of the GFP expression profile in the F1 progeny showed that GFP was detected in all organs examined. The fluorescence intensity differed among the various organs (see Table 2). The expression profile of the rabbit line was comparable to those of the mouse (46) and rat (17). Furthermore, there appeared to be no differences in the GFP expression profile between male and female rabbits in all examined organs. Therefore, this established GFP-transgenic rabbit is a useful tool for various studies such as tissue engineering, regenerative medicine, and developmental biology.

4. GFP-Transgenic Pigs

Recent advances in gene manipulation have allowed the development of a variety of transgenic animals (47–49), and the procedure used for the microinjection of animal zygotes has continuously improved (50, 51). However, the use of classical transgenic technology through microinjection into zygotes is not always suitable with large transgenic animals. In light of the development of new therapeutic strategies (e.g., regenerative medicine and organ transplantation), the pig represents the xenogeneic donor of choice for future organ transplantation in humans, for anatomical and physiological reasons. Therefore, it is thought that transgenic pigs may have the potential to overcome both physiological and immunological barriers that have previously impeded this field. In fact, the use of alpha-1,3-galactosyltransferase gene-knockout pigs as donors has shown marked improvements in xenograft survivals (52–54). The somatic cell nuclear-transfer method has been employed to generate these genetically engineered pigs. It is known that the somatic cell nuclear-transfer technique works better in pigs than in other large animals (10, 55). Taking advantage of GFP as a stable fluorescent bio-probe, it is worth describing the expression profile of GFP-transgenic pigs. The tissues and organs of GFP-transgenic pigs could also provide a useful and helpful marker to monitor tissue fate in the above studies.

4.1. GFP Expression Pattern in the Cloned Pig

Two cloned pig lines have been developed that ubiquitously express GFP (10, 55). GFP cDNA is driven under the CAG promoter in both of these cloned pig lines (35). The *Jinhua* GFP-transgenic pig has a relatively small body size in comparison with common domestic pig strains. The GFP expression pattern was well characterized (summarized in Table 2) (10, 55). Skeletal muscle and the pancreas showed strong GFP expression in the cloned pigs (see Fig. 2a, b), GFP expression in the gastro-intestinal tract and eyes were weak in comparison (see Table 2). GFP expression in the

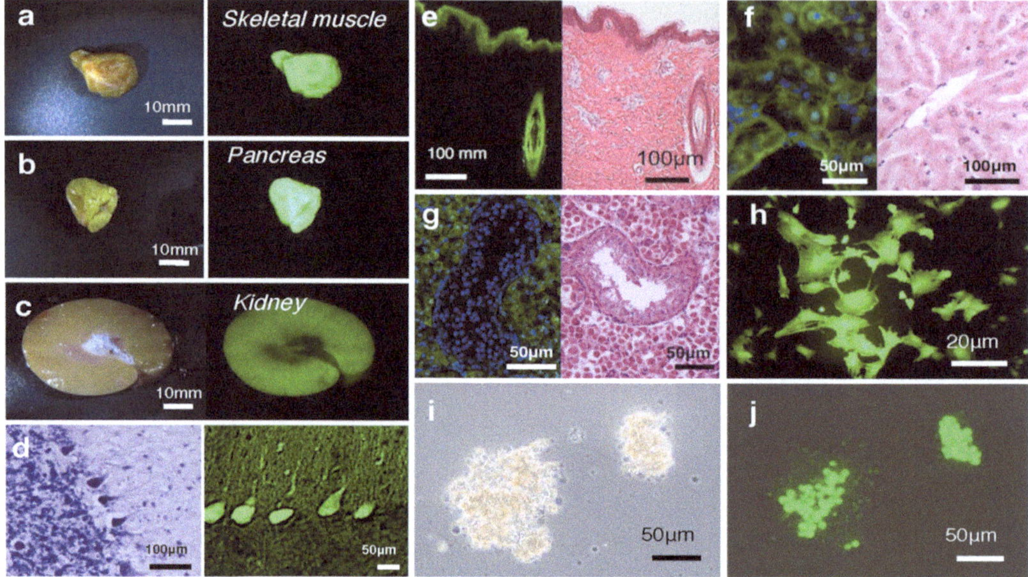

Fig. 2. GFP expression pattern of the GFP-transgenic Jinhua pig. (**a–c**) Representative organs (skeletal muscle (**a**), pancreas (**b**), kidney (**c**)) were removed from GFP-Tg pigs and macroscopically examined under a visible (*left*) or 489-nm excitation (*right*) light. (**b**) Photomicrographs of GFP fluorescence (*right*) and Nissl-staining (*right*) in the brain (cerebellum) of cloned pigs. (**d**) Large cell bodies in the cerebellum expressing GFP (*right panel*). (**e–g**) Microscopic GFP expression in representative tissue sections from cloned pigs: skin (**e**), liver (**f**), and testis (**g**). *Left panels*: 489-nm excitation light, *Right panels*: hematoxylin and eosin staining. (**h**) Strong GFP expression in bone marrow-derived MSCs from the GFP-Tg pig. (**i**, **j**) Langerhans islets isolated from the transgenic pig pancreas were cultured and GFP expression was examined under a visible (**h**) and 489-nm excitation (**i**) light (original magnification 200×).

kidney and liver of cloned pigs appeared intermediate and heterogeneous (see Fig. 2c). Kawarasaki et al. (10) showed extensive microscopic GFP expression patterns in various organs of the cloned pig (see Fig. 2). In the central nervous system (CNS), small-sized and round-shaped GFP-positive cell bodies were found in the olfactory bulb, lateral ventricle, and hippocampus (10). In contrast, GFP-positive large-sized cell bodies were observed in the cerebellum (see Fig. 2d). Both large and small cell bodies were found in the medulla oblongata, and GFP-positive fibers were also found in the olfactory bulb, cerebellum, and medulla oblongata (10).

GFP expression of the skin was predominant in the epidermis (the granular layer and stratum spinosum) and the hair follicle, but less so in the dermis (see Fig. 2e). Expression in skeletal and cardiac muscle appeared modestly heterogeneous, but most of the muscle fibers were GFP-positive. Regarding expression in the liver (see Fig. 2f), parenchymal cells appeared GFP-positive and interstitial cells were GFP-negative. Acinus cells in the pancreas were strongly GFP-positive. In the gastrointestinal tract, GFP was heterogeneously expressed in the epithelium of the stomach, intestine, and colon. Primary and secondary spermatocytes in the testis weakly expressed GFP, although the sperm was GFP-negative (see Fig. 2g). Kawarasaki et al. (10) demonstrated GFP expression of peripheral blood cells from cloned pigs. Leukocytes were GFP-positive, granulocytes exhibited a particularly strong GFP expression, and mononuclear cells were also moderately GFP-positive (10). GFP expression of erythrocytes was negative.

In the cloned pig, sufficient levels of GFP expression were observed in bone marrow-derived MSCs (see Fig. 2h). The cells rapidly proliferated and formed colonies. GFP expression levels were not altered during cell passage. These cells are capable of differentiating readily into osteocytes (10), but poorly into adipocytes (10). These results demonstrate that GFP expression was stable in MSCs from cloned pigs which preferentially differentiated into osteocytes. GFP expression was observed in proliferating hepatocytes in vitro (10). A similar phenomenon was observed in Langerhans islets isolated from the cloned pig pancreas (see Fig. 2i, j). These results suggest that cloned GFP-pigs should be a valuable imaging resource for biomedical science.

5. Notes

GFP-Tg animals are maintained in the hemizygous state. Homozygous conditions remain to be elucidated. Given that there is a multiple copy number of the injected expression construct, it is likely that homozygous conditions may lead to infertility (unstable reproductive activity), lethality in embryonic stages, or premature growth in newborn animals. Thus, it appears prudent to use these imageable transgenic animal resources under hemizygous conditions.

Together with modern advances in optical imaging, the transgenic system provides innovative tools and a new platform on which to further our understanding of in vivo cell biology. Since *Jinhua* GFP-pigs have a mid-size body similar to that of humans, organs from this breed may represent an appropriate resource for experimental xeno-transplantation. Thus, animal resources described in this chapter should provide useful imaging information in research fields such as tissue engineering, experimental regenerative medicine, and transplantation.

Transgenic rats are available from the National Bio Resource Project (NBRP) for the Rat in Japan (http://www.anim.med.kyoto-u.ac.jp/nbr/) or the Rat Resource Research Center (RRRC) in USA (http://www.nrrrc.missouri.edu/).

Acknowledgments

The authors wish to thank Drs. Yoji Hakamata (Nippon Veterinary and Life Science University, Tokyo, Japan) and Masafumi Takahashi (Jichi Medical University, Tochigi, Japan) for contribution to the transgenic rat project, Dr. Tatsuo Kawarasaki (Tokai University, Kumamoto, Japan) for generation of the GFP-transgenic pig, and Drs. Masatsugu Ueda and Ri-ichi Takahashi for creation of the transgenic rat and rabbit (PhoenixBio Co., Ltd., Tochigi, Japan). This study was supported by a research grant from the "Strategic Research Platform" Project for Private Universities: matching fund subsidy from the Ministry of Education, Culture, Sports, Science, and Technology of Japan, and by a grant from the Health and Labor Science Research Grants from the Ministry of Health, Labor, and Welfare.

References

1. Weissleder, R. and Pittet, M. J. (2008) Imaging in the era of molecular oncology. *Nature* **452**, 580–589.
2. Lippincott-Schwartz, J. and Patterson, G. H. (2003) Development and use of fluorescent protein markers in living cells. *Science* **300**, 87–91.
3. Murakami, T. and Kobayashi, E. (2005) Color-engineered rats and luminescent LacZ imaging: a new platform to visualize biological processes. *J Biomed Opt* **10**, 41204.
4. Hoffman, R.M. (2005) The multiple uses of fluorescent proteins to visualize cancer in vivo. *Nat Rev Cancer* **10**, 796–806.
5. Lyon, M. F., Rastan, S., and Brown, S. D. M. (eds) (1996) *Genetic Variants and Strains of the Laboratory Mouse* 3rd edn (Oxford University Press, Oxford).
6. Gondo, Y. (2008) Trends in large-scale mouse mutagenesis: from genetics to functional genomics. *Nat Rev Genet* **9**, 803–10.
7. Daley, G. Q. (2010) Stem cells: roadmap to the clinic. *J Clin Invest* **120**, 8–10.
8. Ronaghi, M., Erceg, S., Moreno-Manzano, V., and Stojkovic, M. (2010) Challenges of stem cell therapy for spinal cord injury: human embryonic stem cells, endogenous neural stem cells, or induced pluripotent stem cells? *Stem Cells* **28**, 93–99.
9. Takahashi, R., Kuramochi, T., Aoyagi, K., Hashimoto, S., Miyoshi, I., Kasai, N., Hakamata, Y., Kobayashi, E., and Ueda, M. (2007) Establishment and characterization of CAG/EGFP transgenic rabbit line. *Transgenic Res* **16**, 115–120.
10. Kawarasaki, T., Uchiyama, K., Hirao, A., Azuma, S., Otake, M., Shibata, M., Tsuchiya, S., Enosawa, S., Takeuchi, K., Konno, K., Hakamata, Y., Yoshino, H., Wakai, T., Ookawara, S., Tanaka, H., Kobayashi, E., and Murakami, T. (2009) Profile of new green fluorescent protein transgenic Jinhua pigs as an imaging source. *J Biomed Opt* **14**, 054017.
11. Zan, Y., Haag, J. D., Chen, K. S., Shepel, L. A., Wigington, D., Wang, Y. R., Hu, R., Lopez-Guajardo, C. C., Brose, H. L., Porter, K. I., Leonard, R. A., Hitt, A. A., Schommer, S. L., Elegbede, A. F., and Gould, M. N. (2003) Production of knockout rats using ENU mutagenesis and a yeast-based screening assay. *Nat Biotechnol* **21**, 645–651.
12. Zhou, Q., Renard, J. P., Le Friec, G., Brochard, V., Beaujean, N., Cherifi, Y., Fraichard, A., and Cozzi, J. (2003) Generation of fertile cloned rats by regulating oocyte activation. *Science* **302**, 1179.
13. Geurts, A. M., Cost, G. J., Freyvert, Y., Zeitler, B., Miller, J. C., Choi, V. M., Jenkins, S. S., Wood, A., Cui, X., Meng, X., Vincent, A., Lam, S., Michalkiewicz, M., Schilling, R., Foeckler, J., Kalloway, S., Weiler, H., Ménoret, S., Anegon, I., Davis, G. D., Zhang, L., Rebar, E. J., Gregory, P. D., Urnov, F. D., Jacob, H. J., and Buelow, R. (2009) Knockout rats via embryo microinjection of zinc-finger nucleases. *Science* **325**, 433.
14. Yamaguchi, Y., Kubo, T., Murakami, T., Takahashi, M., Hakamata, Y., Kobayashi, E., Yoshida, S., Hosokawa, K., Yoshikawa, K., and Itami, S. (2005) Bone marrow cells differentiate into wound myofibroblasts and accelerate the healing of wounds with exposed bones when combined with an occlusive dressing. *Br J Dermatol* **152**, 616–622.
15. Shimizu, T., Sekine, H., Yang, J., Isoi, Y., Yamato, M., Kikuchi, A. Kobayashi, E., and Okano, T. (2006) Polysurgery of cell sheet grafts overcomes diffusion limits to produce thick, vascularized myocardial tissues. *FASEB J* **20**, 708–710.

16. Francis, J. S., Olariu, A., Kobayashi, E., and Leone, P. (2007) GFP-transgenic Lewis rats as a cell source for oligodendrocyte replacement. *Exp Neurol* **205**, 177–189.
17. Hakamata, Y., Tahara, K., Uchida, H., Sakuma, Y., Nakamura, M., Kume, A., Murakami, T., Takahashi, M., Takahashi, R., Hirabayashi, M, Ueda, M., Miyoshi, I., Kasai, N., and Kobayashi, E. (2001) Green fluorescent protein-transgenic rat: a tool for organ transplantation research. *Biochem Res Commun* **286**, 779–785.
18. Inoue, H., Ohsawa, I., Murakami, T., Kimura, A., Hakamata, Y., Sato, Y., Kaneko, T., Okada, T., Ozawa, K., Francis, J., Leone, P., and Kobayashi, E. (2005) Development of new inbred transgenic strains of rats with LacZ and GFP. *Biochem Biophys Res Commun* **329**, 289–296.
19. Sekine, H., Shimizu, T., Yang, J., Kobayashi, E., and Okano, T. (2006) Pulsatile myocardial tubes fabricated with cell sheet engineering. *Circulation* **114(1 Suppl)**, I87–93.
20. Sekine, H., Shimizu, T., Kosaka, S., Kobayashi, E., and Okano, T. (2006) Cardiomyocyte bridging between hearts and bioengineered myocardial tissues with mesenchymal transition of mesothelial cells. *J Heart Lung Transplant* **25**, 324–332.
21. Sekine, H., Shimizu, T., Hobo, K., Sekiya, S., Yang, J., Yamato, M., Kurosawa, H., Kobayashi, E., and Okano, T. (2008) Endothelial cell coculture within tissue-engineered cardiomyocyte sheets enhances neovascularization and improves cardiac function of ischemic hearts. *Circulation* **118**(14 Suppl), S145–52.
22. Miyashita, H., Hakamata, Y., Kobayashi, E., and Kobayashi, K. (2004) Characterization of hair follicles induced in implanted, cultured rat keratinocyte sheets. *Exp Dermatol* **13**, 491–498.
23. Mothe, A. J., Kulbatski, I., van Bendegem, R. L., Lee, L., Kobayashi, E., Keating, A., and Tator, C. H. (2005) Analysis of green fluorescent protein expression in transgenic rats for tracking transplanted neural stem/progenitor cells. *J Histochem Cytochem* **53**, 1215–1226.
24. Inoue, H., Murakami, T., Ajiki, T., Hara, M., Hoshino, Y., and Kobayashi, E. (2008) Bioimaging assessment and effect of skin wound healing using bone-marrow-derived mesenchymal stromal cells with the artificial dermis in diabetic rats. *J Biomed Opt* **13**, 064036.
25. Ajiki, T., Takahashi, M., Inoue, S., Sakuma, Y., Oyama, S., Kaneko, T., Hakamata, Y., Murakami, T., Kume, A., Kariya, Y., Hoshino, Y., and Kobayashi, E. (2003) Generation of donor hematolymphoid cells after rat-limb composite grafting. *Transplantation* **75**, 631–636.
26. Inoue, S., Tahara, K., Kaneko, T., Ajiki, T., Takeda, S., Sato, Y., Hakamata, Y., Murakami, T., Takahashi, M., Kaneko, M., and Kobayashi, E. (2004) Long-lasting donor passenger leukocytes after hepatic and intestinal transplantation in rats. *Transpl Immunol* **12**, 123–131.
27. Sakuma, Y., Sato, Y., Inoue, S., Kaneko, T., Hakamata, Y., Takahashi, M., Murakami, T., and Kobayashi, E. (2004) Lympho-myeloid chimerism achieved by spleen graft of green fluorescent protein transgenic rat in a combined pancreas transplantation model. *Transpl Immunol* **12**, 115–122.
28. Takeuchi, K., Sereemaspun, A., Inagaki, T., Hakamata, Y., Kaneko, T., Murakami, T., Takahashi, M., Kobayashi, E., and Ookawara, S. (2003) Morphologic characterization of green fluorescent protein in embryonic, neonatal, and adult transgenic rats. *Anat Rec A Discov Mol Cell Evol Biol* **274**, 883–886.
29. Hara, M., Murakami, T., and Kobayashi, E. (2008) In vivo bioimaging using photogenic rats: fate of injected bone marrow-derived mesenchymal stromal cells. *J Autoimmun* **30**, 163–171.
30. Horie, M., Sekiya, I., Muneta, T., Ichinose, S., Matsumoto, K., Saito, H., Murakami, T., and Kobayashi, E. (2009) Intra-articular Injected synovial stem cells differentiate into meniscal cells directly and promote meniscal regeneration without mobilization to distant organs in rat massive meniscal defect. *Stem Cells* **27**, 878–887.
31. Sato, Y., Igarashi, Y., Hakamata, Y., Murakami, T., Kaneko, T., Takahashi, M., Seo, N., and Kobayashi, E. (2003) Establishment of Alb-DsRed2 transgenic rat for liver regeneration research. *Biochem Biophys Res Commun* **311**, 478–481.
32. Misawa, R., Ise, H., Takahashi, M., Morimoto, H., Kobayashi, E., Miyagawa, S., and Ikeda, U. (2006) Development of liver regenerative therapy using glycoside-modified bone marrow cells. *Biochem Biophys Res Commun* **342**, 434–440.
33. Sato, Y., Endo, H., Ajiki, T., Hakamata, Y., Okada, T., Murakami, T., and Kobayashi, E. Establishment of Cre/LoxP recombination system in transgenic rats. *Biochem Biophys Res Commun* **319**, 1197–1202.
34. Sato, Y., Ajiki, T., Inoue, S., Hakamata, Y., Murakami, T., Kaneko, T., Takahashi, M., and Kobayashi, E. (2003) A novel gene therapy to

the graft organ by a rapid injection of naked DNA I: long-lasting gene expression in a rat model of limb transplantation. *Transplantation* **76**, 1294–1298.

35. Niwa, H., Yamamura, K., and Miyazaki, J. (1991) Efficient selection for high-expression transfectants with a novel eukaryotic vector. *Gene* **108**, 193–199.
36. Kimura, A., Ohmori, T., Ohkawa, R., Madoiwa, S., Mimuro, J., Murakami, T., Kobayashi, E., Hoshino, Y., Yatomi, Y., and Sakata, Y. (2007) Essential roles of sphingosine 1-phosphate/S1P1 receptor axis in the migration of neural stem cells toward a site of spinal cord injury. *Stem Cells* **25**, 115–124.
37. Kimura, A., Ohmori, T., Kashiwakura, Y., Ohkawa, R., Madoiwa, S., Mimuro, J., Shimazaki, K., Hoshino, Y., Yatomi, Y., and Sakata, Y. (2008) Antagonism of sphingosine 1-phosphate receptor-2 enhances migration of neural progenitor cells toward an area of brain. *Stroke* **39**, 3411–3417.
38. Karp, J. M. and Leng Teo, G. S. (2009) Mesenchymal stem cell homing: the devil is in the details. *Cell Stem Cell* **4**, 206–216.
39. Salem, H. K. and Thiemermann, C. (2010) Mesenchymal stromal cells: current understanding and clinical status. *Stem Cells* **28**, 585–596.
40. Studeny, M., Marini, F. C., Champlin, R. E., Zompetta, C., Fidler, I. J., and Andreeff, M. (2002) Bone marrow-derived mesenchymal stem cells as vehicles for interferon-beta delivery into tumors. *Cancer Res* **62**, 3603–3608.
41. Studeny, M., Marini, F. C., Dembinski, J. L., Zompetta, C., Cabreira-Hansen, M., Bekele, B. N., Champlin, R. E., and Andreeff, M (2004) Mesenchymal stem cells: Potential precursors for tumor stroma and targeted-delivery vehicles for anticancer agents. *J Natl Cancer Inst* 96, 1593–1603.
42. Wang, H. J., Lin, A. X., Zhang, Z. C., and Chen, Y. F. (2001) Expression of porcine growth hormone gene in transgenic rabbits as reported by green fluorescent protein. *Anim Biotechnol* **12**, 101–110.
43. Chesne, P., Adenot, P. G., Viglietta, C., Baratte, M., Boulanger, L., and Renard, J. P. (2002) Cloned rabbits produced by nuclear transfer from adult somatic cells. *Nat Biotechnol* **20**, 366–369.
44. Chrenek, P., Vasicek, D., Makarevich, A. V., Jurcik, R., Suvegova, K., Parkanyi, V., Bauer, M., Rafay, J., Batorova, A., and Paleyanda, R. K. (2005) Increased transgene integration efficiency upon microinjection of DNA into both pronuclei of rabbit embryos. *Transgenic Res* **14**, 417–428.
45. Al-Gubory, K. H. and Houdebine, L. M. (2006) In vivo imaging of green fluorescent protein-expressing cells in transgenic animals using fibred confocal fluorescence microscopy. *Eur J Cell Biol* **85**, 837–845.
46. Okabe, M., Ikawa, M., Kominami, K., Nakanishi, T., and Nishimune, Y. (1997) 'Green mice' as a source of ubiquitous green cells. *FEBS Lett* **407**, 313–319.
47. Hunter, C. V., Tiley, L. S., and Sang, H. M. (2005) Developments in transgenic technology: applications for medicine. *Trends Mol Med* **11**, 293–298.
48. Houdebine, L. M. (2007) Transgenic animal models in biomedical research. *Methods Mol Biol* **360**, 163–202.
49. Niemann, H., Tian, X. C., King, W. A., and Lee, R. S. (2008) Epigenetic reprogramming in embryonic and foetal development upon somatic cell nuclear transfer cloning. *Reproduction* **135**, 151–163.
50. Tesson, L., Cozzi, J., Menoret, S., Remy, S., Usal, C., Fraichard, A., and Anegon, I. (2005) Transgenic modification of the rat genome. *Transgenic Res* **14**, 531–546.
51. Dann, C. T. and Garbers, D. L. (2008) Production of knockdown rats by lentiviral transduction of embryos with short hairpin RNA transgenes. *Methods Mol Biol* **450**, 193–209.
52. Chen, G., Qian, H., Starzl, T., Sun, H., Garcia, B., Wang, X., Wise, Y., Liu, Y., Xiang, Y., Copeman, L., Liu, W., Jevnikar, A., Wall, W., Cooper, D. K., Murase, N., Dai, Y., Wang, W., Xiong, Y., White, D. J., and Zhong, R. (2005) Acute rejection is associated with antibodies to non-Gal antigens in baboons using Gal-knockout pig kidneys. *Nat Med* **11**, 1295–1298.
53. Kuwaki, K., Tseng, Y. L., Dor, F. J., Shimizu, A., Houser, S. L., Sanderson, T. M., Lancos, C. J., Prabharasuth, D. D., Cheng, J., Moran, K., Hisashi, Y., Mueller, N., Yamada, K., Greenstein, J. L., Hawley, R. J., Patience, C., Awwad, M., Fishman, J. A., Robson, S. C., Schuurman, H. J., Sachs, D. H., and Cooper, D. K. (2005) Heart transplantation in baboons using alpha1,3-galactosyltransferase gene-knockout pigs as donors: initial experience. *Nat Med* **11**, 29–31.
54. Yamada, K., Yazawa, K., Shimizu, A., Iwanaga, T., Hisashi, Y., Nuhn, M., O'Malley, P., Nobori, S., Vagefi, P. A., Patience, C., Fishman, J., Cooper, D. K., Hawley, R. J., Greenstein, J., Schuurman, H. J., Awwad, M., Sykes, M., and Sachs, D. H. (2005) Marked prolongation of porcine renal xenograft survival in baboons through the use of alpha1,3-galactosyltransferase

gene-knockout donors and the cotransplantation of vascularized thymic tissue. *Nat Med* **11**, 32–34.

55. Brunetti, D., Perota, A., Lagutina, I., Colleoni, S., Duchi, R., Calabrese, F., Seveso, M., Cozzi, E., Lazzari, G., Lucchini, F., and Galli, C. (2008) Transgene expression of green fluorescent protein and germ line transmission in cloned pigs derived from in vitro transfected adult fibroblasts. *Cloning Stem Cells* **10**, 409–420.
56. Takahashi, M., Hakamata, Y., Murakami, T., Takeda, S., Kaneko, T., Takeuchi, K., Takahashi, R., Ueda, M., and Kobayashi, E. (2003) Establishment of lacZ-transgenic rats: a tool for regenerative research in myocardium. *Biochem Biophys Res Commun* **305**, 904–908.
57. Hakamata, Y., Murakami, T., and Kobayashi, E. (2006) "Firefly rats" as an organ/cellular source for long-term in vivo bioluminescent imaging. *Transplantation* **81**, 1179–1184.

Chapter 13

The Use of Fluorescent Proteins for Developing Cancer-Specific Target Imaging Probes

Thomas E. McCann, Nobuyuki Kosaka, Peter L. Choyke, and Hisataka Kobayashi

Abstract

Target-specific imaging probes represent a promising tool in the molecular imaging of human cancer. Fluorescently-labeled target-specific probes are useful in imaging cancers because of their ability to bind a target receptor with high sensitivity and specificity. The development of probes relies upon preclinical testing to validate the sensitivity and specificity of these agents in animal models. However, this process involves both conventional histology and immunohistochemistry, which require large numbers of animals and samples with costly handling. In this chapter, we describe a novel validation tool that takes advantage of genetic engineering technology, whereby cell lines are transfected with genes that induce the target cell to produce fluorescent proteins with characteristic emission spectra, thus enabling their easy identification as cancer cells in vivo. Combined with multicolor fluorescence imaging, this can provide rapid validation of newly-developed exogenous probes that fluoresce at different wavelengths. For example, the plasmid containing the gene encoding red fluorescent protein (RFP) was transfected into cell lines previously developed to either express or not express specific cell surface receptors. Various antibody-based or ligand-based optical-contrast agents, with green fluorophores were developed to concurrently target cancer cells and validate their positive and negative controls, such as the β-D-galactose receptor, HER1, and HER2 in a single animal/organ. Spectrally-resolved multicolor fluorescence imaging was used to detect separate fluorescence emission spectra from the exogenous green fluorophore and RFP. Here, we describe the use of "co-staining" (matching the exogenous fluorophore and the endogenous fluorescent protein to the positive control cell line) and "counter-staining" (matching the exogenous fluorophore to the positive control and the endogenous fluorescent protein to the negative control cell line) to validate the sensitivity and specificity of target-specific probes. Using these in vivo imaging techniques, we are able to determine the sensitivity and specificity of target-specific optical contrast agents in several distinct animal models of cancer in vivo, thus exemplifying the versatility of our technique, while reducing the number of animals needed to conduct these experiments.

Key words: Cancer, Spectral fluorescence imaging, Fluorescent probe, Fluorescent protein, Sensitivity, Specificity, Molecular imaging, Molecular targeting

Robert M. Hoffman (ed.), *In Vivo Cellular Imaging Using Fluorescent Proteins: Methods and Protocols*, Methods in Molecular Biology, vol. 872, DOI 10.1007/978-1-61779-797-2_13, © Springer Science+Business Media, LLC 2012

1. Introduction

As more novel imaging probes are developed, it is necessary to optimize a method to efficiently assess their effectiveness at identifying pathological processes in vivo. The utilization of optically-detectable agents is promising for clinical applications because of their high sensitivity and specificity. Fluorescently-labeled probes provide the advantage of high signal-to-background ratios, the ability to utilize multiple fluorophores to differentiate even subtle cellular differences, and decreased exposure to ionizing radiation. The application of using fluorescently-labeled probes in the identification and clinical monitoring of cancers shows promise for the future. However, novel probes must undergo sufficient preclinical assessment before these probes can be utilized clinically. The ideal probe binds to its target receptor with high affinity resulting in high sensitivity and specificity, has high signal-to-background ratio, and any probe that remains unbound is quickly removed from the body with low systemic toxicity. Therefore, a preclinical model must be able to monitor binding efficiency, clearance, sensitivity, and specificity efficiently.

An example of the kind of target-specific optical imaging agent that may play a role in disease in the future is one designed to detect ovarian cancer in the peritoneum as an aid to surgery. Using fluorescence imaging in a mouse model with disseminated peritoneal cancer, the agent was able to localize tumor nodules as small as 100 μm in diameter, which were invisible to the naked eye (1). In order to validate new imaging probes such as this, however, it is important to document that the areas of fluorescence actually correspond to sites of disease. Both the "true positives" and the "true negatives" need to be assessed to determine the sensitivity and specificity of the probe. The traditional method of doing this is to use large numbers of animals, painstakingly remove tumor implants, as well as normal regions and correlate these with the images. In addition to being time consuming, expensive, and necessitating large numbers of animals, the analysis is incomplete because it is difficult to sample more than a small percentage of the total lesions. There are also always uncertainties regarding the reliability of the co-localization of the histologic specimen to the image. When the tumor nodules are too small and too closely clustered to be individually defined by direct visualization, it is almost impossible to correlate each positive nodule in the fluorescence image to its respective histologic counter part. In order to overcome this problem, we developed an alternative approach using genetically-engineered probes that constitutively express fluorescent proteins in the cancer cells (2).

In this chapter, we describe a method to assess the sensitivity and specificity of novel exogenous imaging probes by comparing them with endogenously-expressed fluorescent proteins. We used "co-staining" (matching the exogenous fluorophore and the endogenous fluorescent protein to the positive control cell line) and "counter-staining" (matching the exogenous fluorophore to the positive control and the endogenous fluorescent protein to the negative control cell line) to validate the sensitivity and specificity of target-specific probes.

2. Materials

2.1. Plasmid

The RFP (DsRed2)-expressing plasmid was purchased from Clontech Laboratories, Inc. (Mountain View, CA).

2.2. Antibodies and Ligands

1. Trastuzumab was purchased from Genentech, Inc. (San Francisco, CA).
2. Daclizumab was purchased from Hoffmann-La Roche, Inc. (Nutley, NJ).
3. Galactosamine serum albumin (GmSA) was purchased from Sigma Chemical (St. Louis, MO).

2.3. Fluorophores

1. Rhodamine Green (RhodG) NHS was purchased from Invitrogen Corporation (Carlsbad, CA).
2. ICG-sulfo-OSu was purchased from Dojindo Molecular Technologies (Rockville, MD).
3. Alexa680-NHS was purchased from Invitrogen Corporation (Carlsbad, CA).

2.4. Molecular Probes

1. GmSA conjugated to RhodG (3).
2. Trastuzumab conjugated to RhodG (4).
3. Daclizumab conjugated to indocyanine green (ICG) (5).
4. Trastuzumab conjugated to Alexa680 (6).

2.5. In Vivo Imaging

1. Female nude mice (National Cancer Institute Animal Production Facility, Frederick, MD, USA).
2. Maestro In Vivo Imaging System (CRi Inc., Woburn, MA, USA).
3. FluorVivo real-time color fluorescence imaging system (INDEC Biosystems, Santa Clara, CA).

3. Methods

3.1. Co-staining Method

3.1.1. Concept

In co-staining, a target cell line is labeled with an endogenous fluorescent protein. Next, the probe of interest is conjugated to an exogenous fluorophore that is optically distinguishable from the transfected fluorescent protein, and then injected into an animal

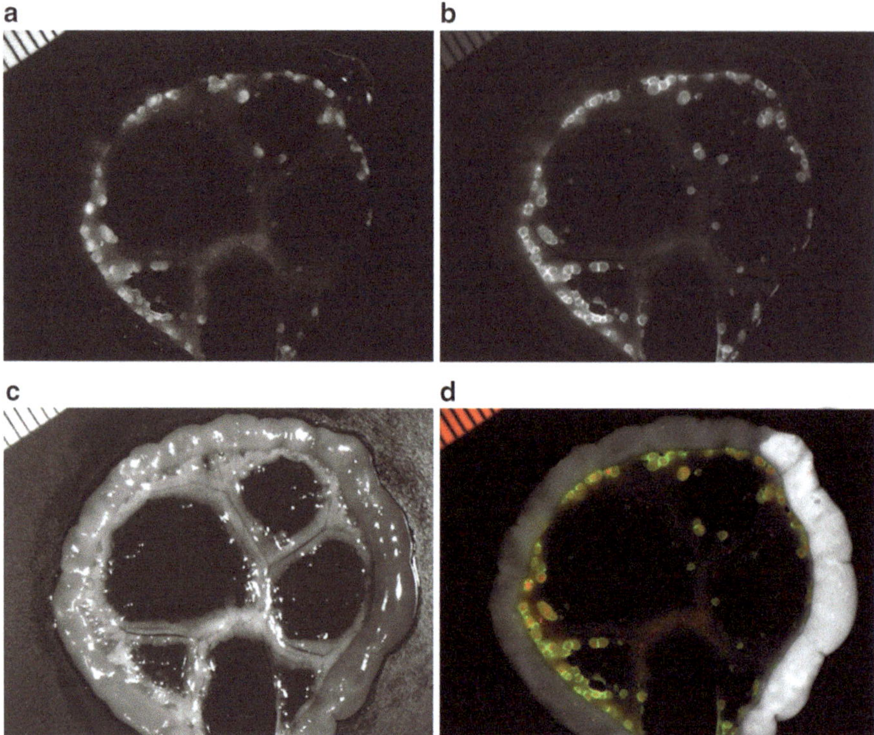

Fig. 1. Co-staining in a peritoneal-dissemination model of ovarian cancer. Spectral fluorescence images of a SHIN3-RFP tumor-bearing mouse which was injected with GmSA-RhodG. Unmixed spectral fluorescence images display (**a**) SHIN3-RFP tumors in the RFP spectrum and (**b**) the tested imaging probe (GmSA-RhodG) in the RhodG spectrum. (**c**) White light image of gut and mesentery. (**d**) Overlay of spectral images illustrates the co-localization of both the RFP and RhodG fluorescence.

bearing tumors derived from a fluorescent-protein-labeled cancer cell line. Since multicolor spectral fluorescence imaging enables the detection of these two distinct fluorescent wave lengths, two fluorescent images, one from the endogenous fluorescent protein and another from the exogenous fluorophore, can be obtained and analyzed (see Fig. 1) (see Note 1). For analysis, true-positive results are tumor foci that show two distinct fluorescence-emission patterns. False positives are those tumor foci that demonstrate exogenous fluorescence only, but not endogenous fluorescence. False negatives are tumor foci that have endogenous fluorescence only, but no exogenous fluorescence. To determine the true negative value, however, it is necessary to randomly choose regions of interest (ROIs) over normal tissues negative for endogenous fluorescence emission and determine whether exogenous fluorescence is present. Once these values are determined, it is possible to determine sensitivity and specificity of a given probe. However, this method is most accurate at determining sensitivity, rather than specificity. This is because specificity estimation requires the knowledge of the true-negative value, and the requirement to place ROIs over "normal" tissue is sometimes difficult because normal tissue is often not validated and this may lead to inaccurate determination of specificity.

3.1.2. Tumor Model and Probe (Peritoneal Dissemination Model)

Here, we describe one method of co-staining using a β-D-galactose receptor-positive cell line (SHIN3) and a newly-developed β-D-galactose receptor-targeted optical probe GmSA-RhodG in a peritoneal-dissemination model of ovarian cancer (3).

1. The red fluorescent protein (RFP:DsRed2) plasmid was transfected into SHIN3 cells (human ovarian cancer cell line) which is positive for the β-D-galactose receptor, and then cloned to establish a cell line stably expressing RFP (SHIN3-RFP).
2. The tested imaging probe, D-galactosamine serum albumin (GmSA) which binds to the β-D-galactose receptor, is conjugated with a green fluorophore, rhodamine green (GmSA-RhodG).
3. An intraperitoneal-dissemination model of ovarian cancer is established 14 days after intraperitoneal injection of 2×10^6 SHIN3-RFP cells, suspended in 200 μL PBS, into female nude mice.
4. Three hours after intraperitoneal injection of 500 pmol GmSA-RhodG diluted in 300 μL PBS, mice are sacrificed with carbon dioxide and then subjected to fluorescence imaging.

3.1.3. In Vivo Fluorescence Imaging (Spectral Fluorescence Imaging)

Spectral fluorescence imaging is a method that enables the unmixing of two or more fluorescent signals by their distinct spectral fluorescence patterns. This has several advantages over conventional fluorescence imaging. First, it can separate and eliminate autofluorescence from the target fluorescence signal efficiently by differentiating their distinct spectral fluorescence patterns. Second, fluorophores/fluorescent proteins having similar, but distinct, emission spectra can be unmixed. Thus, multiplexed use of several fluorophore/fluorescent proteins is feasible with this method. Also, low-intensity target fluorescence can be detected and enhanced after spectral unmixing. Therefore, the brightness of target fluorescence matters less than conventional fluorescence imaging. The procedure using spectral fluorescence imaging, with co-staining to validate GmSA as a probe to identify SHIN3 tumors, in the intraperitoneal-dissemination model, is described below.

1. After euthanasia, the abdominal cavity is exposed, and the mesentery is spread out on a nonfluorescent plate.
2. Close-up spectral fluorescence imaging of the mesentery is performed with an in vivo spectral-fluorescence imager (Maestro, CRi, Waltham, MA). Spectral-fluorescence image data are obtained in which each pixel has its own distinct spectral pattern in a range of acquisition wavelengths. In this particular experiment (RFP and RhodG), parameters of acquisition are as follows: a band-pass filter from 445 to 490 nm and a long pass filter over 515 nm were used for excitation and emission light, respectively. The tunable filter was automatically

stepped in 10 nm increments from 500 to 800 nm while the camera captured images at each wavelength interval with constant exposure. Note that these parameters need to be optimized depending on the combination of fluorescent proteins and fluorophores used.

3. To unmix spectral fluorescence data into its component spectral fluorescences, it is necessary to first create a standard library that represents each fluorophore of interest and a library that represents autofluorescence (see Note 2). In our sample, an RFP spectrum is obtained using RFP transfected SHIN3 cells resuspended in solution, a RhodG spectrum is established using GmSA-RhodG in solution, and an autofluorescence spectrum of the gut is created using a nude mouse without any fluorophore.

4. All spectral-fluorescence data are unmixed with this spectral library, and finally unmixed fluorescence images of RhodG and RFP are created and then subjected to image analysis.

3.1.4. Image Analysis

ROIs were drawn on two unmixed fluorescence images of either endogenous fluorescent protein or exogenous fluorophore, to determine sensitivity and specificity as described in Subheading 3.1.1. In this experiment, ROIs are drawn over both the lesions, depicted by RFP spectral fluorescence (standard reference for cancer foci) and in the surrounding adjacent areas (standard reference for non-cancerous foci). The average fluorescence intensity of each ROI is calculated, both on the RFP and the RhodG spectral unmixed images. Then, additional ROIs are drawn on the nodules depicted only by the RhodG spectral fluorescence to count the number of false-positive lesions. The average fluorescence intensities of false-positive foci are calculated, both on the RFP and the RhodG unmixed images. The number of foci positive for both RhodG and RFP, negative for both RhodG and RFP, and positive only for RhodG or RFP, are counted. Finally, sensitivity and specificity are calculated based on these values, as described in Subheading 3.1.1.

3.2. Counter-Staining Method

3.2.1. Concept

In counter-staining, two or more cancer cells lines are utilized to determine sensitivity and specificity of a given probe. One cell line must express the target receptor of interest that will bind to a testing probe and the other cell line must not express the receptor of interest. The cell line that does not express the target receptor of interest is labeled with an endogenous fluorescent protein. Then, the probe of interest is conjugated with an exogenous fluorophore that is optically distinguishable from the transfected fluorescent protein. The conjugated probe is then injected into an animal bearing tumors derived from both cell lines. Next, spectral fluorescence imaging is used to distinguish the endogenous fluorescent protein and the exogenous probe (see Fig. 2). For analysis, true

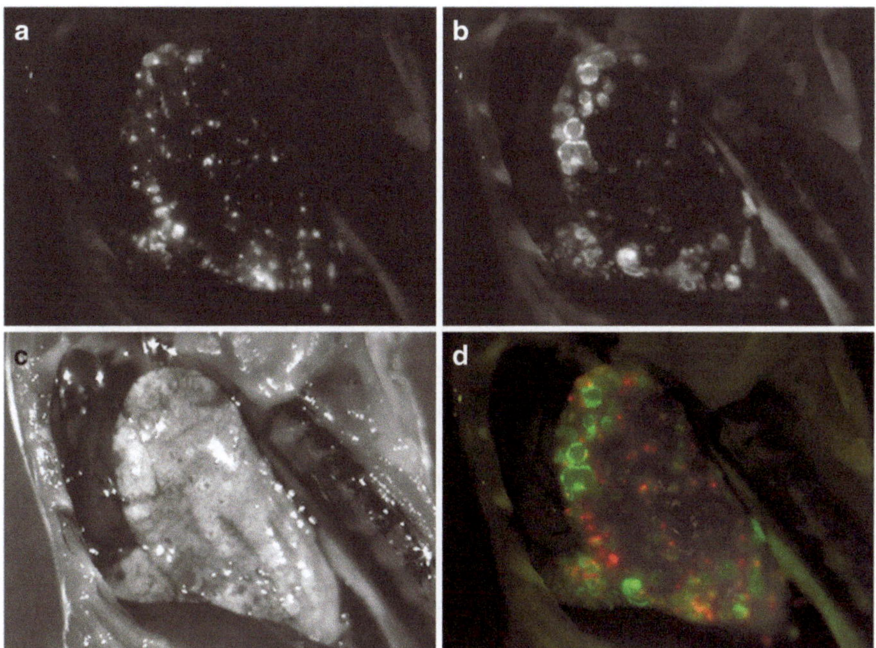

Fig. 2. Counter staining in a model of lung metastasis. Spectral-fluorescence images of a mouse bearing 3T3-HER2 (HER2+) and Balb3T3-RFP (HER2−) tumors in a lung metastasis model that was injected with RhodG-conjugated trastuzumab (anti-HER2). (**a**) Unmixed RFP spectral image localizes RFP tumors. (**b**) Unmixed RhodG spectral image identifies 3T3-HER2 tumors labeled with trastuzumab-RhodG. (**c**) White light images of the left lung. (**d**) Two-color spectral-fluorescence overlay localizes HER2+ and HER2− tumors which are clearly differentiated from each other and do not co-localize.

positive results are tumor foci that fluoresce at the spectra of the exogenous probe. False positives are the tumor foci that fluoresce at both the exogenous-probe and the endogenous-protein fluorescence spectra. True negatives are tumor foci that fluoresce at only the spectra that represent the endogenous fluorescent protein. False negatives are represented by tumor foci that form nodules visible with white light imaging or microscopic imaging but have no exogenous or endogenous fluorescence emission. In contrast to co-staining, this method is most accurate in determining specificity, because this can determine the true-negative value more clearly than the co-staining method. However, to determine the false-negative value, white light images serve as the gold standard and small tumors may be invisible on white light images leading to an overestimation of specificity.

3.2.2. Tumor Model and Probe (Metastatic Lung Cancer Model)

Here, we demonstrate a counter-staining method using a HER-2 receptor-positive cell line (NIH3T3 cells transfected with the HER-2 gene) and a HER-2 receptor-negative cell line (Balb3T3) in a lung metastasis model (7).

1. NIH3T3 cells were transfected with the HER-2 gene and cloned to establish a cell line stably over-expressing HER-2 receptors

(3T3-HER2). The red fluorescent protein (RFP:DsRed2) plasmid was transfected into Balb3T3 cells, which are negative for the HER-2 receptor and cloned to establish a cell line stably expressing RFP (Balb3T3-RFP).

2. The imaging probe to be tested, trastuzumab which binds to the HER2 receptor, is conjugated with a green fluorophore, rhodamine green (trastuzumab-RhodG).
3. A metastatic lung tumor model is established 12 days after intravenous injection of 2×10^6 Balb3T3/RFP cells and 19 days post-injection of 2×10^6 3T3/HER-2^+ cells suspended in 200-μL PBS into female nude mice.
4. 24 h after tail vein injection of 50 μg trastuzumab-RhodG, diluted in 200 μL PBS in mice with lung metastatic tumors, the mice are sacrificed with carbon dioxide and then subjected to fluorescence imaging.

3.2.3. In Vivo Fluorescence Imaging

1. Immediately after sacrifice, the pleural cavity is exposed, and the mouse is placed on a nonfluorescent plate.
2. Spectral-fluorescence images were obtained using an in vivo spectral-fluorescence imager as described in Subheading 3.1.3. Then fluorescence images of RhodG and RFP are unmixed based on the respective emission spectra of RhodG and RFP.

3.2.4. Image Analysis

Two unmixed fluorescent images are obtained representing either the endogenous RFP or exogenous RhodG. Detection of RhodG-positive tumors, RFP-positive tumors, and double-negative tumors are performed using magnified fluorescence composite images and stereoscopic microscope images. Finally, sensitivity and specificity were determined as described above in Subheading 3.2.1.

3.3. Multi-excitation

In the method described above, we utilized a single excitation method for in vivo spectral fluorescence imaging. Using this method, we are limited by the range of excitation light. By using a blue excitation light (445–490 nm), we are limited to fluorophores that are excitable in that range. When a green fluorophore and RFP are used, although both fluorophores are excitable by blue light, the RFP is excited less efficiently than the green fluorophore, increasing "false positive" tumor nodules in the co-staining analysis. For instance, when calculating the sensitivity of GmSA-RhodG using co-staining, three foci positive for GmSA-RhodG were not confirmed by RFP and were determined to be false positives (3). It is difficult to determine, but this may have resulted from RhodG being excited with better efficiency than RFP by the blue light (see Fig. 3) (see Note 3).

Ideally, we would like to be able to excite each fluorophore with different wavelengths of light that excite each fluorophore

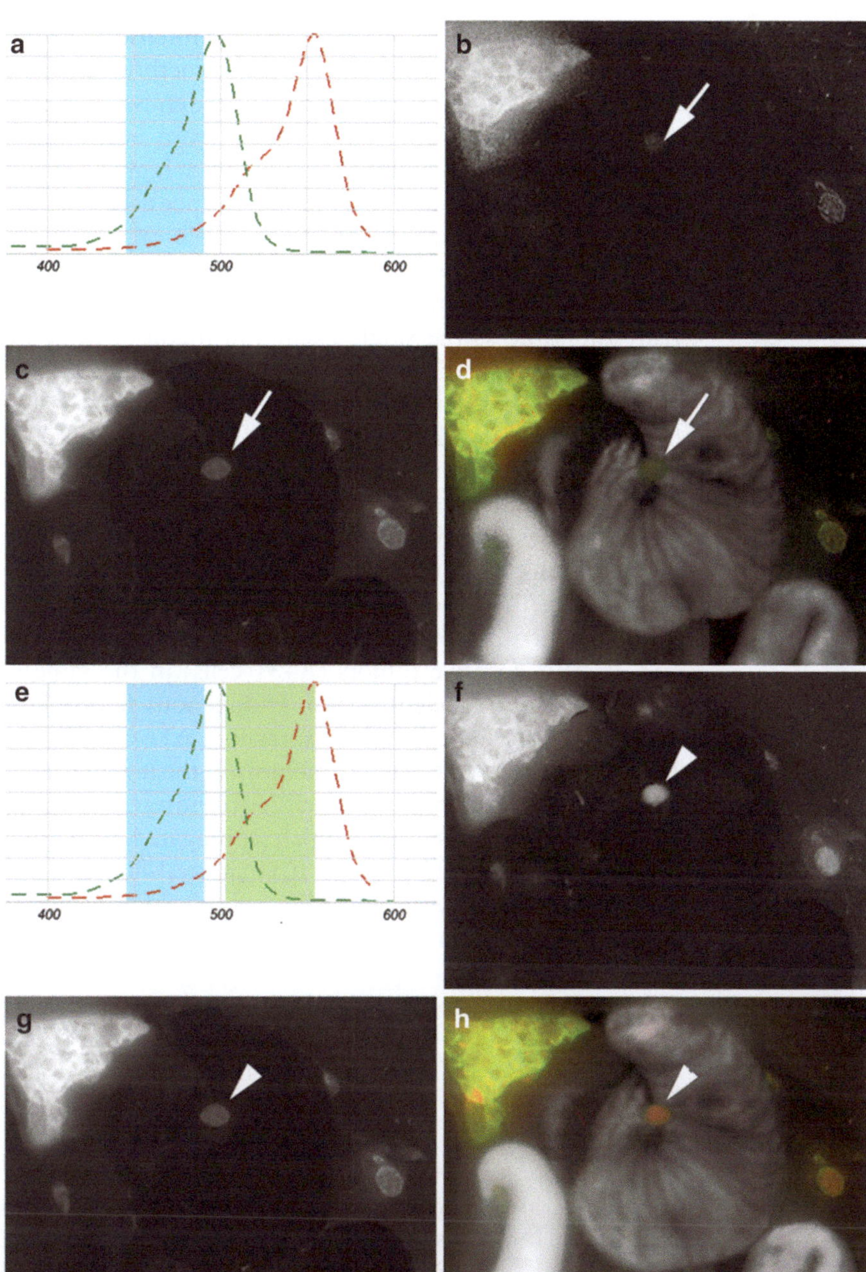

Fig. 3. Comparison of single- and multi-excitation in a peritoneal-dissemination model of ovarian cancer. (**a**) Schematic of RFP (*red*) and RhodG (*green*) absorbance curves, and range of single blue excitation light (*blue*). The blue light adequately excites RhodG, but only suboptimally excites RFP. (**b–d**) Spectral fluorescence images taken with a single excitation blue light using a co-staining method in a SHIN3-RFP-tumor-bearing mouse receiving GmSA-RhodG. (**b**) RFP spectrum identifies endogenous expression of RFP by SHIN3 cells but is unable to recognize tumor nodules (*arrows*) that are visible on the RhodG spectrum (**c**). (**d**) Two-color overlay using single excitation light. (**e**) Schematic of multi-excitation (*blue* and *green*) light on the absorbance curves of RFP (*red*) and RhodG (*green*) demonstrating more efficient excitation of RFP by green light. (**f–h**) Spectral-fluorescence images, taken with multiple-excitation filters in the same mouse, demonstrates the ability to identify the previously-invisible tumor nodules on the RFP spectrum (**f**, *arrow head*) that are still present on the RhodG spectrum (**g**). (**h**) Two-color overlay using multiple excitation filters.

with equal efficiency, resulting in the most intense emission possible for each fluorophore. Multiple excitation methods enable us to use many more fluorophores that span a wider emission spectra. Using this method, the fluorescent signals from each fluorophore/ fluorescent protein increase, leading to a higher signal-to-noise ratio on unmixed images. Furthermore, since each pixel has two or more spectral patterns excited by multiple excitation lights, this method can unmix each fluorophore's spectrum more efficiently (8). The procedure for multiple-excitation image acquisition is essentially the same as the single excitation method, except the filter settings are changed during acquisition of fluorescence imaging. Currently, several filter settings can be used for excitation during one image acquisition with the Maestro spectral-fluorescence imager (CRi) (e.g., blue, green, and red excitation).

3.4. Other Tumor Models

3.4.1. Subcutaneous Transplant Model

The strategies described above may also be used in subcutaneous-xenograft models of tumors. For example, we applied the counter-staining method to a subcutaneous-transplant model with interleukin-2α-receptor (IL-2Rα)-positive tumors (ATAC4 cells) and IL-2Rα-negative tumors (A431 cells) in the same mice (5). A431 cells were labeled with an endogenous fluorophore (RFP) and cloned to establish stable expression. ATAC4 cells and A431-RFP cells were injected subcutaneously in the left and right dorsum of female nude mice, respectively. After intravenous injection of an exogenous probe, daclizumab conjugated to ICG, ATAC4 tumors were depicted by only ICG spectral-fluorescence. A431-RFP was depicted with only RFP spectral fluorescence (counter-staining) (see Fig. 4).

3.4.2. Two Tumor Peritoneal Dissemination Model

Co-staining and counter-staining methods can be performed simultaneously in a single mouse model. This has been demonstrated using our animal model of "co-incident" peritoneal cancer implants (6). Using two cancer cell lines, SHIN3 and SKOV3, which both express the β-D-galactose receptor (D-galR), but differ in their expression of another receptor (HER-2), we were able to specifically distinguish between the two cell populations in a single mouse. SHIN3 cells were transfected with a plasmid expressing RFP (D-galR$^+$, HER2$^-$, RFP$^+$). SKOV3 cells (D-galR$^+$, HER2$^+$, RFP$^-$) and SHIN3 cells were co-injected into nude mice to establish a coincident model of intraperitoneal dissemination. After establishing the tumor model, galactosyl serum albumin (GSA) conjugated to RhodG, which targets D-galR, and trastuzumab conjugated to Alexa680, which targets HER-2, were co-injected into the tumor-bearing mice. In vivo multi-excitation spectral imaging was carried out and each fluorescence signal was unmixed. SHIN3 cells were identified by fluorescence in the RFP and RhodG spectrum (co-staining), but no emission in the Alexa680 spectrum. SKOV3 cells were identified by fluorescence in the RhodG

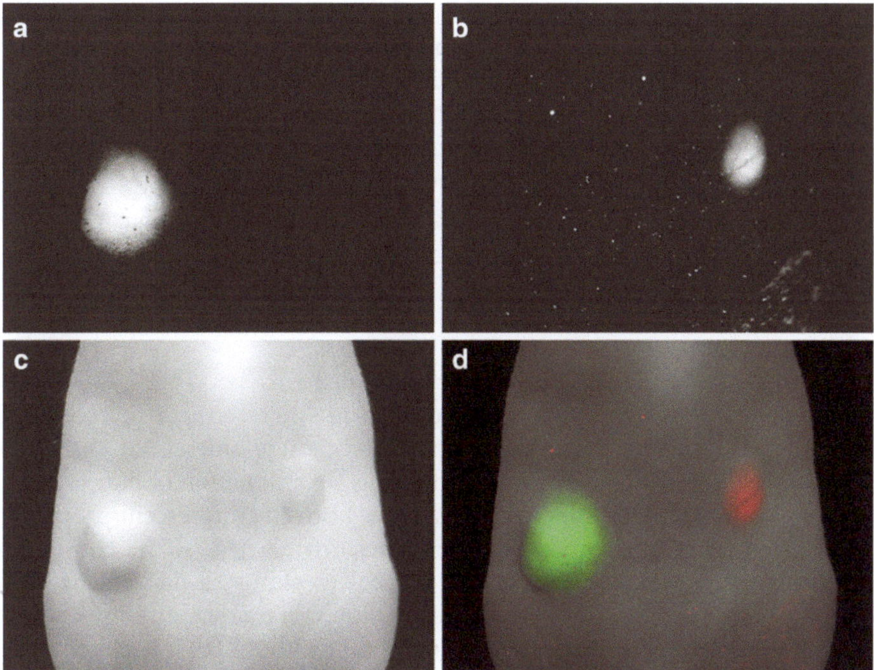

Fig. 4. Subcutaneous-xenograft model identified with counter-staining. Spectral-fluorescence images of a mouse bearing ATAC4 (IL-2Rα+) and A431-RFP (IL-2Rα−) tumors receiving ICG-conjugated daclizumab (anti-IL-2Rα). Unmixed spectral fluorescence images illustrate (**a**) the ICG spectrum demonstrating the localization of ATAC4 tumors labeled with ICG conjugated daclizumab, (**b**) the RFP spectrum localizing A431 cells endogenously expressing RFP. (**c**) White light image, and (**d**) two-color overlay demonstrates that ATAC4 tumor (IL-2Rα+) is depicted only in the ICG spectrum (*green*), while a A431-RFP tumor is depicted only in the RFP spectrum (*red*) and does not show any fluorescence in ICG spectrum.

and Alexa680 spectrum, but no emission in the RFP spectrum (counter-staining). With this method, we were able to verify the multiplexed use of probes to distinguish between subpopulations of cancer cells in vivo. Theoretically, this method can be applied to other mouse models, including subcutaneous-xenograft and lung metastasis models.

3.5. Other Fluorescence Imaging Methods

The in vivo fluorescence imaging methods described above are performed using a spectral-fluorescence imager (Maestro, CRi). Theoretically, these strategies can be applied to other fluorescence imaging devices. For example, counter-staining was demonstrated with an in vivo real-time color fluorescence imager (FluorVivo, Indec) in a co-incident two-tumor peritoneal-dissemination model (SHIN3-RFP and SKOV3) (4). In this study, trastuzumab-RhodG was injected into the co-incident two-tumor model. Fluorescence-guided surgery was then performed. SHIN3-RFP tumors were depicted by their red fluorescence, while SKOV3 tumors were depicted by green fluorescence (labeled fluorescence) in real-time (counter-staining). Furthermore, if emission wavelengths of two fluorescence spectra can be distinguished, it may be possible to perform co-staining or counter-staining with conventional fluorescence imagers which can separate two fluorescent signals only by filter settings. However, it may require higher emission light yields and lower autofluorescence than are currently attainable to assess each tumor nodule accurately.

4. Notes

1. Selection of fluorescent proteins

   In using single-excitation filter settings for co-staining and counter-staining methods, there are several characteristics to consider when choosing a combination of fluorescent proteins and fluorophores. First, both the fluorescent protein and fluorophore must be sufficiently excited by the chosen excitation light. Second, it is ideal to choose a fluorescent protein and fluorophore with emission spectra that are close enough to one another that their spectra will both fall into the predetermined acquisition spectrum. However, it should be possible to differentiate emission signals when both fluorophores have well-characterized emission spectra that are distinct from one another. Spectral fluorescence imaging is also effective at resolving two fluorophores with very similar peak emission spectra that have distinct spectral widths (9). When combining a fluorophore with a narrow spectral width with a fluorophore with a wide spectral width, even if both have the same peak emission wavelength, spectral imaging can be used to effectively differentiate between the two fluorophores.

   This differs from real-time fluorescence imaging where the ideal fluorophores have distinct emission spectra (i.e., colors) with as little overlap as possible, but are still simultaneously excited by a single excitation filter. As fluorophores commonly exhibit considerable spectral overlap, it becomes more difficult to optically differentiate between a fluorescent protein and a fluorophore.

   When using multiple excitation methods, the limitations in fluorophore selection can be reduced (8). However, one must still be careful that unmixed signals do not overlap one another, since this could yield confounding data due to cross-talk between fluorophores.

2. Unmixing procedure

   In the unmixing process, it is necessary to compare an acquired signal relative to a reference spectral library that represents the "pure" fluorophore of interest. However, difficulty in unmixing two target fluorescence spectra correctly is occasionally encountered with spectral libraries, as described above. Fluorophores may undergo processing if taken into cells after binding their target. This may expose the fluorophores to different pH levels or enzymatic changes that may affect the fluorophore's spectral map. A fluorophore may be exposed to different levels of tissue-dependent absorption or scatter that may result in a change in spectral map (10). Thus, it may sometimes be necessary to create a spectral library for a given fluorophore after it is already in the tissue of interest. In that case, it is acceptable to obtain a reference spectral library from an excised tumor nodule labeled with the fluorophore of interest.

After determining a spectral library, it is also important to confirm that there is no crosstalk between endogenous fluorescent proteins and exogenous fluorophores. In the above-described SHIN3-RFP and GmSA-RhodG experiment, mice bearing SHIN3-RFP tumors need to be sacrificed and imaged without addition of the exogenous fluorophore conjugated probe, and confirmation that a lack of fluorescence signal on RhodG spectral unmixed images must be made. Also, mice bearing SHIN3 (no RFP) tumors also need to be sacrificed and imaged after GmSA-RhodG administration, in order to confirm lack of fluorescence signal in RFP spectral unmixed images.

3. Weak fluorescence from fluorescent protein-labeled tumor nodules

   Another possibility that may lead to a "false positive" is the unstable expression of fluorescent proteins in a transfected cell population. This causes weak fluorescence from tumor nodules, especially small tumor nodules that cannot be detected by spectral-fluorescence imaging. This may be an unavoidable limitation when using an endogenously-expressed fluorescent protein. However, using brighter fluorescent proteins and/or obtaining a new clone expressing a fluorescent protein more stably can minimize this limitation. Furthermore, in some circumstance, it is also acceptable to disregard very tiny tumor nodules from image analysis, although it depends on the purpose of the experiment. Even though cancer cells may express fluorescent proteins heterogeneously, larger tumor nodules containing exponentially more cancer cells, will naturally have enough fluorescent protein expression to be detected by spectral-fluorescence imaging.

References

1. Hama, Y., Urano, Y., Koyama, Y., Kamiya, M., Bernardo, M., Paik, R.S., et al. (2006) In vivo spectral fluorescence imaging of submillimeter peritoneal cancer implants using a lectin-targeted optical agent. *Neoplasia* **8(7)**, 607–612.
2. Kobayashi, K., Hama, Y., Koyama, Y., Barrett, T., Urano, Y., Choyke, P. (2007) Whole-body multicolor spectrally resolved fluorescence imaging for development of target-specific optical contrast agents using genetically engineered probes. *Proc SPIE* **6449**, 644914.

3. Hama, Y., Urano, Y., Koyama, Y., Choyke, P.L., Kobayashi, H. (2007) d-galactose receptor-targeted in vivo spectral fluorescence imaging of peritoneal metastasis using galactosamine-conjugated serum albumin-rhodamine green. *J Biomed Opt* **12(5),** 051501.
4. Longmire, M., Kosaka, N., Ogawa, M., Choyke, P.L., Kobayashi, H. (2009) Multicolor in vivo targeted imaging to guide real-time surgery of HER2-positive micrometastases in a two-tumor coincident model of ovarian cancer. *Cancer Sci* **100(6),** 1099–1104.
5. Ogawa, M., Kosaka, N., Choyke, P.L., Kobayashi, H. (2009) In vivo molecular imaging of cancer with a quenching near-infrared fluorescent probe using conjugates of monoclonal antibodies and indocyanine green. *Cancer Res* **69(4)**, 1268–1272.
6. Kosaka, N., Ogawa, M., Longmire, M.R., Choyke, P.L., Kobayashi, H. (2009) Multi-targeted multi-color in vivo optical imaging in a model of disseminated peritoneal ovarian cancer. *J Biomed Opt* **14(1),** 014023.
7. Koyama, Y., Hama, Y., Urano, Y., Nguyen, D.M., Choyke, P.L., Kobayashi, H. (2007) Spectral fluorescence molecular imaging of lung metastases targeting HER2/neu. Clin *Cancer Res* **13(10),** 2936–2945.
8. Koyama, Y., Barrett, T., Hama, Y., Ravizzini, G., Choyke, P.L., Kobayashi, H. (2007) In vivo molecular imaging to diagnose and subtype tumors through receptor-targeted optically labeled monoclonal antibodies. *Neoplasia* **9(12),** 1021–1029.
9. Neher, R., Neher, E. (2004) Optimizing imaging parameters for the separation of multiple labels in a fluorescence image. *J Microsc* **213(Pt 1)**, 46–62.
10. Mansfield, J.R., Gossage, K.W., Hoyt, C.C., Levenson, R.M. (2005) Autofluorescence removal, multiplexing, and automated analysis methods for in-vivo fluorescence imaging. *J Biomed Opt* **10(4)**, 41207.

Chapter 14

In Vivo Imaging of the Developing Mouse Embryonic Vasculature

Irina V. Larina and Mary E. Dickinson

Abstract

Live confocal microscopy of vital fluorescent markers, expressed in mouse embryonic tissues, is a powerful and exciting method to study mammalian embryonic development. This chapter discusses imaging approaches to visualize and characterize dynamic changes of the yolk-sac vasculature and blood flow in mouse embryos. We describe static embryo-culture protocols, which allow maintaining early mouse embryos on the imaging stage for over 24 h. We also describe vital fluorescent-reporter lineages, which can be used to image the developing vasculature and characterize hemodynamics by tracking individual blood cells. Imaging approaches described in this chapter can be used to analyze cardiovascular defects in mutant animals and can provide insights into understanding how genetic signaling pathways and physiological inputs regulate development.

Key words: Live imaging, Fluorescent proteins, Mouse embryo culture, Cell tracking, Mammalian, confocal microscopy

1. Introduction

Cardiovascular researchers have long been interested in understanding how embryonic circulation is initiated and the role it plays in regulating different aspects of cardiovascular development. In experiments performed in avian embryos, Thoma observed that blood vessels can adapt to changes in blood flow during development (1). We now know that the physical forces imparted by blood flow are necessary and sufficient for vessel remodeling (2). However, many open questions remain about the signaling pathways that relate changes in blood flow to changes in cellular morphology. Through the use of advanced imaging methods, we are beginning to gain insights into these challenging problems. The use of tagged

Robert M. Hoffman (ed.), *In Vivo Cellular Imaging Using Fluorescent Proteins: Methods and Protocols*, Methods in Molecular Biology, vol. 872, DOI 10.1007/978-1-61779-797-2_14, © Springer Science+Business Media, LLC 2012

fluorescent proteins under control of specific regulatory elements has significantly increased the ability to visualize and analyze tissues of interest, with subcellular resolution, using confocal and multiphoton microscopy in different animal models (3–5). However, mammalian embryos develop in utero, which limits imaging access to the embryo. To overcome this limitation, live mouse embryo manipulation techniques and static embryo-culture protocols have been optimized to allow imaging of mammalian development (6). These methods have been successfully used in combination with confocal microscopy of transgenic embryos with fluorescent markers (2, 7, 8) as well as optical coherence tomography (OCT) imaging of early embryonic development and cardiodynamics (9, 10).

In this chapter, we described approaches for dynamic in vivo imaging of the mouse embryonic cardiovascular system. Mouse embryo manipulation and live embryo-culture techniques allow one to maintain the embryos on the imaging stage for over 24 h. Confocal microscopy, combined with high-contrast fluorescent protein reporter lines, enables the visualization of the forming and remodeling embryonic vasculature with subcellular resolution and detailed hemodynamic analysis.

2. Materials

2.1. Embryo-Dissection Medium

1. DMEM/F12 (cat. #11330-032, Invitrogen, Grand Island, NY).
2. Fetal bovine serum (FBS) (Invitrogen, Grand Island, NY), stored in aliquots of 5 mL at –30°C.
3. 100× penicillin/streptomycin (Invitrogen, Grand Island, NY), stored in aliquots of 0.5 mL at –30°C.

2.2. Rat-Serum Extraction

1. Adult male Sprague Dawley rats (Charles River Laboratories, Wilmington, MA).
2. Ether (J.T. Baker, Phillipsburg, NJ).
3. Vacutainer blood collection tubes (BD Biosciences, Franklin Lakes, NJ).
4. Vacutainer blood collection sets (REF 367283, BD Biosciences, Franklin Lakes, NJ).
5. 0.45-μm filter (Nalgene).

2.3. Embryo-Culture Medium

1. DMEM/F12 (cat. #11330-032, Invitrogen, Grand Island, NY).
2. Rat serum (the preparation procedure is described below), stored in aliquots of 1 mL at –80°C.
3. 100× penicillin/streptomycin (Invitrogen, Grand Island, NY), stored in aliquots of 0.5 mL at –30°C.

2.4. Live Static Embryo-Culture

1. Dissection/culture medium (the preparation procedure is described above).
2. Glass-bottom culture dishes (MatTek, Ashland, MA) or Lab-tek 2-well chamber slides (Nunc, Rochester, NY).

3. Methods

3.1. Rat-Serum Extraction

Usually 30 rats provide about 50–100 mL of serum (see Notes 1, 2).

1. The rat is anesthetized with ether.
2. Blood is collected from the rat through the dorsal aorta, exposed by abdominal incision, into a Vacutainer blood collection tube using a Vacutainer blood collection set.
3. After the extraction, the tube is placed on ice for up to 40 min during blood extraction from other rats.
4. The tubes with blood are centrifuged at 1,300 × g for 20 min, and the supernatant (blood serum) is collected by pooling. If the supernatant looks pink in some tubes, it is because of the lysis of the red blood cells. We recommend discarding those tubes.
5. The serum is centrifuged again at 1,300 × g for 10 min to eliminate the remaining blood cells. The supernatant is collected by pooling.
6. The serum should be heat-inactivated at 56°C for 30 min, with the lid unscrewed, to allow ether evaporation.
7. For further ether evaporation, the tubes are left overnight at 4°C with lids unscrewed.
8. The serum is filtered using a 0.45-μm filter.
9. The serum can be stored in aliquots of 1 mL at −80°C for up to 1 year.

3.2. Live Static Embryo-Culture

1. For timed pregnancies, mating pairs are set overnight and checked for the presence of a vaginal plug every morning. The presence of the plug is taken as 0.5 dpc.
2. The dissection medium should be freshly prepared (approximately 50 mL per pregnant female) with 89% DMEM/F-12, 10% FBS, and 1% 100× penicillin/streptomycin.
3. The dissection medium should be pre-heated to 37°C.
4. The dissection station should also be preheated and maintained at 37°C using custom-made heater or a conventional heater, and a temperature controller.
5. At the desired stage, the female is sacrificed and the embryos are dissected out of the uterus, with the yolk-sac intact, under a dissection microscope in dissection medium.

6. Freshly-dissected embryos are transferred to the 37°C 5% CO_2 incubator for recovery for at least 30 min.
7. If embryo-culture experiments are limited to a few hours (e.g., for live hemodynamic or cardiodynamic analysis), embryos can be imaged in the DMEM/F-12 medium prepared for embryo dissection as described above. However, for longer imaging sessions (such as visualization of vascular-plexus formation and remodeling and endothelial cell tracking), the culture medium should be supplemented with rat serum: two parts DMEM/F12, one part rat serum and 1% 100× penicillin/streptomycin.
8. For microscopic imaging, embryos are cultured in glass-bottom culture dishes (MatTek) which have a glass cover slip attached to the microwell at the bottom of the dish. Alternatively, Lab-tek 2-well chambers, with glass cover-slip bottoms (Nunc), can be used. The glass-bottom culture dishes use about 3 mL of medium, while the Lab-tek chambers use about 2 mL. The glass cover slip in the bottom of the chambers allows for imaging with inverted microscope (see Notes 3, 4).

3.3. Imaging Developing Embryonic Vasculature

Embryonic vasculature can be visualized using two transgenic markers, Flk1-H2B::EYFP which expresses a nuclear yellow fluorescent protein (YFP) in the endothelial cells (11) and Flk1-myr::mCherry, which expresses a membrane-tethered mCherry fluorescent protein in the embryonic endothelium and endocardium (8). Intercrossing these markers provides a powerful combination, since the EYFP marker allows for the analysis of cell division and the tracking of each endothelial cell, whereas the membrane-targeted mCherry marker reveals cell boundaries and highlights vessel organization. This combination of markers reveals the distribution of individual cells within each vessel segment.

The fluorescence of mCherry and EYFP is first detectable at 7.5 dpc (also known as the early headfold stage) in the blood islands of the yolk-sac. At this stage, the fluorescence is relatively dim and is only detectable in homozygous embryos. At 8.0 dpc, the markers label the vascular plexus as it develops in the embryonic yolk-sac. By 8.5 dpc the vascular plexus is brightly outlined by mCherry with clearly distinguishable YFP-labeled nuclei (see Fig. 1a). Both markers remain in the endothelial cells as the vascular plexus of the embryonic yolk-sac undergoes remodeling into a more mature circulatory system consisting of larger vessels which branch into progressively smaller vessels. Figure 1b shows the remodeling yolk-sac vasculature, at embryonic day 9.5, labeled by the mCherry and the EYFP. Membrane boundaries between endothelial cells can be detected in some cells at 63× magnification (see Fig. 1c). In the embryo proper, the markers are co-expressed from 8.5 dpc until late gestation throughout the vasculature of different embryonic tissues, such as the trunk, skin, brain, and eye. Detailed subcellular

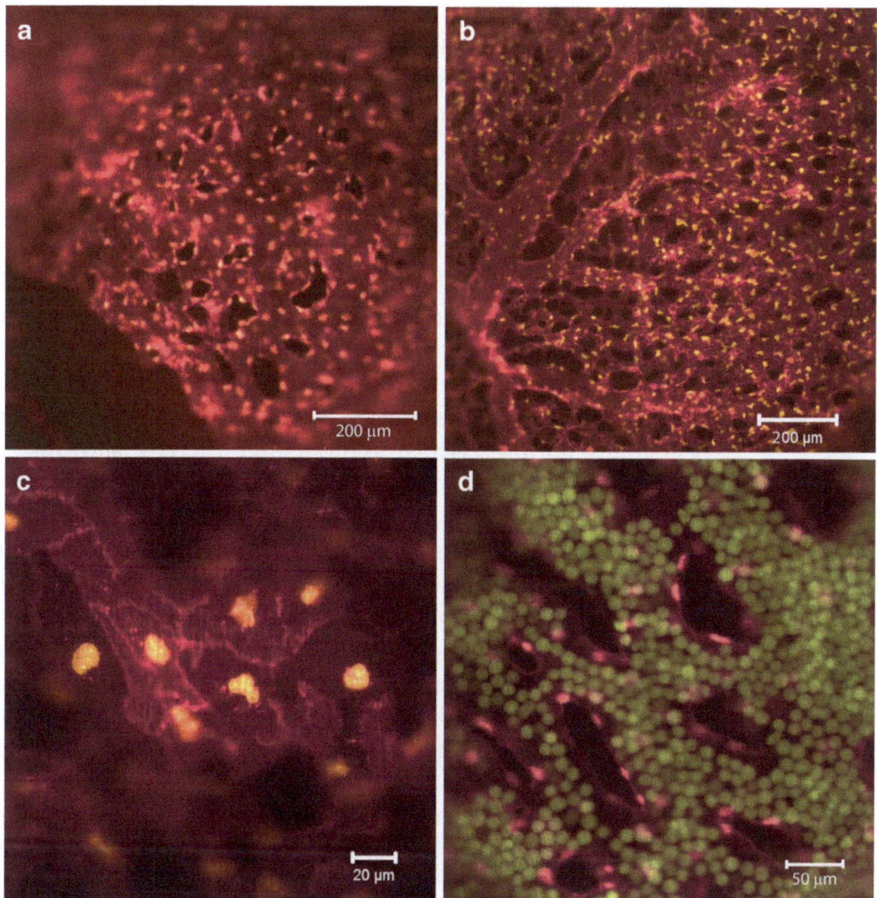

Fig. 1. In vivo imaging of embryonic vasculature in Tg(Flk1-myr::mCherry) × Tg(Flk1-H2B::EYFP) mice. (**a**) Vascular plexus of the embryonic yolk-sac labeled by mCherry (membrane) and EYFP (nuclei) at 8.5 dpc. (**b**) Remodeling yolk-sac vasculature at 9.5 dpc. (**c**) High magnification imaging (63×) of the yolk-sac vasculature at 9.5 dpc. The boundaries between endothelial cells can be distinguished. (**d**) Vascular plexus of Tg(Flk1-myr::mCherry) × Tg(Flk1-H2B::EYFP) × Tg(ε-globin-GFP) mice at 8.5 dpc. Endothelial cells are labeled by the membrane-tethered mCherry and nuclear EYFP. The primitive erythrocytes are marked by GFP for blood flow imaging.

structures can be resolved from high-magnification views of embryonic vessels with outlined membrane boundaries and brightly labeled nuclei. During late gestation, expression of myr::mCherry and H2B::EYFP decreases in major blood vessels, but remains in some smaller vessels throughout postnatal life.

Fluorescent proteins mCherry and EYFP can be combined with other fluorescent labels, for example, green fluorescent protein (GFP). Figure 1d shows an image, acquired with fast scanning confocal microscopy, from an embryonic yolk-sac at 8.5 dpc where the Flk1-driven myr::mCherry and the H2B::EYFP are combined with the GFP expressed in the embryonic blood cells. Even though the excitation and emission spectra of GFP and EYFP overlap significantly, they can be distinguished. In the example shown, the mCherry was excited at 532 nm, while the EYFP and the EGFP were excited at 488 nm. The emission was detected using only two channels: BP 500-525 and BP 560-675. The EGFP fluorescence is detected entirely by the first channel and the EYFP fluorescence is mostly detected by the first channel. However, since its emission spectrum is shifted towards longer wavelengths relative to EGFP, there is some signal detected in the second channel as well. mCherry is only detected by the second channel. This difference can be used to distinguish between the fluorophores. Because the ε-globin-EGFP transgene allows one to track individual blood cells and perform detailed hemodynamic analysis (as discussed later in this chapter), this combination of labels provides an opportunity to study the correlation between hemodynamic and structural changes during vascular remodeling, thereby allowing a better understanding of the role which blood flow and shear stress play in vascular formation.

To visualize dynamic changes during the formation of the vasculature in the embryonic yolk-sac, time-lapse confocal microscopy can be performed on live embryos grown in culture on the microscopic stage. The expression of the above-mentioned fluorescent markers, starting at early stages when the vascular-plexus is forming and continuing through the remodeling process, provides useful insight into the dynamics of vessel formation and maturation. At early stages, confocal time-lapse imaging shows many small sprouts forming, which in turn increases the number of vessel branches. In contrast, at later stages after blood flow begins, many vessel branches regress. By starting the culture at 7.5 dpc, when mCherry is detected only in the region of blood islands, and following the continuous increase in the level of expression throughout the yolk-sac for 24 h, the formation of the entire primitive vascular-plexus in the yolk-sac can be visualized (8).

Analysis of live dynamic changes of endothelial cells during vascular remodeling allows for a better understanding of the process of refinement of the vascular structure in which some vessels regress while those along the major flow trajectories remain. As an example, Fig. 2 shows time-lapse imaging and endothelial cell tracking in Tg(Flk1-myr::mCherry) × Tg(Flk1-H2B::EYFP) embryos. The time-lapse was started at approximately 8.5 dpc; a time when blood flow is establishing and the fluorescent markers are labeling the remodeling vascular-plexus. The images were taken every 5 min. Image processing software Imaris (Imaris 5.0.3, Bitplane) was used for analysis. Figure 2a shows a representative image acquired during the time-lapse. To visualize and analyze the behavior of endothelial cells comprising the vascular-plexus, each detectable EYFP-labeled nucleus in the focal plane was automatically associated with a spot, as shown in Fig. 2b. This procedure was performed in Surpass mode of the Imaris software using "Add new spots" function in the Objects window (we usually set a minimal

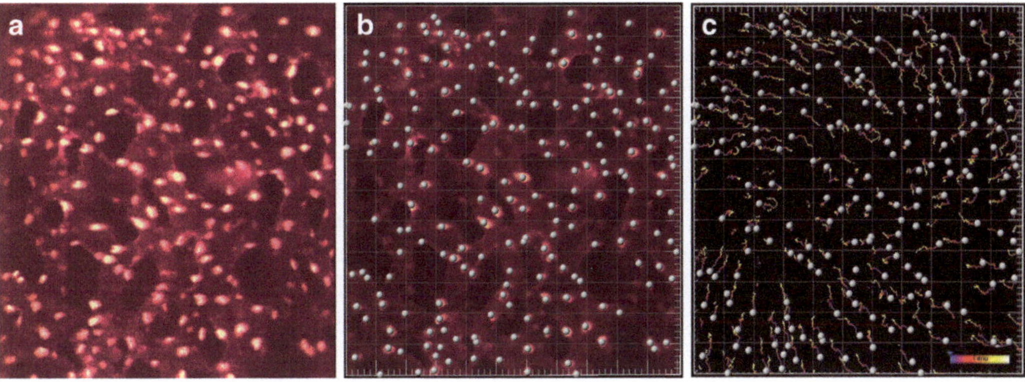

Fig. 2. Imaging vascular remodeling in the Tg(Flk1-myr::mCherry) × Tg(Flk1-H2B::EYFP) embryonic yolk sac at 8.5 dpc. (**a**) A frame from a time lapse acquired from a vascular plexus over a 6-h period. Images are taken every 5 min. (**b**) Spots corresponding to the individual nuclei. (**c**) Cell tracks generated by the analysis software representing movements of individual cells during the time lapse. Different colors correspond to different time points.

diameter of 4 μm for the detected spots). The number of nuclei-associated spots can be adjusted by changing the threshold for spot identification to include or exclude nuclei which are partially out of focus. Once the spots are generated, they can be automatically tracked via the "Tracking" function in the Spots menu (we usually use the Autoregressive Motion algorithm and set the Maximum Distance traveled between the neighboring images as 5–10 μm). Figure 2c shows spots and tracks representing dynamics of the EYFP-labeled nuclei of the endothelial cells over the course of 6 h. The color of the tracks changes from blue to white according to the time scale as shown in the bar. Visual analysis of the generated tracks allows one to appreciate the overall cellular behavior, including divisions (splitting tracks) and apoptosis (disappearing tracks, which can be confirmed by analysis of the original time-lapse) of endothelial cells. Statistical analysis of cellular behavior is performed using the "Statistics" function on the Spot Track group and can include (*x*, *y*) coordinates of each spot at each time frame, track speed, track distance, track displacement, and other parameters.

Because endothelial cell tracking time-lapse images are usually acquired over the course of several hours, one of the major limitations for successful experiments is a sample drift over time caused by embryo growth as well as possible floating and rotation. To prevent excessive movement and floating of the embryo, a part of decidua can be left attached to the embryo during dissection or, alternatively, a strand of a hair can be placed over the embryo to keep it in the same position. However, some drift is unavoidable. If bulk movement of the endothelial cells is observed, but the majority of the cells remain in the imaging plane, the drift can be corrected through using Imaris software. For that purpose, the Drift Correction function should be used under the Spot Track group menu. After the drift is corrected, the spot detection and tracking steps should be repeated as described above.

3.4. Live-Embryonic Hemodynamic Analysis

It has been shown in multiple studies that blood flow plays an important role in the formation and maintenance of the vasculature. In the embryonic yolk sac, shear-stress generated by the circulating blood cells is essential for vascular remodeling (2). Being able to acquire live hemodynamic measurements in wild-type as well as genetically-engineered animals with cardiovascular abnormalities provides a valuable tool to understand molecular mechanisms regulating this relationship. Additionally, blood flow patterns in the embryonic yolk sac are an indication of proper heart function. Therefore, hemodynamic analysis can be used to characterize cardiac defects in mutants and pharmacologically-altered embryos.

A transgenic fluorescent reporter mouse model, useful for hemodynamic analysis in early embryos, was generated by Dyer et al. (12). These animals carry GFP under the control of a ε-globin promoter which drives GFP expression in primitive erythroblasts. Blood cells, brightly labeled with the GFP, are first detected in the blood islands of the embryonic yolk sac prior to the beginning of the heart beat. A few hours after the beginning of the heart beat (about 8.5 dpc), when the plasma flow is stronger, the blood cells depart the blood islands and join the circulation. By using this reporter in live embryo culture, these events can be directly studied.

Classical point-scanning confocal microscopy is too slow to record the movement of the rapidly-moving blood cells. To overcome this limitation, several groups have used a line-scanning approach where, instead of full frame imaging, scanning is performed along a designated line, either along or perpendicular to vessels (7, 13, 14). If the scanning line is set perpendicular to the vessel, the circulating blood cells cross the line, and the number of scans and time it takes for each cell to go across the line can be calculated to determine the velocity. If the scanning line is set parallel to the flow, the blood cells move along the line, and their velocity can also be measured. These techniques provide valuable hemodynamic measurements in live vessels and are still widely used, though they are time-consuming since only a single line recording can be performed at a time.

The recent development of fast scanning, confocal microscopes provides the opportunity for full-frame real-time imaging of rapid dynamic events. If the brightness of the fluorescent marker permits, the Zeiss LSM 5 LIVE confocal microscope allows one to acquire 512×512 pixel images up to 120 frames per second (fps). This is about 100 times faster than using a standard confocal microscope. Since the blood flow in early embryos is in the range of millimeters per second, the dynamics of all blood cells in the

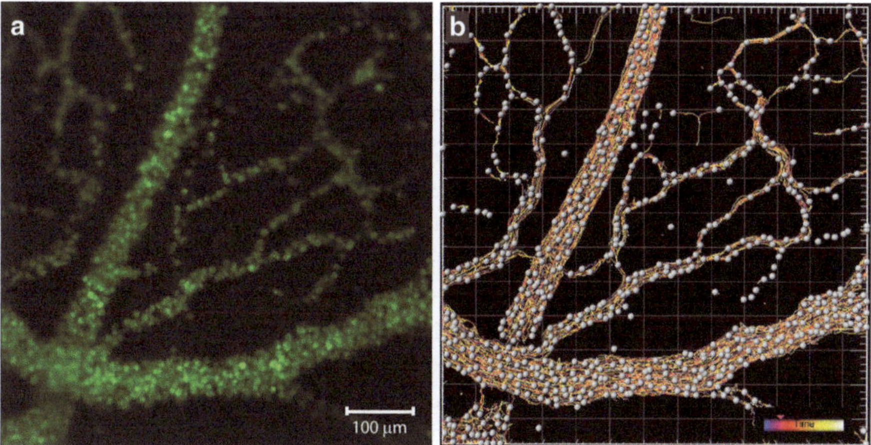

Fig. 3. Blood flow analysis in the embryonic yolk sac of the Tg(ε-globin-GFP) mice at 10.5 dpc using fast scanning confocal microscopy. (**a**) A representative frame from a time series acquired at 30 fps. *Green spots* represent individual circulation blood cells labeled by the GFP. (**b**) Tracks of all detected blood cells during the acquisition time 100 frames. Different colors representing different time points.

field of view can be followed with superior temporal and spatial resolution for detailed hemodynamic analysis.

Figure 3 shows an example of blood flow imaging in cultured ε-globin-GFP embryos using a Zeiss LSM 5 LIVE fast scanning confocal microscope. The imaging is performed at 10.5 dpc, when the yolk-sac vasculature is remodeled into a network of large vessels with high flow and smaller branches with lower flow. An image sequence of 100 frames (512×512 pixels) was acquired at 30 fps using 10× magnification. Figure 3a shows a representative frame from the time-lapse images showing individual blood cells (green dots) circulating through the yolk-sac vasculature. To visualize and analyze the dynamics of blood flow, cell tracking can be performed automatically in using Imaris (Imaris 5.0.3, Bitplane) software via Spot Detection and Spot Tracking functions as discussed above. Figure 3b shows automatically-detected spots for the image in Fig. 3a and all individual cell tracks generated for the time lapse. The color of the tracks changes according to the time scale. Quantitating spot movement can be used for heart rate and blood flow determination in a variety of vessels at different phases of the cardiac cycle.

4. Notes

1. Even though rat serum is commercially-available through different sources, we prefer extracting the rat serum in the lab. Using the commercially-available serum resulted in very inconsistent

results probably caused by traces of anesthetics used during blood extraction as well as variations in serum extraction and handling. On the other hand, home-made rat serum consistently allowed us to maintain embryos in static culture for over 24 h.

2. It is critical to use ether for rat anesthesia because it evaporates easily from the serum, while other anesthetic agents remain in the serum and can affect embryonic growth and survival.
3. It is important to maintain the temperature of the microscope stage at 37°C throughout the duration of the imaging session. This can be achieved with a custom-built incubation system fitted around the microscope. However, we have found that the embryo culture technique is more robust if a commercial incubation system is used (such as from Carl Zeiss, Inc.) since it provides precise control of the temperature and CO_2 and is equipped with a humidifier.
4. As always with time lapse imaging of live embryos, it is important to minimize the exposure of the specimens to the laser light, as repetitive imaging with high intensity lasers can be harmful to embryo growth. According to our observations, time lapse sequences can be acquired for at least 24 h if the images are taken every 5–10 min with minimal laser power. However, due to varying expression of reporter lineages and variations from microscope to microscope, the tolerance of embryos in each individual experiment will need to be determined by comparison to cultured embryos that are not imaged.

Acknowledgments

The study is supported by the National Institutes of Health (R01HL095586) and the American Heart Association (10SDG3830006).

References

1. Thoma, R. (1893) Untersuchungen über die Histogenese und Histomechanik des Gefässsystems. Stuttgart: Ferdinand Enke.
2. Lucitti, J.L., Jones, E.A.V., Huang, C., Chen, J., Fraser, S.E., Dickinson, M.E. (2007) Vascular remodeling of the mouse yolk sac requires hemodynamic force. *Development* **134**, 3317–3326.
3. Hadjantonakis, A.K., Dickinson, M.E., Fraser, S.E., Papaioannou, V.E. (2003) Technicolour transgenics: imaging tools for functional genomics in the mouse. *Nat Rev Genet* **4**, 613–625.
4. Lichtman, J.W., Fraser, S.E. (2001) The neuronal naturalist: watching neurons in their native habitat. *Nat Neurosci* **4** **Suppl**, 1215–1220.
5. Kulesa, P.M. (2004) Developmental imaging: Insights into the avian embryo. *Birth Defects Res C Embryo Today* **72**, 260–266.
6. Jones, E.A.V., Crotty, D., Kulesa, P.M., et al. (2002) Dynamic in vivo imaging of postimplantation mammalian embryos using whole embryo culture. *Genesis* **34**, 228–235.
7. Jones, E.A.V., Baron, M.H., Fraser, S.E., Dickinson, M.E. (2004) Measuring hemodynamic

changes during mammalian development. *Am J Physiol Heart Circ Physiol* **287,** H1561–1569.

8. Larina, I., Shen, W., Kelly, O., Hadjantonakis, A., Baron, M., Dickinson, M. (2009) A membrane associated mCherry fluorescent reporter line for studying vascular remodeling and cardiac function during murine embryonic development. *Anatomical Record* **292**, 333–341.
9. Larina, I.V., Sudheendran, N., Ghosn, M., et al. (2008) Live imaging of blood flow in mammalian embryos using Doppler swept source optical coherence tomography. *Journal of Biomedical Optics* **13,** 0605061–3.
10. Larina, I.V., Ivers, S., Syed, S., Dickinson, M.E., Larin, K.V. (2009) Hemodynamic measurements from individual blood cells in early mammalian embryos with Doppler swept source OCT. *Optics Letters* **34,** 986–988.
11. Fraser, S.T., Hadjantonakis, A-K., Sahr, K.E., et al. (2005) Using a histone yellow fluorescent protein fusion for tagging and tracking endothelial cells in ES cells and mice. *Genesis* **42,** 162–171.
12. Dyer, M.A., Farrington, S.M., Mohn, D., Munday, J.R., Baron, M.H. (2001) Indian hedgehog activates hematopoiesis and vasculogenesis and can respecify prospective neurectodermal cell fate in the mouse embryo. *Development* **128,** 1717–1730.
13. Dirnagl, U., Villringer, A., Einhaupl, K.M. (1992) In vivo Confocal Scanning Laser Microscopy Of The Cerebral Microcirculation. *J Microsc-Oxf* **165,** 147–157.
14. Kleinfeld, D., Mitra, P.P., Helmchen, F., Denk, W. (1998) Fluctuations and stimulus-induced changes in blood flow observed in individual capillaries in layers 2 through 4 of rat neocortex. *Proc Natl Acad Sci USA* **95,** 15741–15746.

Chapter 15

Screening Reef Corals for Novel GFP-Type Fluorescent Proteins by Confocal Imaging

Anya Salih

Abstract

The discovery of multicolored fluorescent proteins (FPs), in reef corals, that are close relatives of the green fluorescent protein (GFP) has led to what is now viewed as the second GFP revolution. Numerous GFP-type proteins, termed "reef FPs," have been cloned from reef organisms and many possess new colors, novel molecular characteristics, protein chemistry and many display unusual photophysical properties. Although some FPs have certain disadvantageous properties, such as the tendency to oligomerize or have slow maturation rates, reef FPs have been developed into versatile probes for cell biology and imaging applications. Screening of natural sources for novel GFP-type proteins continues to be valuable due to the need to expand the range of spectral colors, brightness, monomeric or dimeric states, faster maturation states, and photoactivity. Confocal imaging, coupled with microspectral detection, provides a rapid technique for in vivo characterization of FPs with desirable spectral and photoactive properties.

Key words: Green fluorescent protein, Coral, Reef GFP, Confocal imaging, Micro-spectra, Photoactivation, Photoconversion, Photoswitching, Marine organisms, Fluorescence

1. Introduction

The first green fluorescent proteins (GFPs) were cloned from hydrozoans—the jelly-fish *Aequorea victoria* and the sea pansy *Renilla reniformis* (1, 2). In the last decade, a variety of GFP-type protein homologs have been discovered in hard reef corals belonging to the order Scleractinia and in other anthozoans lacking a hard skeleton (soft corals, sea anemones, zoanthids, corallimorphs), often referred to as "reef FPs" (3–11). They were also discovered in other marine organisms, such as crustaceans and even a basic chordate animal, amphioxus (12, 13). Thus, marine organisms have become a new source of novel and diverse GFP-type proteins.

Robert M. Hoffman (ed.), *In Vivo Cellular Imaging Using Fluorescent Proteins: Methods and Protocols*, Methods in Molecular Biology, vol. 872, DOI 10.1007/978-1-61779-797-2_15, © Springer Science+Business Media, LLC 2012

Reef FPs now provide a multitude of genetically-expressible proteins for use in fluorescence imaging applications. Their colors are significantly more diverse than those of wtGFP, with excitation wavelengths extending from the violet to the orange (400–590 nm) and emission maxima covering almost the full rainbow color palette, from blue to red (440–660 nm). They frequently have high quantum yields, are typically extremely stable, and resistant to photobleaching (13–15). While highly advantageous for conventional imaging methods, reef-FPs' resistance to photobleaching limited their applications in dynamic studies using fluorescence recovery after photobleaching (FRAP) techniques, since the light levels needed to bring about bleaching were phototoxic to cells. The recent discovery of the photoactive fluorescent proteins (PAFPs) in anthozoans, which respond to irradiation, by altering their optical properties, now provides an improvement to FRAP-based techniques (8, 15–17). Their fluorescent state can be precisely controlled (dim/bright; or converted from one color to another) by irradiation. PAFPs enable direct color labeling and selective monitoring of labeled proteins, organelles and cells. PAFP-based imaging is less phototoxic to cells, does not require continuous imaging, and is a much more versatile method for studying dynamic processes in living cells. Another novel application of photoactivatable reef FPs is the recent development of super-resolution imaging by photoactivated localization microscopy (PALM) and related methods. Irradiation of PAFPs is used to generate images that precisely localize single fluorophores to within a few tens of nanometers (18).

GFP-type proteins contribute a surprisingly high fraction to the overall soluble protein content of anthozoan tissues, ranging from 4.5% to 14.7% (Salih unpublished data; 10). In the majority of cases, several FPs co-occur within the same anthozoan organism whose visual color patterns are determined by the more highly-expressed proteins (7, 9–11, 19). These more abundant FPs frequently mask the rare types when tissues are screened for GFP-type proteins using conventional fluorescence spectroscopic methods. The most widely used screening methods includes the use of molecular techniques. For example, four spectrally-distinct GFP-like genes were discovered in the coral *Montastraea cavernosa* in one study using RNA extraction, cDNA synthesis, and cloning of FPs (7). Another study, using similar methods, identified more than ten in the same coral (9). Molecular screening methods are time consuming and require the use of expensive reagents. Moreover, they do not always identify all the GFP-type proteins that may be present and those that occur in lower concentrations or those that have low homology to the previously cloned types may be missed.

Confocal microscopy in vivo imaging of FPs in tissues of marine organisms, coupled with micro-spectral characterization described here, provides a convenient extension beyond spectroscopic and molecular methodologies in the search for novel GFP-type proteins (20–22). Spectral characteristics, photobleaching stability, and photoactive properties (color conversion or photoswitching upon irradiation) can be rapidly identified. Since the confocal technique focuses light from the sample on a very small pinhole, all information not coming from the focal plane is eliminated. In this way, emissions from the less common FPs within the selected focal plane can be visualized without being swamped by the more abundant FPs above and below the focal plane as is the case when using the spectrophotometric techniques described in Subheading 3.4. Confocal micro-spectroscopy can thus provide information on proteins which otherwise may be undetected by conventional spectroscopic methods. Owing to major demands for new fluorescent labels for bioimaging applications, we can expect the arsenal of GFP-type fluorescent probes to continue to be increased with novel types sourced from marine organisms.

2. Materials

2.1. Coral Sample Collection, Dissection, Storage, and RNA Preservation

1. Powerful underwater flashlight with yellow barrier and blue exciter filters (NightSea LLC).
2. 10% magnesium chloride in filtered seawater solution.
3. Sharp forceps.
4. RNA*later*® (Ambion Applied Biosystems, Austin, USA).

2.2. Confocal Imaging and Micro-spectral Detection of FPs

1. Microscope coverslips of appropriate thickness (# 1.5; 170-μm thick).
2. Live-cell chamber slides (e.g., Lab-Tek, Nunc or μ-Slides, Ibidi).
3. Press-to-seal CoverWell perfusion chambers (Grace Bio-Labs, Inc.) which form incubation chambers when pressed to glass coverslips or microscope slides. Used to position dissected coral samples on glass slides for imaging.
4. Mounting medium: Antifade (Molecular Probes, Eugene, OR).
5. Confocal laser scanning microscope system (e.g., Leica TCS SP2 or SP5 AOBS) suited for live-cell microscopy (inverted set-up) equipped with the following: a range of high numerical-aperture (1.2–1.4 NA) oil, glycerol, or water objectives (20×, 40×, 63×, or 100×) and UV (405 nm) and visible light laser lines.

3. Methods

Except for the red chlorophyll fluorescence emanating from the intracellular symbiotic microalgae populating coral tissues, the majority of cyan, green, yellow, orange, and red coral pigments, both the fluorescent and nonfluorescent types (see Fig. 1a–e), are GFP homologs, ranging from ~20% to 95% homology to jelly-fish-derived wtGFP. Initial screening of corals for GFP-type proteins can be done during daylight visual assessment by snorkeling, by SCUBA diving or during night-time surveys using powerful underwater flashlights fitted with a blue light excitation filter.

When searching for GFP-like proteins with specific properties, it is important to consider coral biology (4, 23). GFPs that have stable expression at higher temperatures (>37°C) can be screened from corals growing in shallow reef-flat pools where trapped seawater becomes heated and temperatures at low tide regularly exceed 40°C. Photostable and photoconvertible GFPs can be obtained from corals from such high-stress, shallow habitats, where photoprotective GFPs are most required. Far red-shifted GFPs can

Fig. 1. Coral tissue colors due to GFP-type fluorescent and chromophoric proteins. (**a**) Corals in daylight. (**b**) Fluorescence of the same corals at blue light illumination. (**c**) Acroporiid corals pigmented by GFP-type proteins including the blue CPs. (**d**) CPs expressed at a wound site of *Acropora nobilis*. (**e**) Purple-pink CPs expressed in coral polyps.

be sourced from corals from light-limited depths (30–40 m) or from under deeply shaded reef overhangs, where red morphs are abundant and GFPs channel light into chlorophyll absorption of symbionts in order to amplify photosynthesis (4).

3.1. Coral Collection, Maintenance, Dissection, and Sample Preservation

1. Coral fragments were collected from a number of sites of the Great Barrier Reef, Australia: One Tree Island (23°30′S, 152°06′E), Heron Island (23°26′S, 151°55′E), Ribbon Reef no. 15-072 (15°30′, 145°46′), and from Osprey Reef (13°53.244′, 146°33.435′) in the Coral Sea. Sampling was from depths of 0.1–40 m in shallow water at low tide, by snorkeling and diving with SCUBA gear (see Note 1).
2. Selection of corals is by visual assessment of their coloration in daylight and during night-time screening for fluorescence by illumination with strong blue light from a diving flashlight mounted with a blue filter (NightSea Inc.) to excite FPs (see Note 2).
3. Collected corals are maintained in flowing seawater tanks until analysis or transported to marine aquaria at the university laboratory. Samples can also be frozen in liquid N_2.
4. Prior to dissection of tissues, corals are anesthetized by injection of 10% $MgCl_2$ (~50% of body weight) or by adding $MgCl_2$ to seawater in a container housing the coral for 30 min.
5. Sharp forceps are used to excise soft coral tissue from the calcium-carbonate skeleton.
6. Immerse excised tissues selected for cloning of FPs into RNA*later*® (Albion) solution that stabilizes RNA (see manufacturer's protocol for details). Care should be taken to remove as much of the coral skeleton as possible, since it can interfere with subsequent RNA isolation.
7. Molecular techniques for cloning and characterization of FPs are not detailed here (3, 6–8). Briefly, total RNA is isolated from tissue samples using an RNAqueous kit (Ambion) and amplified cDNA is prepared from it using the SMART protocol (24). The complete cDNA coding sequences for GFP-like proteins can be obtained as described in ref. 25. RNA*later*®-stored samples can be used for microscopic studies of FP cellular localization, since this reagent preserves cell structure and FP fluorescence, although cell swelling or shrinkage does occur, especially after prolonged storage.

3.2. Digital Photography of Coral Fluorescence

Fluorescence of corals in daylight is photographed with any digital camera (e.g., Nikon Coolpix 4500). When photographying underwater, an underwater housing (e.g., Ikelite) for the camere is used (see Figs. 1a, b and 2a, b) (26). A yellow blue light barrier filter (NightSea LLC) is fitted to the dome port of the underwater housing and an underwater strobe (Ikelite DS50) is fitted with a blue

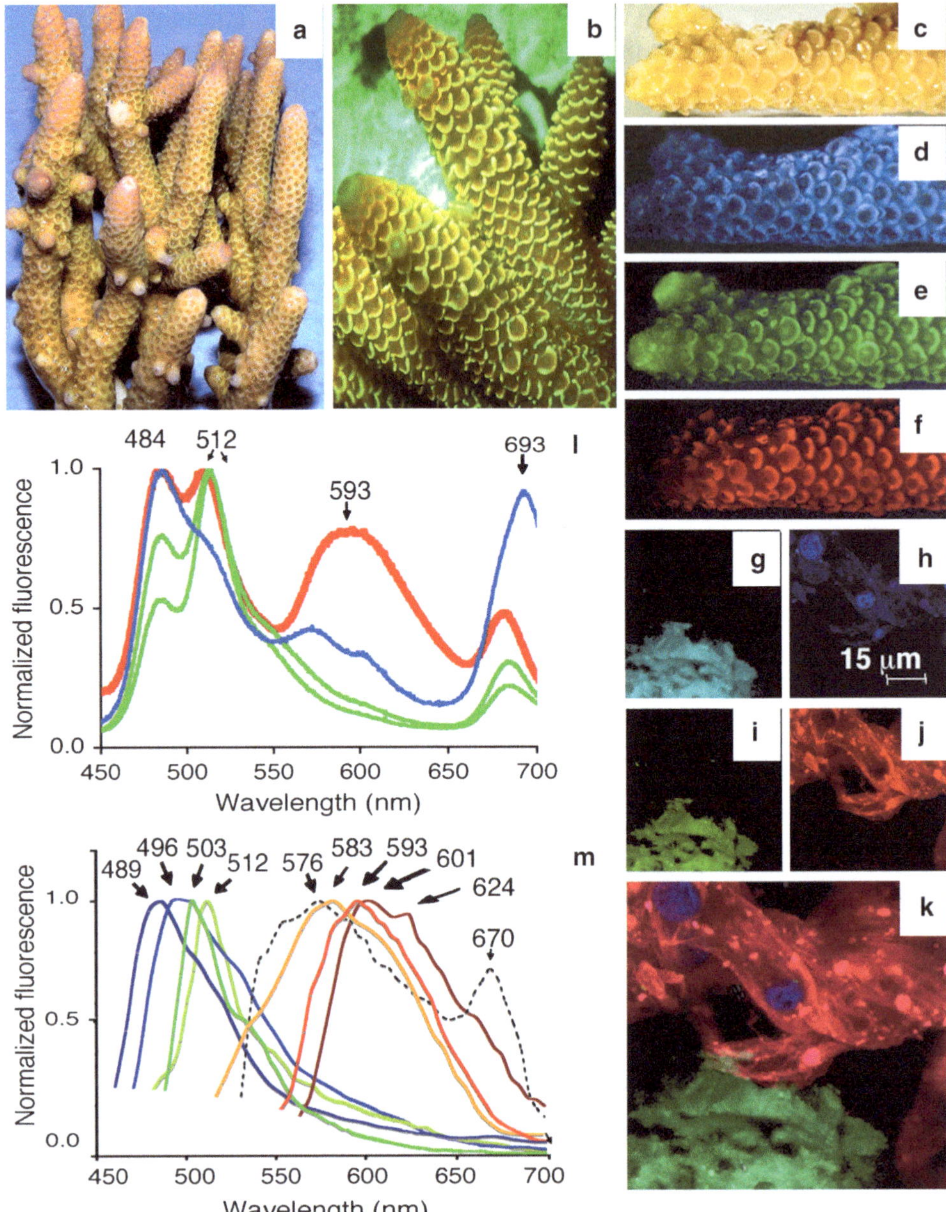

Fig. 2. Diversity of fluorescent proteins in the coral *Acropora millepora*. (**a**) Colony coloration in daylight. (**b**) Fluorescent colors photographed at blue light illumination. Imaging of the sample using a dissection microscope in (**c**) transmitted light, (**d**) by UV 360–390 nm, (**e**) by blue (440–490 nm), and (**f**) by green (530–540 nm) illumination. Confocal imaging of tissue at excitation (**g**) by 405 nm laser line—cyan FPs localized in ectodermal cells, (**h**) by 488 nm laser line—red emissions from symbiotic algae in endodermal cells, (**i**) green FP emissions co-locolized with cyan FPs, (**j**) by 561 nm laser line—red amilRFP emissions in the endodermal cells, and (**k**) super-imposed image from 4 PMTs. (**l**) Fluorescence emission spectra of the above-mentioned cyan, green, and red FPs, as well as symbionts' chlorophyll. (**m**) Diverse fluorescence emission spectra of FPs from the same tissue obtained by micro-spectroscopic confocal imaging. *Dotted line* non-GFP autofluorescence mainly due to photosynthetic pigments.

light exciter filter (NightSea LLC) as the fluorescence excitation source (see Note 3) (26, 27).

3.3. Epifluorescence Microscopic Examination

1. Whole coral samples can be examined for FPs by any dissection epifluorescence microscopes using appropriate fluorescence filter sets. The majority of coral samples are multi-pigmented and tissues show different fluorescence at excitation by UV, blue, or green light (11, 22). The following filters can be used: for cyan FPs—BP 340–380 (UV), 400 LP 425 (e.g., for Dapi, Hoechst, AMCA); for green FPs—BP 450–490 (blue) 510 LP 515 or BP 440–520 (blue) 505 BP 497–557 (e.g., for GFP, Alexa488, Cy2); for red FPs—BP 515–560 (green) 580 LP 590 (e.g., for dsRed, Alexa546, Cy3) (see Fig. 2c–f).
2. Excised tissues, mounted on slides under a coverslip, can be examined using any compound fluorescence microscope. Tissues can also be mounted on Live-cell chamber slides or within press-to-seal CoverWell perfusion chambers or microscope slides. Seawater is used for mounting. If bleaching is observed, an antifade reagent can be added to seawater.

3.4. Spectral Characterization of Fluorescent Proteins

1. The in vivo excitation and emission spectra of FPs of corals immersed in seawater are obtained using a fluorescence spectrophotometer (Varian Cary Eclipse, Varian Inc., Palo Alto, CA, USA). The 1-m optical probe enables investigation of spectral properties without the need for protein extraction and analysis in a cuvette or a microplate. The tip of the probe is placed in direct contact with the coral tissue or separated by a rubber ring mounted on the probe tip to avoid abrasion by the coral surface. The emission spectra is collected from 300 to 800 nm; excitation spectra from 200 to 650 nm. FP emissions of coral tissues can be measured with an Ocean Optics fluorospectrophotometer and fiber-optic probe, which is especially useful for field-work, since it is compact and easy to transport. Similar to the above-mentioned Cary Eclipse, this optical probe can allow fluorescence emission spectral characterization without the need for protein extraction (28).
2. To resolve FP spectral variants that are present at the microcellular level, the Ocean Optics fluorescence spectrometer is fitted to an epifluorescence microscope (e.g., Nikon 50i Epifluorescence microscope). Pigmented tissues are dissected, mounted on glass slides and FP emissions determined by imaging with 20× or 40× objectives with excitation by an Hg arc lamp with UVA filter sets (ex 330–360 nm, FT395, LP420). By limiting the field of view with an iris diaphragm, spot measurements of fluorescence at single-cell resolution can be made and data collected by a PC-type computer and Ocean Optics software. This analysis can

distinguish GFP-like proteins from non-GFP pigments, aid in resolving the number of GFP spectral variants present in a sample, and enable selection of specimens for further analysis. In Fig. 2l, three FPs in cells of the coral *Acropora millepora*—cyan 484 nm, green 512 nm, and red 593 nm, that were identified by this method, correlate to spectral variants that have been subsequently cloned from this species (amilCFP—ex 441 nm/em 489 nm; amilGFP—ex 503 nm/em 512 nm and amil RFP—ex 560 nm/em 593 nm) (11, 22).

3. Absorption characteristics of FPs and CPs are analyzed in a solution made from strongly pigmented tissues dissected from anesthetized or frozen corals in which a single protein dominates. Tissues are homogenized in 50 mM phosphate buffer (pH 7), extracted for 48 h at 4°C, and centrifuged. Protein absorbance spectra at 1 nm resolution are recorded on Cary 1 absorption spectrophotometer (Varian, Germany).

3.5. Characterization of Multicolor FPs by Confocal Imaging

1. Confocal microscopes from Leica (Leica Microsystems, Heidelberg) and other confocal microscopes with spectrophotometric detection, such as the Zeiss Meta, or Olympus FluoView 1000, are suitable and are easily adaptable to the below-mentioned formats.
2. To image emissions from the less common FPs, that can be swamped by the more abundant FPs (as occurs when using spectroscopic techniques in Subheading 3.4), it is important to use the confocal pinhole set to 1 Airy unit.
3. Leica TCS SP2 or Leica TCS SP5 confocal upright or inverted microscopes (Leica Microsystems, Heidelberg, Germany) are used for single- and multi-photon imaging and micro-spectrometry. Leica TCS SP2 has the following laser lines: argon laser with lines 458, 476, 488, 514 nm and HeNe with 543 nm line. Leica TCS SP5 has the following lasers: diode 405 laser; argon laser with lines 458, 476, 488, 496, 514 nm; DPSS laser with 561 nm line; HeNe lasers with 594 and 633 nm lines.
4. Tissue pieces are mounted on a well slide and sealed under a glass coverslip in seawater from live or frozen corals or in RNA*later*® for fixed samples using nail varnish (see Note 4). Larger samples with skeletal material can be viewed on Live-cell chamber slides (e.g., Lab-Tek, Nunc or μ-Slides, Ibidi). Alternately, coral samples can be put on slides or coverslips in required medium using adhesive "press-to-seal" CoverWell perfusion chambers (Grace Bio-Labs, Inc.).
5. Water immersion 20× or 63× objectives are particularly suited for imaging coral tissues.
6. To image various FPs enable the *xy* or *xyz* scan mode, set the 405 laser line excitation for cyan FPs; blue excitation of 458 or 488 nm laser lines for green FPs; and 514, 543 or 561 nm laser

lines for the excitation for red FPs. Define emission bands at 460–480 nm for cyan FPs into PMT1; 500–540 nm for green and yellow FPs to PMT2; 570–620 nm for red FPs to PMT 3 (see Fig. 2g–k). Each emission band should be set at 5 nm to the red side of the excitation line to avoid interference from reflected light (28).

7. Screening for far red FPs can be done in PMT4 at excitation 488 nm/emission 620–670 nm. To distinguish chlorophyll fluorescence emanating from algal endosymbionts, set emissions from 675 to 700 nm for PMT5. Algal chloroplasts are clearly distinguishable and can be separated from other cellular structures containing red FPs.

8. For non-AOBS confocal microscopes, an appropriate beam splitter should be selected for each excitation line. If dual excitation is required, select the double dichroic DD 488/543 for FP emissions at excitation wavelengths 488 and 543 nm and detection bandwidths of 500–530 and 555–700 nm, respectively. The triple dichroic TD 488/543/633 beam splitter can also be used for the above emitters, with detection bandwidths of 500–535, 555–620, and 650–750 nm, respectively.

9. The Leica TCS SP5 system is more appropriate since it combines three tunable devices—the Acousto Optical Tuneable Filters (AOTF) for fast selection and stepless attenuation of laser lines; the Acousto Optical Beam Splitter (AOBS®) for excitation–emission separation and beam splitters are not required; and the SP-Detector (spectrophotometer detector).

10. Optimal scan speed is 400 Hz; image size 512 × 512 pixels or at 1,024 × 1,024 pixels for high resolution, publication quality images. Use "glow-over/under" pseudo-coloring to adjust the brightness so that the maximum dynamic range of the PMT is made available. PMT gain should be set so that the brightest pixels are just slightly under saturated and the offset such that the darkest pixels are just above dark values.

11. Image averaging improves the signal-to-noise ratio of an image and is done by line (e.g., six lines) or by frame (e.g., 3–7 frames), by scanning a number of times and averaging the brightness of each pixel across scans.

12. To scan for the presence of various FPs deeper in tissues, collect a *z* series which can be reconstructed into 3D views from image stacks. Complete 3D models of the specimen can be rendered and examined from any direction using Leica software. By using the "zoom" function, subcellular resolution of multi-colored FPs can be improved.

13. 3D sections down to a depth of ~200 μm can be made. The Glycerol Leica objective PL APO 63× 1.3 HCX CS permits deeper subsurface imaging of FPs due to its better refractive index match than water or oil immersion objectives and reduces

spherical aberrations enabling an extra-long working distance of 280 µm.

14. To test for cross-talk between FP spectral variants, especially those with closely matching spectral emissions, the "sequential scan" of the microscope can be used and their emissions (e.g., orange and red emitters) can be collected separately.
15. To test multiphoton excitation of FPs, a Coherent Verdi-Mira titanium sapphire laser or the Chameleon™-Ultra 690–1,040 nm 2.5 W laser can be used. Optimal FP excitation is at 800–950 nm, even for red FPs.

3.6. Spectral Characterization of FPs by Confocal Micro-spectrometry

1. Samples are mounted on slides as described in Subheading 3.5.
2. Spectral imaging is with *xyλ* scan mode. Select an appropriate dichroic beam splitter for the chosen excitation line on a Leica TCS SP2 or any other non-AOBS confocal microscope system requiring beam splitters. Be aware that most beam splitters will introduce spectral artifacts such as those shown in Fig. 1 in the spectral study of corals by Ainsworth et al. (29) (see Note 5).
3. Since Leica TCS SP5 is equipped with an AOBS, the beam path is optimized automatically for the selected wavelength and no dichroic filters are required. Spectral artifacts introduced by beam splitters are thus avoided. Define the detection bandwidth for the selected PMT. To improve the signal-to-noise ratio, average 2–3 frames (see Note 6).
4. Alternately, 2-photon excitation can be used for which no dichroic filters are required (see Note 7).
5. Once λ series are collected, select the "Quantify" in Leica software, define regions of interests (ROIs) in areas showing distinct spectral FP variants or mixtures. Data can be imported into Excel spread-sheet as ASCII text file.
6. The above analyses will enable a fuller complement of GFP-like proteins to be identified than by using conventional spectroscopic techniques (see Fig. 2m). Confocal micro-spectral imaging demonstrated that all FPs isolated from the coral *A. millepora* by cloning (11) can be spectroscopically characterized as described here. Moreover, this method resolved the presence of additional spectral variants not previously cloned from this species (see Fig. 2m).

3.7. Screening for Kindling and Photoswitching FPs

1. Corals and other anthozoans are frequently pigmented with pink and purple-blue proteins belonging to the GFP family (see Fig. 1c–e). These colorful proteins are non-fluorescent, absorbing light mainly in the green wave-band. First discovered in sea anemones, where they color tentacle tips pink (28), many of these proteins are photoactivatable and acquire

red fluorescence on irradiation by green light. They are called kindling fluorescent proteins (KFPs) (30, 31) and have attracted considerable attention as photoswitchable imaging probes (15–17). Screening for KFPs is by sampling visibly blue, pink, mauve, or purple parts of corals: tips of branching corals (e.g., Pocilloporidae, Acroporidae); tentacle tips of large polyped or fleshy species (e.g., Gonioporiidae, Faviidae, Trachyphyliidae); or "scarred" tissue areas that often express large quantities of KFPs (see Fig. 1d) (28).

2. Excised tissues are mounted on slides or on glass coverslips as described in Subheading 3.5.
3. Kindling is by irradiation with green light, either in epifluorescent or confocal modes. For confocal imaging of kindling, irradiation is with 514, 543, or 561 nm laser lines set at 30–50% laser power, depending on the objective used. Emission wavelengths are set at 570–650 nm. Kindling is more rapid at higher magnification or using a higher zoom option (see Fig. 3a–b).
4. To study kindling kinetics, collect sequential images in *xyt* mode for a total time of 30 seconds (or as required) at 256 × 256 scanning format at 675 ms/frame scan rate. When KFPs are present in cells at low concentration, kindling requires a longer scanning time or scanning at a higher zoom setting (taking care not to bleach the sample). Figure 3b shows kindling of KFP by confocal green laser line irradiation—the protein was cloned from *A. millepora* (11, 28).
5. KFPs are usually quenched partially or fully (i.e., switched off) by blue light, either in epifluorescence or laser irradiation modes (31). In confocal mode, use 458 or 488 nm laser lines at 40% or 20% intensity, respectively, to test for quenching while performing an *xyt* scan (see Note 8).
6. Coral tissues contain other types of photoswitching fluorescent proteins similar to Dronpa (32). To screen tissues for such proteins, look for instances when fluorescence is reversibly quenched by irradiation by switching-off the laser, followed by switching-on by another laser line, while performing an *xyt* scan.
7. The most common switching-on wavelengths are in the UVA band. A 405 laser is the most appropriate for confocal activation. The commonest quenching wavelengths are in blue waveband and 488 nm works best (see Fig. 3f).
8. Green tissues of *A. millepora* exposed to pulses of 405 nm laser at 20% intensity while imaging green emissions (500–520 nm) at 3% intensity of the 488 nm laser line showed photoswitching events of various magnitudes indicating the presence of photoswitching green FPs (see Fig. 3f).

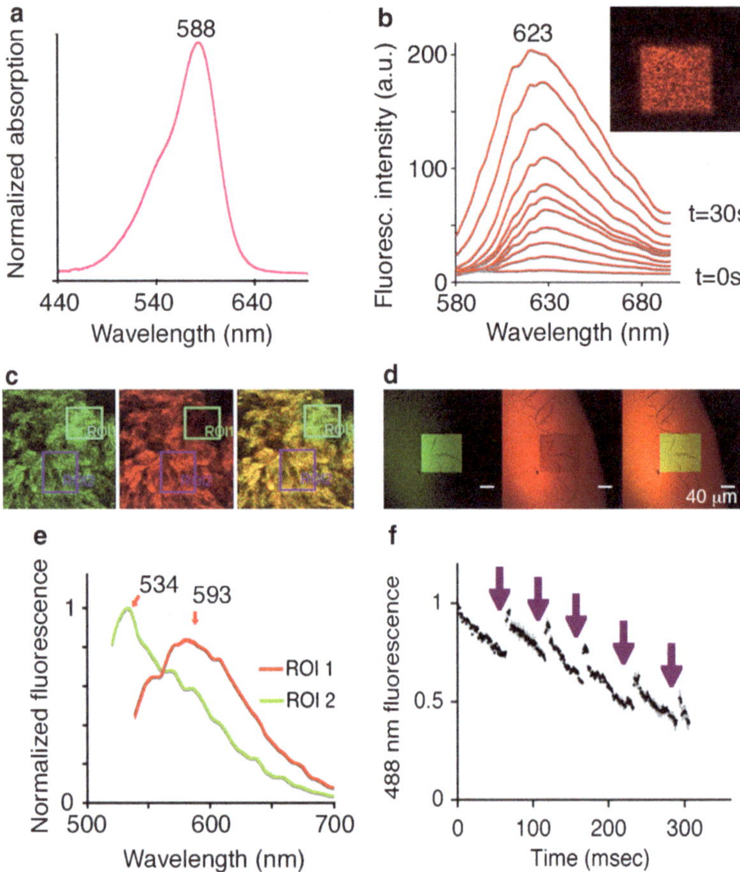

Fig. 3. Photoactive fluorescent proteins of *A. millepora*. (**a**) Absorption of purple-blue CP extracted from tissues in phosphate buffer. (**b**) Kindling by a 561 nm laser line of cloned and purified KFP from this species. (**c**) Confocal imaging in two PMTs showing green and red emissions as well as superimposed image of the two. ROI 1—cells irradiated by 514 nm at high zoom convert from red to green color due to the presence of amilRFP. ROI 2—cells lack the photoconverting amilRFP and their red color is due to other red FPs. (**d**) Photoconversion of purified and dried amilRFP from yellow-green to red fluorescence by a 514 nm laser line. (**e**) Red spectrum—ROI 1 un-converted amilRFP and green spectrum—ROI 2 converted form at excitation at 514 nm obtained by micro-spectroscopic confocal imaging. (**f**) Photoswitching of green FPs in live tissues during imaging at 488 nm and brief irradiation by a 405 nm laser. *Arrows* 405 nm laser pulses.

3.8. Screening for Photoconvertible FPs

1. These protocols relate to screening coral samples for the presence of photoconvertible FPs that switch their fluorescence from one color to another upon irradiation at a specific wavelength. Well-known examples of the group include the green-to-red converters such as Kaede (33) and EosFP (8). An example presented here is a novel red-to-yellow FP named amilRFP, discovered in the coral *A. millepora* (28). AmilRFP photoconverts from red (emission max = 593 nm) to yellow (emission max = 538 nm) with bright fluorescence upon irradiation with green light in a process somewhat similar to that reported for the red-to-green conversion of DsRed after multiphoton irradiation (28).

2. To test for the presence of red-to-yellow photoconverters in coral tissues, select a sample that strongly expresses red fluorescence using spectroscopic or microscopic procedures described above. In the case of *A. millepora*, tissues express a mixture of FPs, including cyan, green, and red emitters. Cells with predominantly red FPs have excitation at 560 nm and emission at 593 nm (11, 22).
3. Dissect out red fluorescent tissue pieces in seawater, mount on a slide in a drop of seawater, as described previously and use confocal microscopy to characterize the localization of red FPs in cells by green light excitation (e.g., 514 or 561 nm laser lines) and image emission at 570–620 nm (see Note 9).
4. Photoconversion using wide-field fluorescence microscopy can be performed by green epifluorescent irradiation using the filter set for DsRed and a low-power objective (e.g., 10–20×) for imaging red fluorescence and switching to a high-power objective (e.g., 40× or 100×) for photoconversion of a small area.
5. If using a confocal microscope lacking FRAP or photoactivation wizard modules, photoconversion can be performed manually by changing the 514 nm laser power from low (imaging mode) to high (photoconversion mode) and back to low (postphotoconversion mode) intensity during an *xyt*-series scan, while simultaneously imaging in two channels: channel 1 for green-yellow at 500–545 nm and channel 2 for red fluorescence at 570–620 nm. Alternately, use the zoom function to zoom to a specific area and photoconvert it.
6. A rapid increase of yellow and a decrease of red emission indicate photoconversion. If no photoconverting proteins are present, only a decrease of fluorescence intensity during irradiation in both PMTs occurs, due to photobleaching.
7. When using Leica SP5 software, activate the FRAP wizard. Use a fast confocal scanning mode as it produces a more rapid photoconversion—e.g., 256 × 256 imaging format. Scanning is at 675 ms per frame, which insures rapid photoconversion. Higher resolution imaging, at fewer frames per second, will result in slower photoconversion.
8. Set imaging intensity of the 514 nm laser line to 5% and image emissions in two channels at 520–545 and 570–620 nm. Set the number of pre-photoconversion scans to ten or more. Set the photoconversion 514 nm laser power to 20–40% intensity. Collect several frames for a post-conversion series. Avoid very strong laser power as this can cause protein bleaching instead of photoconversion. When the red-to-yellow FPs are present, a 3–12-fold increase of yellow fluorescence will be observed.
9. Spectrally characterize the photoconverted and non-converted ROIs by performing $xy\lambda$ scans at 514 nm excitation to capture

emissions of the converted yellow and the un-converted red forms, respectively.

10. A similar method is appropriate for testing for the presence of other types of photoconverters such as the red-to-green converting proteins, by UV irradiation, including EosFP and Kaede. The fluorescence of EosFP changes irreversibly from green (emission max. 516 nm) to red (emission max. 581 nm) upon illumination with a wavelength of approximately 400 nm. Use two lasers 488 and 561 nm to image the green and red forms, respectively, and the 405-nm laser to induce photoconversion.
11. Since novel photoconverters will be found in coral tissues, their presence can be investigated by irradiation by various laser lines and running a *xyt* series while confocally imaging emission in different wavebands, as described in the above protocols.
12. Color quantification and fluorescence data analysis can be done using the intensity quantification tool of the proprietary software bundled with the Leica TCS SP5 confocal microscope. It provides relative fluorescence intensities in different PMTs in the selected ROIs, along a free-form line or a straight line. Other software can also be used, such as open source, public domain Java image processing program ImageJ http://rsbweb.nih.gov/ij/, or Adobe Photoshop.

4. Notes

1. Note that a valid coral collection permit is required for collections on reefs in Australia and many other countries.
2. When visually-selecting corals for the presence of FPs, it should be noted that some FPs produce strong color effects (34). Other FPs absorb wavelengths that are transmitted well by seawater or emit at wavelengths to which human eyes are least sensitive. Cyan and blue-green FPs (emission max. 460–500 nm) are invisible at ambient daylight illumination. Screening corals for such proteins requires excitation with specific wavelengths in the dark. When present in high concentrations, cyan and green FPs impart a gray or white coloration to coral tissues due to light scattering and reflection by FPs in dense micro-granule layers (4, 19, 20). FPs emitting in orange or red wavebands, are usually invisible in daylight in shallow water, but in deeper water where there is little competing down-welling light, they become brightly visible.

3. Refer to http://www.nightsea.com/ for information and equipment for photographic and video imaging of coral fluorescence. During daylight photography of fluorescence the shutter speed of the camera greater than ~1/125 s would reduce the influence of the ambient light. An underwater strobe should be set to maximum power enabling the lens aperture to be sufficiently small to reduce the influence of sunlight. This technique also applies to film-based cameras. The inability to synchronize the strobe at faster shutter speeds would limit daylight fluorescence photography to deeper water, unless a very powerful strobe is used. Images of coral fluorescence in the dark are taken with a similar camera set-up using an appropriate shutter speed.
4. Take care that the acetone in the nail varnish does not touch the sample as acetone can affect FP fluorescence properties.
5. The reflection filter in the Leica RT 30/70 is suitable for collection of a spectral series with most laser lines as it does not distort the collected spectrum by introducing spectral gaps in place of laser excitation lines. The drawback is that it can introduce reflected light, particularly from coverslips. Alternately, select the dichroic RSP 500 for imaging at the excitation wavelength of 488 nm and emission at 500–600 nm (not suitable for imaging far-red FPs). When using double-dichroic filters (e.g., DD 488/543 for FP emissions at excitation wavelengths of 488 and 543 nm) or triple-dichroic (e.g., TD 488/543/633) beam splitters, significant spectral artifacts are produced, as seen in the study by Ainsworth et al. (29).
6. We found that PMT 2 is best for a λ series. Ensure there is a minimum of 5 nm separating the laser excitation line and the start of the detection bandwidth in order to reduce reflected light from the laser being detected. Set the λ band-width to 10 nm and λ step size to 5 nm. Setting the bandwidth to less than 10 nm can cause significant spectral artifacts in both the Leica SP2 and SP5 models.
7. However, 2-photon systems will have poor spectral resolution above 700 nm due to a filter blocking 2-photon excitation from the detector.
8. The property of reversible on- and off-switching of red fluorescence by green (on) and blue (off) irradiation can be used to distinguish the presence of KFPs in the sample from other photoactive red emitters or green FPs exhibiting redding.
9. Take care not to expose samples to high irradiation by green (or other wavelengths) during sample examination or when attempting to focus since even brief pulses of epifluorescence or laser light can induce photoconversion. Use low intensity white transmitted light.

References

1. Chalfie, M., Tu, Y., Euskirchen, G., Ward, W. W., Prasher, D. C. (1994) Green fluorescent protein as a marker for gene expression *Science* **263** (5148), 802–805.
2. Chalfie M. (1995) Green fluorescent protein. *Photochem Photobiol* **62** (4), 651–656.
3. Matz, M. V., Fradkov, A. F., Labas, Y. A., Savitsky, A. P., Zaraisky, A. G., Markelov, M. L., and Lukyanov, S. A. (1999) Fluorescent proteins from nonbioluminescent Anthozoa species. *Nat Biotechnol* **17**, 969–973.
4. Salih, A., Larkum, A., Cox, G., Kuhl, M., and Hoegh-Guldberg, O. (2000) Fluorescent pigments in corals are photoprotective *Nature* **408**, 850–853.
5. Dove, S. G., Hoegh-Guldberg, O., and Ranganathan, R. (2000) Major colour patterns of reef-building corals are due to a family of GFP-like proteins. *Coral Reefs* **19**, 197–204.
6. Labas, Y. A., Gurskaya, N. G., Yanushevich, Y. G., Fradkov, A. F., Lukyanov, K. A., Lukyanov, S. A. and Matz, M. V. (2002) Diversity and evolution of the green fluorescent protein family *Proc Natl Acad Sci USA* **99**, 4256 – 4261.
7. Kelmanson, I. V., and Matz, M. V. (2003) Molecular basis and evolutionary origins of color diversity in great star coral Montastraea cavernosa (Scleractinia: Faviida) *Mol Biol Evol* **20** (7), 1125–1133.
8. Wiedenmann, J., Ivanchenko, S., Oswald, F., Schmitt, F., Rocker, C., Salih, A., Spindler, K. D., and Nienhaus, G. U. (2004) EosFP, a fluorescent marker protein with UV-inducible green-to-red fluorescence conversion *Proc Natl Acad Sci USA* **101** 15905–15910.
9. Kao H. T., Sturgis, S., DeSalle, R., Tsai, J., Davis, D., Gruber, D. F., and Pieribone, V. A. (2007) Dynamic regulation of fluorescent proteins from a single species of coral *Marine Biotechnology* 9(6), 733–746.
10. Oswald, F., Schmitt, F., Leutenegger, A., Ivanchenko, S., D'Angelo, C., Salih, A., Maslakova, S., Bulina, M., Schirmbeck, R., Nienhaus, G. U., Matz, M. V., and Wiedenmann, J. (2007) Contributions of host and symbiont pigments to the coloration of reef corals *FEBS J* **274** (4), 1102–1109.
11. Alieva, N. O., Konzen, K. A., Field, S. F., Meleshkevitch, E. A., Hunt, M. E., Beltran-Ramirez, V., Miller, D. J., Wiedenmann, J., Salih, A., and Matz, M. V. (2008) Diversity and Evolution of Coral Fluorescent Proteins *PLoS* **3** (7) e2680.
12. Deheyn, D. D., Kubokawa, K., McCarthy, J. K., Murakami, A., Porrachia, M., Rouse, G. W. and Holland, N. D. (2007) Endogenous green fluorescent protein (GFP) in amphioxus. *Biol Bull* **213**, 95–100.
13. Shagin, D. A., Barsova, E.V., Yanushevich, Y.G., Fradkov, A.F., Lukyanov, K.A., Labas, Y.A, et al. (2004) GFP-like proteins as ubiquitous metazoan superfamily: evolution of functional features and structural complexity. *Mol Biol Evol* **21**, 841–850.
14. Verkhusha, V. V. and Lukyanov, K. A. (2004) The molecular properties and applications of Anthozoa fluorescent proteins and chromoproteins *Nature Biotech* **22** (3), 289–296.
15. Nienhaus, G.U., and Wiedenmann, J. (2009) Structure, dynamics and optical properties of fluorescent proteins: perspectives for marker development *Chemphyschem* **13**, 10(9-10), 1369–79.
16. Lukyanov, K. A., Chudakov, D. M., Lukyanov, S., and Verkhusha, V. V. (2005) Innovation: photoactivatable fluorescent proteins (Review). *Nat Rev Mol Cell Biol* **6**, 885–891.
17. Prescott, M. and Salih, A. (2008) Genetically encoded fluorescent proteins: properties and applications in the life sciences. Fluorescent Applications in Biotechnology and Life Sciences John Wiley & Sons, Inc., Hoboken, New Jersey. USA & Canada 55–77.
18. Hofmann, M., Eggeling, C., Jakobs, S., and Hell, S. H. (2005) Breaking the diffraction barrier in fluorescence microscopy at low light intensities by using reversibly photoswitchable proteins. *Proc Natl Acad Sci USA* **102**, 17565–17569.
19. Cox, G. and Salih, A. (2006) Fluorescent characteristics of fluorochromatophores in corals. Focus on Multidimensional Microscopy. Ed. Cheng, P.C, Hwang, P.P., Wu, J.L., Wang, G. & H. Kim. World Scientific Publishing Co. Volume **3**, 35–44.
20. Salih, A., Cox, G., and Larkum, A.W.D (2003) Cellular organization and spectral diversity of GFP-like proteins in coral cells studied by single and multi-photon imaging and microscpectroscopy. Multiphoton microscopy in biomedical Sciences III, Ed. A. Pereasamy & P.T. So. *SPIE Proceedings*, **496**, 194–200.
21. Salih, A., Larkum, A., Cronin, T., Wiedenmann, J., Szymczak, R., and Cox. G. (2004) Biological properties of coral GFP-type proteins provide clues for engineering novel optical probes and biosensors. *SPIE Proceedings* Vol. **5329**: Genetically Engineered and Optical Probes for Biomedical Applications II: 61–72.

22. Cox, G., Matz, M. and Salih, A. (2007) Fluorescence lifetime imaging of coral fluorescent proteins. *Microscopy Res Techniques* **70**, 243–251.
23. Adam M. Gilmore, A. M., Larkum, A.W.D., Salih, A., Itoh, S., Shibata, S., Bena, C., Yamasaki, H., Papina, M. and Woesik, R.V. (2003) Simultaneous time resolution of the emission spectra of fluorescent proteins and zooxanthellar chlorophyll in reef-building corals. *Photochemistry and photobiology* 77(5), 515–523.
24. Zhu, Y. Y., Machleder, E. M., Chenchik, A., Li, R., Siebert, P. D. (2001) Reverse transcriptase template switching: a SMART approach for full-length cDNA library construction. *Biotechniques* **30**(4), 892–897.
25. Matz, M., Shagin, D., Bogdanova, E., Britanova, O., Lukyanov, S., Diatchenko, L., and Chenchik, A. (1999) Amplification of cDNA ends based on template-switching effect and step-out PCR. *Nucleic Acids Res* **27**, 1558–1560.
26. Mazel, C. H. (2005) Underwater fluorescence photography in the presence of ambient light. *Limnol. Oceanogr. Methods* **3**, 499–510.
27. Baird, A., Salih, A., and Trevor-Jones, A. (2006) Fluorescence census techniques for the early detection of coral recruits. *Coral Reefs* **23**, 73–76.
28. Salih, A., Wiedenmann, J., Matz, M., Larkum, A.W., and Cox, G. (2006) Photoinduced activation of GFP-like proteins in tissues of reef corals. *Progress in Biomedical Optics and Imaging* (7) **21**, 60980B.1–60980B.12.
29. Ainsworth, T.D., Hoegh-Gudlberg, O., and Leggat, W (2008) Imaging the fluorescence of marine invertebrates and their associated flora *Journal of Microscopy* **232**, 197–199.
30. Lukyanov, K.A., Fradkov, A. F., Gurskaya, N. G. , Matz, M. V., Labas, Y. A., Savitsky, A. P., Markelov, M. L., Zaraisky, A. G., Zhao, X., Fang, Y., Tan, W., and Lukyanov, S. A. (2000) Natural animal coloration can be determined by a nonfluorescent green fluorescent protein. *J Biol Chem* **275**, 25879–25882.
31. Chudakov, D.M., Belousov, V.V., Zaraisky, A.G., Novoselov, V.V., Staroverov, D.B., Zorov, D. B., Lukyanov, S., and Lukyanov, K.A. (2003) Kindling fluorescent proteins for precise in vivo photolabeling. *Nature Biotechnology* **21**, 191–194.
32. Habuchi, S., Ando, R., Dedecker, P., Verheijen, W., Mizuno, H., Miyawaki, A., and Hofkens, J. (2005) Reversible single-molecule photoswitching in the GFP-like fluorescent protein Dronpa. *Proc Natl Acad Sci* USA **102**, 9511–9516.
33. Ando, R., Hama, H., Yamamoto-Hino, M., Mizuno, H., and Miyawaki, A. (2002). An optical marker based on the UV-induced green-to-red photoconversion of a fluorescentprotein. *Proc Natl Acad Sci USA*, **99**, 12651–12656.
34. Mazel, C. H., and E. Fuchs (2003) Contribution of fluorescence to the spectral signature and perceived color of corals. *Limnol. Oceanogr* **48**, 390–401.

Chapter 16

What Does It Take to Improve Existing Fluorescent Proteins for In Vivo Imaging Applications?

Marc Zimmer

Abstract

Although fluorescent proteins are ubiquitously used as genetic tracers and imaging agents, there is significant room for improvement. This chapter discusses how new improved fluorescent proteins can be designed. It focuses on the design of far-red and infrared fluorescent proteins, since the currently-available red fluorescent proteins are not optimal for in vivo applications.

Key words: Fluorescent proteins, Phytochrome, Bilirubin, DsRed, mKate

1. Introduction

Green fluorescent protein (GFP) was first isolated from the jellyfish, *Aequorea victoria*, in 1962 (1). However, it was its cloning in the early 1990s (2) that led to the birth of fluorescent protein imaging techniques (3). The main reason that fluorescent proteins (FPs) have become such popular in vivo imaging tools is that the chromophore is an inherent part of the protein that is formed by an autocatalytic cyclization of the Ser65Tyr66Gly67 triad. This means that the FPs are genetic markers that do not require the addition of any additional chemicals for fluorescence.

Wild-type GFP is rarely used in imaging, enhanced GFP (EGFP) and other GFP mutants are much more useful, as are GFP-like proteins derived from species other than *A. victoria*. The main drawback to wild-type GFP that has led to the search and development of new GFP mutants, as in vivo imaging agents, is that it is difficult to distinguish GFP fluorescence from background fluorescence, creating the need for brighter (brightness = fluorescence quantum yield × peak molar extinction coefficient) red-shifted

Robert M. Hoffman (ed.), *In Vivo Cellular Imaging Using Fluorescent Proteins: Methods and Protocols*, Methods in Molecular Biology, vol. 872, DOI 10.1007/978-1-61779-797-2_16, © Springer Science+Business Media, LLC 2012

mutants. Furthermore, the chromophore formation is slow (on the order of hours) and can be incomplete, especially at higher temperatures. In the last 15 years numerous orange, red, and far-red FPs have been developed, but none of them match the overall utility of EGFP and many are cytotoxic (4). For deep-tissue imaging, mKate2 and tdKatushka (5) are the most useful of the Fps due to their long-wavelength excitation and emission. However, they are not as bright as tdTomato or as photostable as mCherry (4). Therefore, there is a continued need to understand FPs and to find new FPs for in vivo imaging.

There are three different approaches to finding new and improved FPs for biomedical applications,

- Rationally-designing mutants.
- Randomly mutating known FPs and selecting for properties of interest.
- Fishing for new FPs in bioluminescent and fluorescent organisms.

Attempts using a combination of the approaches are the most efficient.

2. Rationally Designing New Mutants

In order to improve the properties of GFP, it is important to understand the photophysics and chromophore chemistry of the protein. Several hundred papers have been published that examine aspects of the chromophore formation, FP brightness, and the relationship between the absorption/emission and the protein surrounding the chromophore. Yet our knowledge is far from complete and rationally-designed wavelength mutations are still beyond our grasp.

BFP, EGFP, and YFP are good examples of rationally-designed GFP mutants. Knowing that the GFP chromophore is formed from the S65Y66G67 residues, these residues are obvious candidates for mutation. Therefore even before the crystal structure of GFP was known, Tsien created the Y66H mutant. He chose to mutate Tyr 66 as it is the central residue in the chromophore and he mutated it to a His as he correctly presumed that a residue with an aromatic side-chain is required for fluorescence. The resultant blue fluorescent protein (BFP) was the first wavelength mutation of GFP (6). BFP has a much lower fluorescence quantum yield than GFP ($\Phi_{fl}=0.20$ vs. 0.80). It has been suggested that this is due to the fact that His66 (BFP-chromophore) forms fewer hydrogen bonds with the surrounding protein than Tyr66 (GFP-chromophore) does. Also the smaller imadazole ring (His66) in BFP may have more conformational freedom than the larger phenol (Tyr66), which leads to more intersystem crossing (7). Recently, two new BFPs (Azurite and A5) with enhanced brightness and photostability were developed (8, 9). The methodology applied to find the brighter BFP mutants was based on the concept that replacing the residues surrounding the chromophore with bulkier amino acids would constrain the chromophore's motion and thereby increase the proteins brightness.

GFP has two excitation peaks due to the neutral and anionic forms of the chromophore. Tsien found that the S65T GFP mutant has only one excitation peak, a sixfold increased brightness and a fourfold increase in the rate of oxidation of chromophore. It is the basis of the most commonly used FP, enhanced green fluorescent protein (EGFP) (10, 11). Gly67 is strictly conserved in all FPs, it is required for chromophore formation.

In the first rationally-designed mutant based on the crystal structure of GFP-S65T, Tsien and co-workers decided to mutate T203 into a tyrosine so that it could π stack with the phenolic group of the tyrosine in the chromophore (12). The resultant yellow fluorescent protein, YFP, is red-shifted by 16 nm relative to GFP-S65T and does indeed have a π stacking interaction between the chromophore and Tyr203 (13). It is significantly brighter than GFP.

In general mutations of the chromophore-forming residues lead to large wavelength shifts, while changing the protein matrix around the chromophore corresponds to spectral fine tuning.

The green, blue, cyan, and yellow fluorescent proteins developed by the late 1990s were the first colors in the rainbow palette of FP's but a very important color, red, was still missing. Since infrared wavelengths are minimally absorbed by hemoglobin, water, and lipids (14, 15), imaging agents that emit in the 650–900 nm range would be ideally suited for in vivo imaging of deep tissue in live animals. A search for red bioluminescent (16–18) or fluorescent protein (5, 19, 20) has been continuing unabated for the last 15 years.

3. Fishing for New FPs in Bioluminescent and Fluorescent Organisms

Groups all over the world tried mutating GFP to form a red GFP mutant. This strategy was not very successful and we would have to wait until 2008 before a red mutant of *A. victoria* GFP was created by rational design (21). Other groups took to oceans and forests to look for red bioluminescent organisms. They were no more successful. It took a conceptual shift to find red fluorescent proteins. Lukyanov and Labas made the breakthrough (22). Thinking that fluorescent proteins did not necessarily have to be associated with other chemiluminescent proteins, they decided to look for organisms that were red fluorescent but were not bioluminescent. In a Moscow aquarium they found corals containing the first "red" fluorescent protein, DsRed. The chromophore in DsRed has undergone a second oxidation to form an acylimine that extends the conjugation of the chromophore—hence the red shift.

There are 15 red-emitting FPs (λ_{emiss} > 600 nm), most are variants of the naturally occurring FPs, Dsred, eqFP578, and eqFP611 (23). The biophysical properties of the red fluorescent proteins obtained from corals and other organisms can be improved by rational mutations, such as those that increase the π-stacking interactions with the chromophore, preferentially stabilize the excited state with polar interactions (24, 25) or use unnatural amino acids (26).

Genes for FPs have been found in over 150 species from four different phyla: Cnidaria, Arthropoda, Chordata, and Ctenophora. The large phylogenetic distances between the FP-containing phyla suggests that new FPs with interesting new properties may ultimately be found in many other Metazoan lineages (27). A tyrosine is the central chromophoric residue in all natural-occurring FPs; perhaps it is required for their, as of yet unknown, function.

4. Randomly Mutating Known FPs and Selecting for Properties of Interest

EqFP611 has the most red-shifted emission (611 nm) and the largest Stokes shift (52 nm) of all naturally-occurring FPs. However, like most other naturally occurring red FPs eqFP611 is tetrameric. This is a problem in applications involving the use of FPs as fusion tags to study motility, localization, and intermolecular binding. By selectively making single-point mutations to residues located in the interface between the tetramers, a monomeric form of eqFp611 was developed. Unfortunately monomerization led to a loss of fluorescence. After multiple rounds of random and multisite-directed mutagenesis, a bright red monomeric FP was obtained—mRuby (19). It was four residues shorter than wild-type eqFP611 and had 28 mutations. DsRed also required 33 mutations before it was monomerized to form the first monomeric red FP (mRFP1) (28). Clearly, at this stage, it is difficult to know the structural and chemical consequences of each of these mutations. This makes it difficult to rationally design mutants and makes a combined approach much more practical.

5. Infrared Fluorescent Proteins

The main reason GFP-like proteins are used in in vivo whole body imaging is that the chromophore is autocatalytically formed from the protein itself and no external factors are needed for fluorescence.

Once expressed GFP-like FPs fold into their characteristic β-barrel shape, the chromophore is formed and fluorescence may be observed. Bioluminescent proteins are also used in in vivo imaging applications, however, they require the addition of an external substrate before light emission is observed, e.g., luciferase/luciferin. To date no infrared in vivo imaging agents based on bioluminescent proteins or GFP-like molecules have been reported. Perhaps the limited conjugational variations of the 2-imino-5-(4-hydroxybenzylidene)-imidazolinone chromophore and/or the size of the β-barrel in the GFP-like proteins are not sufficient for an infrared imaging agent. An alternative protein/chromophore system, such as the one described below may be required.

The first infrared fluorescent protein was recently reported. It was engineered from bacterial phytochrome (29). The protein has excitation and emission maxima of 684 and 708 nm, respectively. This contrasts with 595 and 663 nm for AQ14, an unstable variant from *Actinia equine* (30) that is the GFP-like protein with the highest peak-emission wavelength (31). The infrared-fluorescent protein (IFP) is based on a bacterio-phytochrome from *Deinococcus radiodurans*, which with a single D207H mutation, is red fluorescent (32). The bacterio-phytochromes are part of the phytochrome super-family that are found in plants, fungi, and bacteria, where they are responsible for light-regulating responses, such as seed germination, flowering, and phototaxis (33–35). The phytochromes form homo-dimeric complexes with a linear tetrapyrrole. The chromophore is auto-catalytically bound to a positionally-conserved cysteine. The bacterio-phytochromes are excellent candidates for in vivo imaging agents, since they bind biliverdin IXα, which is ubiquitous in all aerobic organisms as the initial intermediate in heme catabolism. These proteins are non-toxic, and have negligible fluorescence on their own. In vivo, the bacteriophytochromes use light to control gene expression.

The fluorescent truncated bacterio-phytochromes were developed by removing the PHY and histidine kinase-related domains. These modifications convert the light energy to chemical energy used in signaling; by optimizing the protein to prevent dimerization; and by limiting the biliverdins conformational freedom, especially around the D ring. They are not GFP-like proteins and do not have an autocatalytically-formed chromophore that is an integral part of the protein backbone. Instead an ubiquitously present chromophore autocatalytically binds the photoprotein. This means that if sufficient biliverdin IXα is present, it will bind the IFP, making it autofluorescent. Despite their low quantum yield, the IFPs have been shown to be suitable for whole body mammalian imaging (29).

References

1. Shimomura, O., Johnson, F. H., and Saiga, Y. (1962) Extraction, purification and properties of aequorin, a bioluminescent protein from the luminous hydromedusan, *Aequorea.*, *J. Cell. Comp. Physiol.* **59**, 223–229.
2. Prasher, D. C., Eckenrode, V. K., Ward, W. W., Pendergast, F. G., and Cormier, M. J. (1992) Primary structure of the Aequorea victoria green fluorescent protein, *Gene* **111**, 229–233.
3. Zimmer, M. (2009) GFP: from jellyfish to Nobel prize and beyond, *Chem. Soc. Rev. DOI:* 10.1039/b904023d.
4. Davidson, M. W., and Campbell, R. E. (2009) Engineered fluorescent proteins: innovations and applications, *Nature Methods* **6**, 713–717.
5. Shcherbo, D., Murphy, C. S., Ermakova, G. V., Solovieva, E. A., Chepurnykh, T. V., Shcheglov, A. S., Verkhusha, V. V., Pletnev, V. Z., Hazelwood, K. L., Roche, P. M., Lukyanov, S., Zaraisky, A. G., Davidson, M. W., and Chudakov, D. M. (2009) Far-red fluorescent tags for protein imaging in living tissues, *Biochemical Journal* **418**, 567–574.
6. Heim, R., Prasher, D. C., and Tsien, R. Y. (1994) Wavelength mutations and posttranslational autoxidation of green fluorescent protein., *Proc. Natl. Acad. Sci. USA* **91**, 12501–12504.
7. Wachter, R. M., King, B. A., Heim, R., Kallio, K., Tsien, R. Y., Boxer, S. G., and Remington, S. J. (1997) Crystal structure and photodynamic behavior of the blue emission variant Y66H/Y145F of green fluorescent protein., *Biochemistry* **36,** 9759–9765.
8. Mena, M. A., Treynor, T. P., Mayo, S. L., and Daugherty, P. S. (2006) Blue fluorescent proteins with enhanced brightness and photostability from a structurally targeted library, *Nature Biotechnology* **24,** 1569–1571.
9. Treynor, T. P., Vizcarra, C. L., Nedelcu, D., and Mayo, S. L. (2007) Computationally designed libraries of fluorescent proteins evaluated by preservation and diversity of function. *Proc Natl Acad Sci USA* **104**, 48–53.
10. Heim, R., Cubitt, A., and Tsien, R. Y. (1995) Improved green fluorescene, *Nature* **373**, 663–664.
11. Cubitt, A. B., Heim, R., Adams, S. R., Boyd, A. E., Gross, L. A., and Tsien, R. Y. (1995) Understanding, improving and using green fluorescent proteins. *TIBS* **20**, 448–455.
12. Ormoe, M., Cubitt, A. B., Kallio, K., Gross, L. A., Tsien, R. Y., and Remington, S. J. (1996) Crystal structure of the Aequorea victoria green fluorescent Protein, *Science* **273**, 1392–1395.
13. Wachter, R. M., Elsiger, M. A., Kallio, K., Hanson, G. T., and Remington, S. J. (1998) Structural basis of spectral shifts in the yellow emission variants of green fluorescent protein, *Structure* **6**, 1267–1277.
14. Jobsis, F. F. (1977) Noninvasive, infrared monitoring of cerebral and myocardial oxygen sufficiency and circulatory parameters, *Science* **198**, 1264–1267.
15. Weissleder, R., and Ntziachristos, V. (2003) Shedding light onto live molecular targets, *Nature Medicine* **9,** 123–129.
16. Branchini, B. R., Ablamsky, D. M., Murtiashaw, M. H., Uzasci, L., Fraga, H., and Southworth, T. L. (2007) Thermostable red and green light-producing firefly luciferase mutants for bioluminescent reporter applications, *Analytical Biochemistry* **361**, 253–262.
17. Caysa, H., Jacob, R., Muther, N., Branchini, B., Messerle, M., and Soling, A. (2009) A red-shifted codon-optimized firefly luciferase is a sensitive reporter for bioluminescence imaging, *Photochemical & Photobiological Sciences* **8**, 52–56.
18. Fischer, A. J., and Lagarias, J. C. (2004) Harnessing phytochrome's glowing potential. *Proc Natl Acad Sci USA* **101,** 17334–17339.
19. Kredel, S., Oswald, F., Nienhaus, K., Deuschle, K., Roecker, C., Wolff, M., Heilker, R., Nienhaus, G. U., and Wiedenmann, J. (2009) mRuby, a Bright Monomeric Red Fluorescent Protein for Labeling of Subcellular Structures, *Plos One* **4,** e4391.
20. Suto, K., Masuda, H., Takenaka, Y., Tsuji, F. I., and Mizuno, H. (2009) Structural basis for red-shifted emission of a GFP-like protein from the marine copepod Chiridius poppei, *Genes Cells* **14,** 727–737.
21. Mishin, A. S., Subach, F. V., Yampolsky, I. V., King, W., Lukyanov, K. A., and Verkhusha, V. V. (2008) The first mutant of the Aequorea victoria green fluorescent protein that forms a red chromophore, *Biochemistry* **47,** 4666–4673.
22. Matz, M. V., Fradkov, A. F., Labas, Y. A., Savitsky, A. P., Zaraisky, A. G., Markelov, M. L., and Lukyanov, S. A. (1999) Fluorescent proteins from nonbioluminescent Anthozoa species, *Nature Biotech.* **17**, 969–973.
23. Day, R. N., and Davidson, M. W. (2009) The fluorescent protein palette: tools for cellular imaging, *Chemical Society Reviews* **38**, 2887–2921.

24. Marques, M. A. L., Lopez, X., Varsano, D., Castro, A., and Rubio, A. (2003) Time-dependent density-functional approach for biological chromophores: The case of the green fluorescent protein - art. no. 258101, *Physical Review Letters* **9025,** 8101–8101.
25. Abbyad, P., Childs, W., Shi, X. H., and Boxer, S. G. (2007) Dynamic Stokes shift in green fluorescent protein variants. *Proc Natl Acad Sci USA* **104,** 20189–20194.
26. Kent, K. P., Oltrogge, L. M., and Boxer, S. G. (2009) Synthetic Control of Green Fluorescent Protein, *Journal of the American Chemical Society* **131**, 15988-+.
27. Haddock, S. H. D., Moline, M. A., and Case, J. F. (2010) Bioluminescence in the Sea, *Annual Review of Marine Science* **2**, 443–493.
28. Campbell, R. E., Tour, O., Palmer, A. E., Steinbach, P. A., Baird, G. S., Zacharias, D. A., and Tsien, R. Y. (2002) A monomeric red fluorescent protein, *Proc Natl Acad Sci USA* **99,** 7877–7882.
29. Shu, X. K., Royant, A., Lin, M. Z., Aguilera, T. A., Lev-Ram, V., Steinbach, P. A., and Tsien, R. Y. (2009) Mammalian Expression of Infrared Fluorescent Proteins Engineered from a Bacterial Phytochrome, *Science* **324**, 804–807.
30. Shkrob, M. A., Yanushevich, Y. G., Chudakov, D. M., Gurskaya, N. G., Labas, Y. A., Poponov, S. Y., Mudrik, N. N., Lukyanov, S., and Lukyanov, K. A. (2005) Far-red fluorescent proteins evolved from a blue chromoprotein from Actinia equina, *Biochemical Journal* **392,** 649–654.
31. Nienhaus, G. U., and Wiedenmann, J. (2009) Structure, Dynamics and Optical Properties of Fluorescent Proteins: Perspectives for Marker Development, *Chemphyschem* **10**, 1369–1379.
32. Wagner, J. R., Zhang, J. R., von Stetten, D., Guenther, M., Murgida, D. H., Mroginski, M. A., Walker, J. M., Forest, K. T., Hildebrandt, P., and Vierstra, R. D. (2008) Mutational analysis of Deinococcus radiodurans bacteriophytochrome reveals key amino acids necessary for the photochromicity and proton exchange cycle of phytochromes, *Journal of Biological Chemistry* **283,** 12212–12226.
33. Rockwell, N. C., Su, Y. S., and Lagarias, J. C. (2006) Phytochrome structure and signaling mechanisms, *Annual Review of Plant Biology* **57,** 837–858.
34. Quail, P. H. (2002) Phytochrome photosensory signalling networks, *Nature Reviews Molecular Cell Biology* **3,** 85–93.
35. Zimmer, M. (2006) Non-retinal chromophoric proteins, in *Cis-trans Isomerizaion in Biochemistry* (Duvage, C., Ed.), pp 77–94, Wiley-VCH, Weinheim.

Chapter 17

In Vivo Imaging of Oligonucleotide Delivery

Fumitaka Takeshita, Ryou-u Takahashi, Jun Onodera, and Takahiro Ochiya

Abstract

RNA interference (RNAi) has rapidly become a powerful tool for drug-target discovery and therapeutics. Cancer is an important application for RNAi therapeutics, since abnormal gene regulation is thought to contribute to the pathogenesis and maintenance of the metastatic phenotype of cancer. Many oncogenic genes present enticing therapeutic target possibilities for RNAi. Small interfering RNA (siRNA) and microRNA (miRNA) are potent and specific examples of RNAi are able to silence tumor-related genes and multiple oncogenic pathways and appear to be a rational approach to inhibit tumor growth. In subsequent in vivo studies, an appropriate animal model must be developed for a better evaluation of gene-silencing effects on tumors. How to evaluate the effect of siRNA and miRNA in an in vivo therapeutic model is also important. Bioluminescence imaging is an optical imaging method that can evaluate RNAi in vivo.

Key words: siRNA, MicroRNA, Cancer, Delivery, Imaging, Luciferase, Oligonucleotides

1. Introduction

RNAi can effect posttranscriptional gene silencing. The introduction into an organism of double-stranded RNA (dsRNA) corresponding to a transcribed sequence results in degradation of the corresponding mRNA (1–6). With RNAi, dsRNA blocks gene expression in a sequence-specific manner. When introduced into cells, dsRNA is processed by the RNase III family nuclease Dicer into siRNA, 21-basepair dsRNA with two overhanging bases at each 3′ terminus. The double-stranded siRNA is passed to the RNA-induced silencing complex (RISC), an RNA–nuclease complex, which is activated as it unwinds the duplex and incorporates one of the antisense strands. The RISC then selectively degrades

Robert M. Hoffman (ed.), *In Vivo Cellular Imaging Using Fluorescent Proteins: Methods and Protocols*, Methods in Molecular Biology, vol. 872, DOI 10.1007/978-1-61779-797-2_17, © Springer Science+Business Media, LLC 2012

RNA containing the sequence complementary to the incorporated antisense strand.

Antisense oligonucleotide drugs were used prior to the discovery of RNAi and several antisense molecules are currently in late-stage preclinical or clinical development (7). Although researchers continue to explore and develop antisense reagents for therapeutic use by morpholino oligomers, a fourth class of oligonucleotide-based compounds, consisting of siRNAs, has recently become widely used for gene knockdown in vitro and in vivo.

Another group of catalytically-active RNA molecules (ribozymes) has also been considered for therapeutic use. However, only a few ribozymes have turned out to be efficient compounds in clinical trials. RNAi is effective because siRNA is highly specific for the target gene, and the single-strand RNA molecule incorporated into the RISC is used to recognize multiple copies of the target RNA. Therefore, an extremely small amount of siRNA can generate reliable gene suppression, making toxicity less of a concern. Furthermore, effective antisense oligonucleotide sequences are determined empirically, resulting in uncertain efficacies.

MicroRNA (miRNA), an endogenously-expressed form of siRNA, approximately 22 nucleotides in length, also works for gene silencing. It is estimated that there are over 1,000 miRNAs in humans. It is believed that a single miRNA can regulate several hundred genes. Current understanding of the molecular mechanism of any disease, including cancer (8, 9), would be incomplete without factoring in the functional significance of miRNA. Mis-expression of miRNAs has been observed in various types of cancers and is also associated with the clinical outcome of cancer patients. Consistently, miRNAs have been implicated in the regulation of various cellular processes that are often deregulated during tumor development and progression (10, 11), suggesting that these miRNAs might be targets for cancer therapy.

The most direct way for molecules to correct expression of altered genes and miRNAs is treatment by RNA oligonucleotides. For this purpose, an in vivo delivery system is a key issue. Here, we describe an in vivo imaging method of delivery of oligonucleotides such as siRNA and miRNA.

2. Materials

2.1. Cell Lines and Medium

1. PC-3M-luc cells (Xenogen Corp., Alameda, CA).
2. Cell culture medium: RPMI 1640 medium (Invitrogen Corp., Carlsbad, CA) supplemented with 10% heat-inactivated fetal bovine serum (Equitech-Bio, Kerrville, TX) and 0.2 mg/ml zeocin (Invitrogen Corp.).

2.2. Oligonucleotide Delivery Mixture

1. Oligonucleotide delivery system: atelocollagen (12–15) for local use AteloGene™ #1390 and systemic use AteloGene™ #1391 (Koken, Tokyo).
2. Oligonucleotides: 5–10 and 20–40 μM oligonucleotide solutions for local and systemic delivery in vivo.

2.3. In Vivo Imaging

1. For in vivo imaging with Renilla luciferase: ViviRen (5 mg/kg, Promega).
2. For in vivo imaging with firefly luciferase: D-luciferin (150 mg/kg, Xenogen).
3. Data analysis: LivingImage software (version 2.50, Xenogen) (16).

3. Methods

Recent progress in the optical imaging of cancers in animal models presents many potential advantages for recreating the disease process, disease detection, screening, diagnosis, drug development, and treatment evaluation. Fluorescence-based imaging (17–21) and bioluminescence-based imaging (12, 13, 22–29) are well developed and allow specific, highly-sensitive, and quantitative measurements of a wide range of tumor-related parameters in mice.

A major advantage of GFP-labeling is that imaging requires no preparative procedures and hence allows for direct visualization in living tissue. In contrast, luciferase imaging requires exogenous injection of luciferin substrate which can stress the animals. In addition, the intensity of the luciferase signal may sometimes be variable and unstable. Furthermore, RFP imaging is about 1,000 times stronger than that of luciferase in vivo. Therefore, for monitoring the tumor metastasis process at the single-cell level, fluorescence imaging may be the more practical method. In fact, fluorescence-based orthotopic metastatic models have been used to study mechanisms and drug discovery. Here, we have used the bioluminescence signal from the luciferase reporter gene in our metastasis model. Luciferase genes in our tumor cells can function stably over significant periods of time in tumors and in their metastases.

3.1. Preparation of Dual Luciferase Expressing Cells

1. For construction of 3′-UTR-Renilla luciferase plasmid and reporter assays, amplify the segment of 3′-UTR of the Bcl2 gene by PCR using genomic DNA from normal human prostate epithelial cells (PrEC, CT-2555, Lonza Walkersville, Inc., Walkersville, MD).
2. Insert the PCR product into a pGL4.75 [HRuc/CMV] vector (Promega, Madison, WI), using the XbaI site immediately

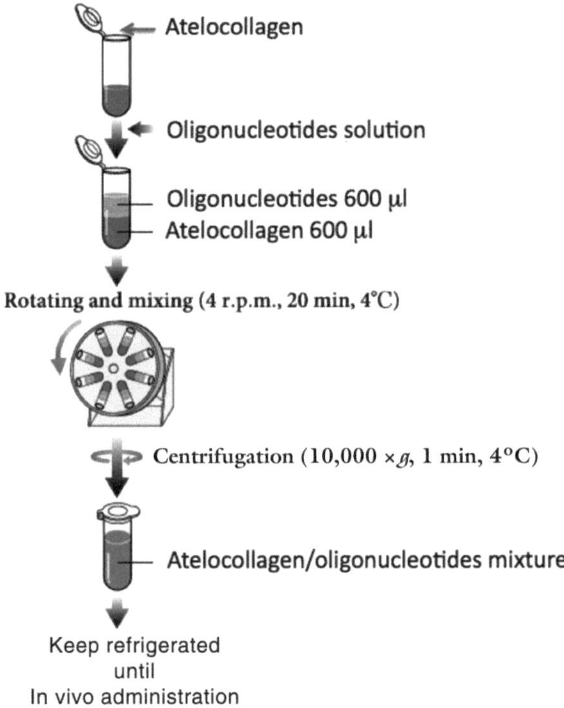

Fig. 1. Preparation of oligonucleotides delivery mixture. Gently add 600 μl of the oligonucleotide solution to 600 μl of the atelocollagen solution. Rotate the mixture solution for 20 min. Set the rotating speed at about 4 rpm when a 20-cm diameter holder is used. After mixing, centrifuge the tube for 1 min at 10,000 x *g* to deform the mixed solution.

downstream from the stop codon of Renilla luciferase (pGL4.75[HRuc/CMV]-Bcl2 3′UTR).

3. For reporter assays, transfect 2 μg pGL4.75[HRuc/CMV]-Bcl2 3′UTR using LipofectAMINE™ 2000 (Invitrogen Corp.) into PC-3M-luc cells.
4. Select stable transfectants in hygromycin (0.2 mg/ml; Invitrogen Corp.) using the Dual-luciferase assay-system (Promega). The intensity of Renilla luciferase is normalized by firefly luciferase. Clones expressing both luciferase genes are named PC-3M-luc/Rluc-Bcl2 3′UTR.

3.2. Oligonucleotides Delivery

1. Gently add 600 μl of the oligonucleotide solution to 600 μl atelocollagen solution.
2. Rotate the mixture solution for 20 min. Set the rotating speed at approximately 4 rpm when using a 20-cm diameter holder (see Note 1).
3. After mixing, centrifuge the tube for 1 min at 10,000 × *g* to deform the mixed solution (see Note 2). The procedure is shown in Fig. 1.

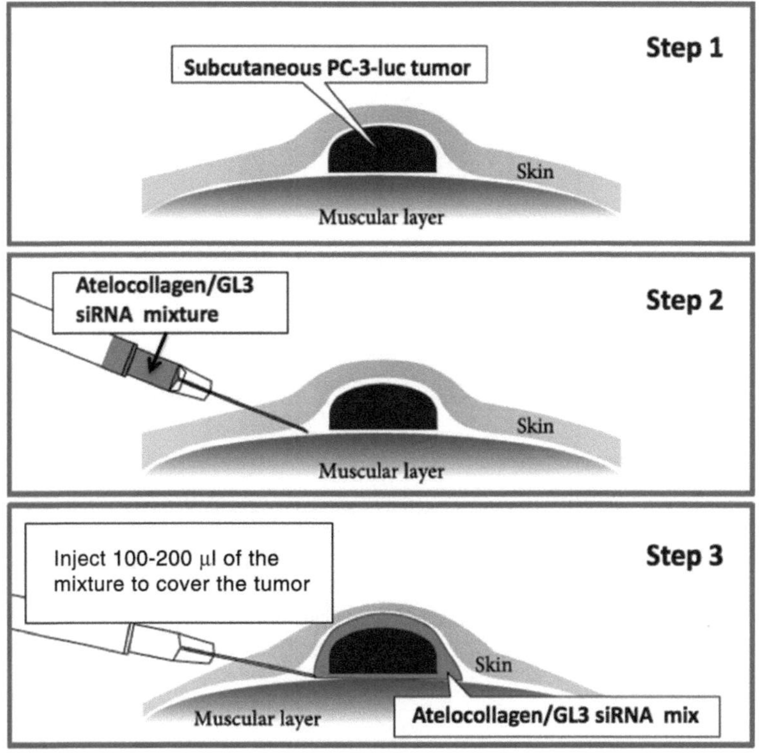

Fig. 2. Administration of the mixture to subcutaneous tumors. Step 1: PC-3-luc subcutaneous tumor; Step 2: insert a 26 G needle subcutaneously approximately 5 mm to the side of the tumor; Step 3: lay the needle parallel to the skin and insert it for 2–3 mm in the direction of the tumor, and then inject the mixture for 20–30 s. It is effective when the mixture is administered so as to cover the whole target site.

3.3. Imaging of Local Delivery of Oligonucleotides

1. Prepare 7- to 10-week-old male athymic nude mice (CLEA Japan, Shizuoka, Japan).
2. To generate a subcutaneous tumor model, the animals are injected with 1×10^6 PC-3-luc cells suspended in 100 ml sterile DPBS.
3. When a tumor develops to 5×5 mm a mixture of GL3 siRNA (specifically knock down firefly luciferase) and atelocollagen are prepared according to the described method.
4. Set the 18-G needle in the disposable syringe and slowly draw the atelocollagen/oligonucleotide mixture (see Note 3).
5. Replace the needle of the syringe with a 26-G injection needle and keep the syringe refrigerated until administration.
6. Insert the injection needle from approximately 5 mm to the side of the subcutaneous tumor with the cut face of the needle turned upward.
7. Lay the needle parallel to the skin and insert it 2–3 mm in the direction of the tumor, and then gently inject 200 μl of the mixture. The procedure is shown in Fig. 2 (see Note 4).

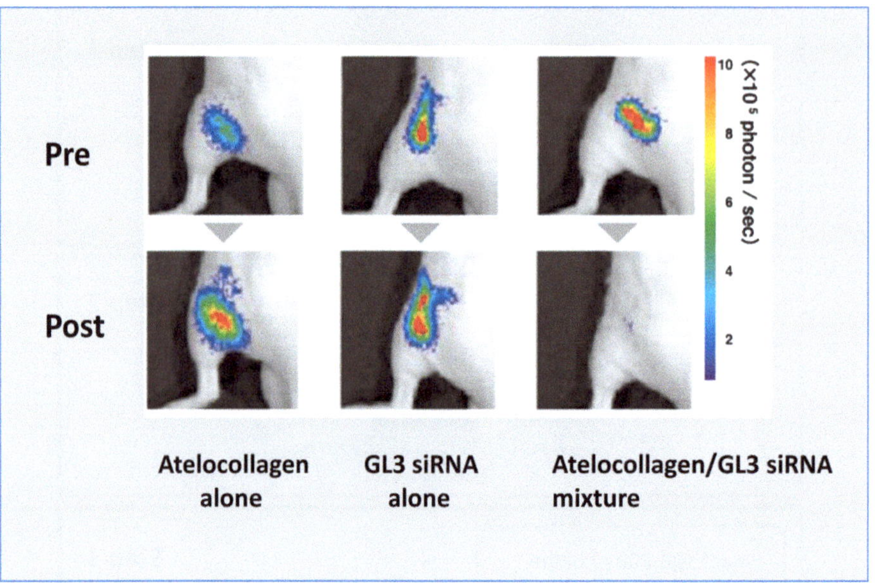

Fig. 3. In vivo imaging of local delivery of oligonucleotides. Firefly luciferase-expressing cells formed subcutaneous tumors. Atelocollagen/GL3 siRNA significantly inhibited the photon count of luciferase as compared to atelocollagen alone and GL3 siRNA alone.

8. 24–48 h after injection (see Note 5), the animals are subjected to bioimaging analysis. An example result is shown in Fig. 3.

3.4. Imaging of Systemic Delivery of Oligonucleotides

1. To generate an experimental metastasis model, the anesthetized animals are injected with 2×10^6 PC-3M-luc cells suspended in 100 ml sterile DPBS into the left heart ventricle (see Note 6).
2. When metastasis develops, a mixture of GL3 siRNA and atelocollagen is prepared according to the above method.
3. Anesthetize the animals, if necessary.
4. Set the 18-G needle in a disposable syringe and slowly draw the atelocollagen/oligonucleotide mixture. In a systemic injection, 100–200 μl of the mixture is used.
5. Replace the syringe needle with the 26-G injection needle and keep the syringe refrigerated until administration.
6. Disinfect the tail of the animal with ethanol.
7. Insert the needle into the vein at a position 1/4 from the tail end.
8. Confirm that the injection needle has entered the vessel and then slowly inject 100–200 μl of the mixture (see Note 7).
9. 24–48 h after injection, the animals are subjected to bioimaging analysis. An example result is shown in Fig. 4.

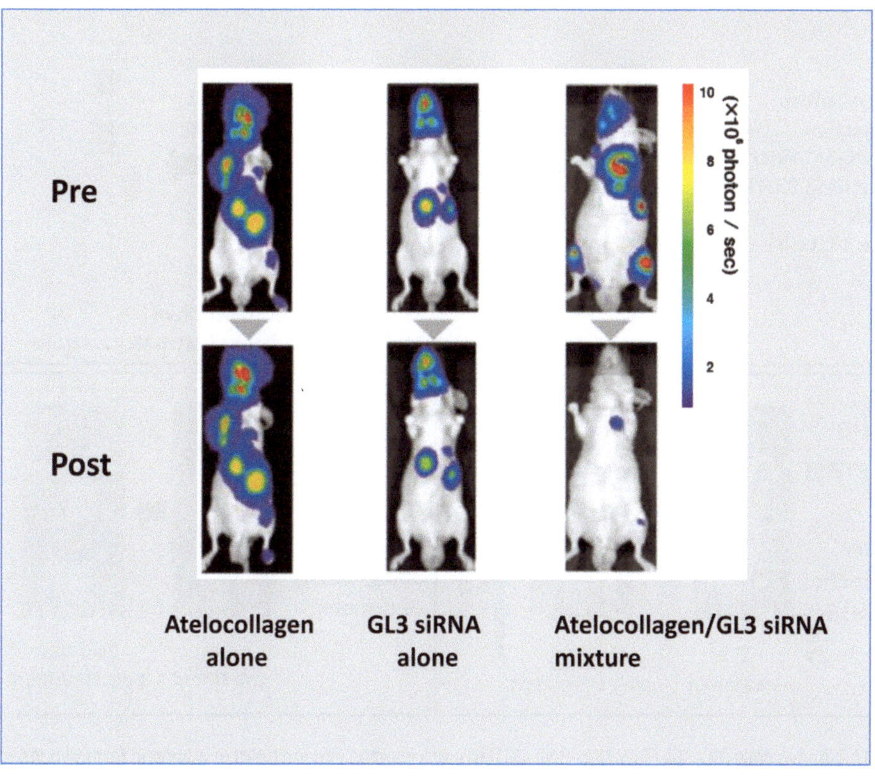

Fig. 4. Firefly-luciferase-expressing cells with formed bone metastasis in mice. Intra-cardiac administration of atelocollagen/GL3 siRNA significantly inhibited the photon count of luciferase as compared to atelocollagen alone and GL3 siRNA alone.

3.5. Dual Luciferase Imaging System for Delivery of Oligonucleotides

1. Seven- to ten-week-old male athymic nude mice (CLEA Japan, Shizuoka, Japan) are anesthetized by exposure to 3% isoflurane on day 0 and subsequent days.
2. On day 0 of the experiments, to generate an experimental metastasis model, the anesthetized animals are injected with 2×10^6 PC-3M-luc/Rluc-Bcl2 3′UTR cells, suspended in 100 ml sterile DPBS, into the left heart ventricle.
3. When metastasis develops, a mixture of miRNA16 and atelocollagen is prepared according to the described method.
4. For systemic injection of the atelocollagen/miRNA mixture, repeat steps 4–8 in Subheading 3.2.
5. For in vivo imaging, the mice are injected with ViviRen (5 mg/kg, Promega) by intravenous tail vein injection and imaged immediately to count the photons from the animal body.
6. After the bioluminescence from Renilla luciferase disappears, the mice are administered D-luciferin (150 mg/kg, Xenogen) by intraperitoneal injection.
7. Ten minutes later, photons from firefly luciferase are counted. An example result is shown in Fig. 5.

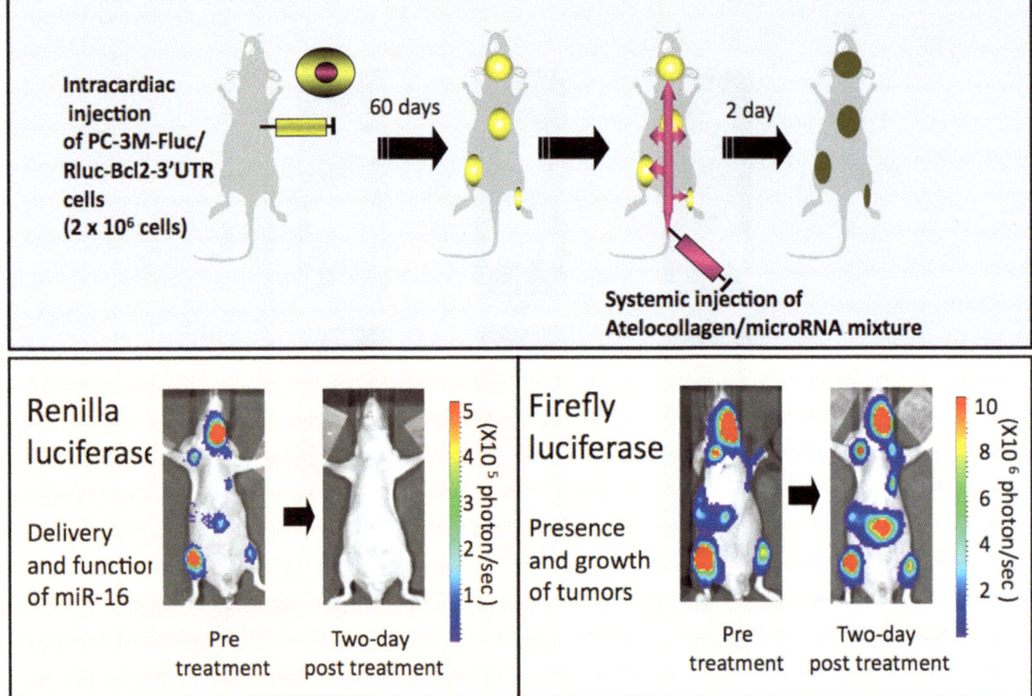

Fig. 5. Dual luciferase cells (PC-3M-Fluc/Rluc-Bcl2-3′UTR) were injected into the heart of mice and formed bone metastases. When the expression of Renilla luciferase from tumor cells could be detected, 50 μg miR-16/atelocollagen was injected intravenously. As can be seen at the bottom of the figure (left side), the photons from the Renilla luciferase were dramatically suppressed one day post-treatment. This result indicates that atelocollagen has the potential for delivering miR-16 throughout the whole body, including bone metastatic sites. Firefly luciferase was not affected, because large-sized tumors like these were not inhibited by a single treatment of miR-16 (29).

4. Notes

1. Avoid mixing by vortexing; otherwise, large aggregates may generate and cause a less efficient delivery of the oligonucleotides.
2. After mixing, ensure there are no visible aggregates.
3. Draw the mixture slowly to avoid incorporation of bubbles.
4. Intratumoral injection is also possible for delivery of atelocollagen/siRNA.
5. The effect of siRNA, such as miRNA delivery by atelocollagen, differs depending on the oligonucleotide sequence, expression level of the target oligonucleotides, the difference in target tumor cells, tissues, site of the tumors, etc. Investigating the

most suitable duration after administration for analysis of gene downregulation is recommended.

6. When the needle is correctly positioned into the left ventricle, bright red oxygenated blood influxes into the needle hub (13, 24). In this model, bone metastases developing in the jaws and/or legs of the mice are detected by non-invasive in vivo bioluminescence imaging.
7. For systemic administration of 200 μl of the mixture, taking 20–30 s will enhance delivery efficacy.
8. In drug resistant tumors, sometimes in vivo bioluminescence imaging does not work well since the luciferin substrate and oligonucleotide mixture are not easily taken up by drug-resistant tumor cells (14).
9. Atelocollagen is a highly purified type I collagen that is modified to have low immunogenicity (Koken, Tokyo). Atelocollagen forms nanosize particles when it is mixed with synthetic miRNAs, via electrostatic binding. The nanoparticles are easily incorporated into cells by endocytosis. The atelocollagen/oligonucleotide complex showed high resistance to nucleases. Therefore, the complex is thought to be stable in vivo (12, 29).
10. It is possible to prepare luciferase-expressing cells with a virus vector system. To generate lentiviral-vector particles containing the luciferase gene, an HIV-1 gag–pol expression plasmid, an HIV-1 Rev expression plasmid and a VSV-G envelope protein expression plasmid were used to package the HIV-based expression vector. In brief, four plasmids were co-transfected into 293FT cells. Two days after transfection, the supernatants were cleared from the cellular debris by low-speed centrifugation (10 min, 1,000 × *g*) and filtration through 0.45-μm filters. Aliquots were stored at −80°C.

Acknowledgements

This work was supported in part by a Grant-in-Aid for the Third-Term Comprehensive 10-Year Strategy for Cancer Control, a Grant-in-Aid for Scientific Research on Priority Areas Cancer from the Ministry of Education, Culture, Sports, Science and Technology, and the Program for Promotion of Fundamental Studies in Health Sciences of the National Institute of Biomedical Innovation (NiBio), and a Takeda Science Foundation.

References

1. Bass, B.L. (2000) Double-stranded RNA as a template for gene silencing. *Cell* **101**, 235–238.
2. Mcmanus, M.T. and Sharp, P.A. (2003) Gene silencing in mammals by small interfering RNAs. *Nat Rev Genet* **3**, 737–747.
3. Shankar, P., Manjunath, N., and Lieberman, J. (2005) The prospect of silencing disease using RNA interference. *JAMA* **293**, 1367–1373.
4. Leung, R.K., Whittaker, P.A. (2005) RNA interference: from gene silencing to gene-specific therapeutics. *Pharmacol Ther* **107**, 222–239.
5. Behlke, M.A. (2006) Progress towards in vivo use of siRNAs. *Mol Ther* **13**, 644–670.
6. Dykxhoorn, D.M., Palliser, D., and Lieberman, J. (2006) The silent treatment: siRNAs as small molecule drugs. *Gene Ther* **13**, 541–552.
7. Rayburn, E.R., Wang, H., and Zhang, R. (2006) Antisense-based cancer therapeutics: are we there yet? *Expert Opin Emerg Drugs* **11**, 337–352.
8. Hagan, J.P. and Croce, C.M. (2007) MicroRNAs in carcinogenesis. *Cytogenet Genome Res* **118**, 252–259.
9. Jiang, J., Gusev, Y., Aderca, I., Mettler, T.A., Nagorney, D.M., Brackett, D.J., Roberts, L.R., and Schmittgen, T.D. (2008) Association of microRNA expression in hepatocellular carcinomas with hepatitis infection, cirrhosis, and patient survival. *Clin Cancer Res* **14**, 419–427
10. Bartel, D.P. (2004) MicroRNAs: genomics, biogenesis, mechanism, and function. *Cell* **116**, 281–297.
11. Osaki, M., Takeshita, F., and Ochiya, T. (2008) MicroRNA as biomarkers and therapeutic drugs in human cancer. *Biomarkers* **13**, 658–670.
12. Minakuchi, Y., Takeshita, F., Kosaka, N., Sasaki, H., Yamamoto, Y., Kouno, M., Honma, K., Nagahara, S., Hanai, K., Sano, A., Kato, T., Terada, M., and Ochiya, T. (2004) Atelocollagen-mediated synthetic small interfering RNA delivery for effective gene silencing in vitro and in vivo. *Nucleic Acids Res* **32**:e109.
13. Takeshita, F., Minakuchi, Y., Nagahara, S., Honma, K., Sasaki, H., Hirai, K., Teratani, T., Namatame, N., Yamamoto, Y., Hanai, K., Kato, T., Sano, A., and Ochiya, T. (2005) Efficient delivery of small interfering RNA to bone-metastatic tumors by using atelocollagen in vivo. *Proc Natl Acad Sci USA* **102**, 12177–12182.
14. Honma, K., Iwao-Koizumi, K., Takeshita, F., Yamamoto, Y., Yoshida, T., Nishio, K., Nagahara, S., Kato, K., and Ochiya, T. (2008) *RPN2* gene confers docetaxel resistance in breast cancer. *Nat Med* **14**, 939–948.
15. Takei, Y., Kadomatsu, K., Goto, T., and Muramatsu, T. (2006) Combinational antitumor effect of siRNA against midkine and paclitaxel on growth of human prostate cancer xenografts. *Cancer*, **107**, 864–873.
16. Vooijs, M., Jonkers, J., Lyons, S., and Berns, A. (2002) *Cancer Res* **62,** 1862–1867.
17. Hoffman, R.M. (1999) Orthotopic transplant mouse models with green fluorescent protein-expressing cancer cells to visualize metastasis and angiogenesis. *Cancer and Metastasis Reviews* **17**, 271–277.
18. Hoffman, R.M. (1999) Orthotopic metastatic mouse models for anticancer drug discovery and evaluation: a bridge to the clinic. *Invest New Drugs* **17**, 343–359.
19. Hoffman, R.M. (2002) In vivo imaging of metastatic cancer with fluorescent proteins. *Cell Death Differ* **9**, 786–789.
20. Hoffman, R.M. (2005) Orthotopic metastatic (MetaMouse) models for discovery and development of novel chemotherapy. *Methods Mol Med* **111**, 297–322.
21. Nakanishi, H., Ito, S., Mochizuki, Y., and Tatematsu, M. (2005) Evaluation of chemosensitivity of micrometastases with green fluorescent protein gene-tagged tumor models in mice. *Methods Mol Med* **111**, 351–362.
22. Hennig, R., Ventura, J., Segersverd, R., Ward, E., Ding, X.Z., Rao, S.M., Jovanovic, B.D., Iwamura, T., Talamonti, M.S., Bell, R.H. Jr, and Adrian, T.E. (2005) LY293111 improves efficacy of gemcitabine therapy on pancreatic cancer in a fluorescent orthotopic model in athymic mice. *Neoplasia* 7, 417–425.
23. Contag, P.R., Olomu, I.N., Stevenson, D.K., and Contag, C.H. (1998) Bioluminescent indicators in living mammals. *Nat Med* **4**, 245–247.
24. Rehemtulla, A., Stegman, L.D., Cardozo, S.J., Gupta, S., Hall, D.E., Contag, C.H., and Ross, B.D. (2000) Rapid and quantitative assessment of cancer treatment response using in vivo bioluminescence imaging. *Neoplasia* **2**, 491–495.
25. Jenkins, DE, Oei, Y, Hornig, YS, Yu, S.F., Dusich, J., Purchio, T., and Contag, P.R. (2003) Bioluminescent imaging (BLI) to improve and refine traditional murine models of tumor growth and metastasis. Clin Exp Metastasis **20**, 733–744.

26. Vooijs, M., Jonkers, J., Lyons, S., and Bernes, A. (2002) Noninvasive imaging of spontaneous retinoblastoma pathway-dependent tumors in mice. *Cancer Res* **62**, 1862–1867.

27. Lyons, S.K. (2005) Advances in imaging mouse tumor models in vivo. *J Pathol* **205**,194–205.

28. Laurie, N.A., Gray, J.K., Zhang, J., Leggas, M., Relling, M., Egorin, M., Stewart, C., and Dyer, M.A. (2005) Topotecan combination chemotherapy in two new rodent models of retinobrastoma. *Clin Cancer Res* **11**, 7569–7578.

29. Takeshita, F., Bader, A.G., Osaki, M., Takahashi, R., Yamamoto, Y., Kosaka, N., Kawamata, M., Kelnar, K., Brown, D., and Ochiya, T. (2010) Systemic delivery of miR-16 for RNAi therapy in prostate cancer. *Mol Ther* **18**, 181–187.

Chapter 18

Subcellular Imaging In Vivo: The Next GFP Revolution

Robert M. Hoffman

Abstract

The use of fluorescent proteins to differentially label cancer cells in the nucleus and cytoplasm and high-powered imaging technology have been used to visualize the nuclear–cytoplasmic dynamics of cancer-cell in vivo. Nuclear–cytoplasmic dynamics have been imaged in cancer cells trafficking in both blood vessels and lymphatic vessels as well as during seeding on organs and interacting with stroma in the live animal. Fluorescent proteins have also been used to color code the phases of the cell cycle which can now be followed in vivo. This technology has furthered our understanding of the spread of cancer at the subcellular level. Fluorescent proteins thereby provide the basis for the new field of in vivo cell biology.

Key words: GFP, RFP, Fluorescence, In vivo, Nuclei, Cytoplasm, Nuclear-cytoplasmic dynamics, Apoptosis, Mitosis, Cell-deformation, Imaging, Nude mice

1. Cellular Imaging In Vivo

Chambers et al. (1) stated that knowledge of metastasis had been limited because it had been a hidden process. Chambers' group developed in vivo (intra-vital) microscopy to visualize metastasis using cells labeled with small-molecule fluorescent dyes in nanospheres. However, such small-molecule labeling is not hereditary. Genetic reporters such as green fluorescent protein (GFP) are not only very bright but also heritable and can be used to follow cancer cells longitudinally in mice, even at the subcellular level, the topic of this chapter.

2. Properties of GFP for In Vivo Cellular Imaging

The GFP gene was cloned from the bioluminescent jellyfish *Aequorea victoria* (2) and shown to have great utility for cellular imaging (3, 4). GFP cDNA encodes a 283-amino acid monomeric

Robert M. Hoffman (ed.), *In Vivo Cellular Imaging Using Fluorescent Proteins: Methods and Protocols*,
Methods in Molecular Biology, vol. 872, DOI 10.1007/978-1-61779-797-2_18, © Springer Science+Business Media, LLC 2012

polypeptide with M_r 27,000 (5, 6) that requires no other substrates, or cofactors to fluoresce (7). Bright mutants of the GFP gene have been generated (8–10) and have been humanized for high expression and strong signal (11). Red fluorescent protein (RFP) from the *Discosoma* coral has similar features as GFP, as well as the advantage of longer wavelength emission (12–15).

Initial studies of tumor biology that used stable GFP expression focused on static images of primary tumors and metastases (16, 17). The first use of GFP in vivo for imaging was by Chishima et al. who visualized cancer cells in mice (18). The first use of GFP to observe motility and shape changes of cancer cells in tumors was described by Farina et al. (19).

Chishima et al. (18) and Huang et al. (20) showed that GFP-transduced cancer cells could be imaged in blood vessels. To examine cell behavior during intravasation, Wyckoff et al. (21) have used GFP to view cells in time-lapse images within a single optical section. Using a confocal microscope, it was shown that both metastatic and non-metastatic cells are motile, but metastatic cells show greater orientation toward blood vessels (22).

Farina et al. (19) visualized GFP-expressing cancer cell movement in live rats with a laser scanning confocal microscope. Metastatic and non-metastatic cancer cells were found to differ in their movement. Using multi-photon microscopy and GFP-expressing cancer cells, Wang et al. (23) found major differences in behavior between the non-metastatic and metastatic primary breast cancers in cell motility, and chemotaxis. Goswami et al. (24) have shown that macrophages promote the invasion of GFP-labeled breast cancer cells.

Naumov et al. (25), visualized fine cellular details of cancer cells, expressing GFP, such as pseudopodial projections, even after extended periods of in vivo growth.

Mook et al. (26) visualized initial arrest and division of GFP-expressing colon cancer cells in sinusoids of the liver, due to size restriction, after portal vein (PV) injection.

Al-Mehdi et al. (27) observed the steps in early hematogenous metastasis of cancer cells expressing GFP in subpleural microvessels in perfused mouse and rat lungs. Cancer cells attached to the endothelia of pulmonary precapillary arterioles and capillaries. Colonies of GFP-expressing cancer cells were observed growing entirely within the blood vessels.

3. Origins of Subcellular Imaging In Vivo

To obtain the dual-color cells suitable for subcellular imaging, RFP was expressed on a retrovirus in the cytoplasm of cancer cells, and GFP linked to histone H2B, was expressed in the nucleus.

The dual color cells could report subcellular dynamics such as mitosis and apoptosis.

The cell cycle position of individual living cancer cells, expressing nuclear GFP and cytoplasmic RFP, was readily visualized by the nuclear-cytoplasmic ratio and nuclear morphology. Real-time induction of apoptosis was observed by nuclear size changes and progressive nuclear fragmentation.

Mitotic cells were visualized by whole-body imaging after injection in the mouse ear. Common-carotid-artery injection of dual-color cells and a reversible skin flap enabled the external visualization of the dual-color cells in microvessels in the mouse brain. Extreme cell deformation was visualized in the capillaries. Dual-color cells in various positions of the cell cycle were visualized in excised mouse lungs after tail-vein injection of the dual-color cells (15).

4. Subcellular Imaging of Cancer Cell Deformation in Capillaries

The dual-color cells, with GFP in the nucleus and RFP in the cytoplasm, were used to visualize the cytoplasmic and nuclear dynamics of cells migrating in capillaries. Immediately after the cells were injected in the heart of nude mice, a skin flap on the abdomen was made. Highly-elongated cancer cells, containing elongated nuclei, in capillaries in the skin flap were observed with a color CCD camera. The cytoplasm and nuclei in the capillaries deformed sufficiently to fit the width of these vessels. The average length of the major axis of the cancer cells in the capillaries increased to approximately four times their normal length. The nuclei increased their length 1.6 times in the capillaries. Cancer cells in capillaries greater than 8 μm in diameter could migrate up to 48.3 μm/h (22).

5. Imaging Cancer Cell Trafficking in Blood Vessels

The nuclear and cytoplasmic behavior of cancer cells was observed in real time in larger blood vessels as they trafficked or adhered to the vessel surface in the abdominal skin flap. During extravasation, time-lapse dual-color imaging showed that cytoplasmic processes of the cancer cells exited the vessels first, with nuclei following along the cytoplasmic projections. Both cytoplasm and nuclei underwent deformation during extravasation (28).

6. Imaging Cancer Cell Trafficking in Lymphatic Vessels

Miller et al. (29), von Andrian and Mempel (30) and Halin et al. (31) visualized lymphocyte trafficking, cell migration, and cell–cell interaction in lymph nodes using GFP-labeled cells. Stoll et al. (32) visualized T-cells interacting with dendritic cells in lymph nodes using multicolor fluorescence. Dadiani et al. (33) imaged breast cancer emboli clustering at a lymphatic vessel junction. However, trafficking of cancer cells in lymphatic vessels was not investigated in these studies.

Dual-color cells were injected into the inguinal lymph node of nude mice and imaged trafficking through lymphatic vessels, via a skin flap, in real time, until they entered the axillary lymph node (34).

7. Imaging of Tumor–Stroma Interaction at the Cellular and Subcellular Level

The use of multiple colors of fluorescent proteins can distinguish cancer cells from host cells (35–37). Multicolor fluorescent proteins were also used to distinguish cancer cells from one another (38).

A transgenic mouse expressing GFP in all of its cells, or in specific cells such as endothelial cells, transplanted with cancer cells expressing RFP, enabled the interaction between the cancer cells and host stromal cells to be visualized in real time (35, 36).

To non-invasively image cancer cell/stromal cell interaction at the cellular and subcellular level in live mice in real time, we developed a three-color animal model. The model consists of GFP-expressing mice transplanted with dual-color cancer cells labeled with GFP in the nucleus and RFP in the cytoplasm. An Olympus IV100 Laser Scanning Microscope, with ultra-narrow microscope objectives, was used for three-color whole-body imaging of the dual-color cancer cells interacting with the GFP-expressing stromal cells. Drug response of both cancer and stromal cells in the intact live animal was also imaged in real time. Various in vivo phenomena of cancer-cell–stromal-cell interaction and cellular dynamics were imaged, including mitotic and apoptotic cancer cells, stromal cells interacting with the cancer cells, tumor vasculature, and tumor blood flow. Tumor blood flow was also non-invasively imaged in this model as the GFP-expressing lymphocytes were brightly fluorescent (39).

We demonstrated using GFP and RFP imaging that stromal cells are necessary for metastasis. After splenic injection of cancer cells, splenocytes co-trafficked with the cancer cells to the liver and facilitated metastatic colony formation. Splenic injection of HCT-116-GFP-RFP cancer cells resulted in the formation of liver and distant metastasis in GFP transgenic nude mice. GFP spleen cells were found in the liver metastases that resulted from intrasplenic injection of the cancer cells. When GFP spleen cells and the RFP cancer cells were co-injected in the portal vein (PV), liver metastasis resulted that contained GFP spleen cells, in contrast to when cancer cells were injected in the PV alone where metastasis did not occur. These results demonstrate that stromal cells are necessary for metastasis to form (40).

8. Subcellular Imaging of Cancer-Cell Viability In Vivo and Increased Cancer Cell Viability After Cyclophosphamide Pretreatment

Additional experiments were performed to understand the poor viability of cancer cells after injection in the PV. Apoptosis was readily visualized in the dual-color cells by their altered nuclear morphology. Extensive clasmocytosis (destruction of the cytoplasm) of the HCT-116-GFP-RFP cells occurred within 6 h after injection in the PV. The number of apoptotic cells rapidly increased within the PV within 12 h of injection. In contrast, dual-color mouse mammary tumor (MMT)-GFP-RFP cells injected into the PV mostly survived in the liver of nude mice 24 h after injection. Many surviving MMT-GFP-RFP cells showed invasive figures with cytoplasmic protrusions. The cells grew aggressively and formed colonies in the liver. However, when the host mice were pretreated with cyclophosphamide, the HCT-116-GFP-RFP cells also survived and formed colonies in the liver after PV injection. These results suggest that a cyclophosphamide-sensitive host cellular system attacked the HCT-116-GFP-RFP cells but could not effectively kill the MMT-GFP-RFP cells (41).

In mice pretreated with cyclophosphamide, intravascular proliferation and extravasation of human HT-1080 fibrosarcoma cells occurred along with extravascular colony formation. In the non-pretreated mice, most cancer cells remained quiescent in vessels without extravasation. Cyclophosphamide does not directly affect the cancer cells because cyclophosphamide has been cleared by the time the cancer cells were injected (42).

These results show an "opposite effect" of chemotherapy which enhances critical steps in malignancy rather than inhibiting them, suggesting that certain current approaches to cancer chemotherapy should be reconsidered (42).

9. Subcellular In Vivo Imaging of Cancer Cell Seeding on the Lung in Real Time

We developed a new in vivo mouse model to image single cancer-cell dynamics of metastasis to the lung in real time. Dual-color HT-1080 human fibrosarcoma cells were injected in the tail vein of the mouse. The right chest wall was then opened in order to image metastases on the lung surface directly. Regulating airflow volume with a novel endotracheal intubation method enabled controlling lung expansion adequate for imaging of the exposed lung surface. After each observation, the chest wall was sutured and the air was suctioned in order to reinflate the lung, and thereby keep the mice alive. Observations have been carried out for up to 8 h per session and repeated up to six times per mouse. Seeding and arresting of single cancer cells on the lung, accumulation of cancer-cell emboli, cancer-cell viability, and metastatic colony formation were imaged in real time thereby enabling the cell biology of metastatic targeting to be studied in vivo in real time (43).

10. Subcellular Imaging of Cancer Cell Killing by UV Light

Dual-color 143B human osteosarcoma cells, HT-1080 human fibrosarcoma cells, Lewis lung carcinoma (LLC), and XPA-1 human pancreatic cancer cells were exposed to various doses of UVA, UVB, or UVC, apoptotic and viable cells were quantitated under fluorescence microscopy. UV-induced cancer cell death was wavelength and dose dependent, as well as cell-line dependent in vitro. After UVA exposure, most cells were viable even when the UV dose was increased. With UVB irradiation, cell death only could be observed with irradiation at 50 J/m^2. For UVC, as little as 25 J/m^2 UVC irradiation killed approximately 70% of the 143B dual-color cells. UV-induced cancer cell death varied among the cell lines. UVC exposure also suppressed cancer cell growth in nude mice in a model of minimal residual cancer (MRC). No apparent side effects of UVC exposure were observed (44).

11. Subcellular In Vivo Color-Coded Imaging of Cell Cycle Progression

Miyawaki's group used cell-cycle-specific cell-cycle proteins linked with different color fluorescent proteins to visualize a cell's position in the cell cycle at any time point. These probes label individual G1 phase nuclei red and those in S/G2/M phases green (45).

12. A Simple Microscope for In Vivo Subcellular Imaging

Dual-color cancer cells expressing GFP in the nucleus and RFP in the cytoplasm were injected by a vascular route in an abdominal skin flap in nude mice. The mice are then imaged with the Olympus

MVX10 macroview fluorescence microscope which is a long-working-distance fluorescence microscope with high-numerical aperture objectives was used for variable-magnification imaging in live mice from macro- to subcellular. With the MVX10, the nuclear and cytoplasmic behavior of cancer cells trafficking in blood vessels of live mice was observed. Lung metastases in live mice from the macro- to the subcellular level were imaged by opening the chest wall to expose the lung in live mice as described above. Injected splenocytes, expressing cyan fluorescent protein (CFP), could also be imaged on the lung of live mice (46).

The studies described in this chapter enabled the founding of the field of in vivo cell biology (47). The wide range of instrumentation available as well as a large and diverse collection of cancer cell lines expressing GFP and/or RFP, and transgenic nude mice expressing GFP, RFP, and CFP, enable widespread investigations of in vivo cell biology (36, 48–53).

References

1. Chambers, A.F., Groom, A.C., Macdonald, I.C. (2002) Dissemination and growth of cancer cells in metastatic sites. *Nat Rev Cancer* **2**, 563–572.
2. Prasher, D.C., Eckenrode, V.K., Ward, W.W., Prendergast, F.G., Cormier, M.J. (1992) Primary structure of the *Aequorea victoria* green-fluorescent protein. *Gene* **111**, 229–233.
3. Chalfie, M., Tu, Y., Euskirchen, G., Ward, W.W., Prasher, D.C. (1994) Green fluorescent protein as a marker for gene expression. *Science* **263**, 802–805.
4. Cheng, L., Fu, J., Tsukamoto, A., Hawley, R.G. (1996) Use of green fluorescent protein variants to monitor gene transfer and expression in mammalian cells. *Nat Biotechnol* **14**, 606–609.
5. Cody, C.W., Prasher, D.C., Westler, W.M., Prendergast, F.G., Ward, W.W. (1993) Chemical structure of the hexapeptide chromophore of the Aequorea green fluorescent protein. *Biochemistry* **32**, 1212–1218.
6. Yang, F., Moss, L.G., Phillips, G.N., Jr. (1996) The molecular structure of green fluorescent protein. *Nat Biotechnol* **14**, 1246–1251.
7. Morin, J., Hastings, J. (1971) Energy transfer in a bioluminescent system. *J Cell Physiol* 77, 313–318.
8. Cormack, B., Valdivia, R., Falkow, S. (1996) FACS-optimized mutants of the green fluorescent protein (GFP). *Gene* **173**, 33–38.
9. Crameri, A., Whitehorn, E.A., Tate, E., Stemmer, W.P. (1996) Improved green fluorescent protein by molecular evolution using DNA shuffling. *Nat Biotechnol* **14**, 315–319.
10. Delagrave, S., Hawtin, R.E., Silva, C.M, Yang, M.M., Youvan, D.C. (1995) Red-shifted excitation mutants of the green fluorescent protein. *Biotechnology (N.Y.)* **13**, 151–154.
11. Heim, R., Cubitt, A.B., Tsien, R.Y. (1995) Improved green fluorescence. *Nature* **373**, 663–664.
12. Zolotukhin, S., Potter, M., Hauswirth, W.W., Guy, J., Muzyczka, N. (1996) A 'humanized' green fluorescent protein cDNA adapted for high-level expression in mammalian cells. *J Virol* **70**, 4646–4654.
13. Gross, L.A., Baird, G.S., Hoffman, R.C., Baldridge, K.K., Tsien, R.Y. (2000) The structure of the chromophore within DsRed, a red fluorescent protein from coral. *Proc Natl Acad Sci USA* **97**, 11990–11995.
14. Fradkov, A.F., Chen, Y., Ding, L., Barsova, E.V., Matz, M.V., Lukyanov, S.A. (2000) Novel fluorescent protein from Discosoma coral and its mutants possesses a unique far-red fluorescence. *FEBS Lett* **479**, 127–130.
15. Yamamoto, N., Jiang, P., Yang, M., Xu, M., Yamauchi, K., Tsuchiya, H., et al. (2004) Cellular dynamics visualized in live cells *in vitro* and *in vivo* by differential dual-color nuclear-cytoplasmic fluorescent-protein expression. *Cancer Res.* **64**, 4251–4256.
16. Hoffman, R.M. (2002) Green fluorescent protein imaging of tumour growth, metastasis, and angiogenesis in mouse models. *Lancet Oncol.* **3**, 546–556.
17. Condeelis, J., Segall, J.E. (2003) Intravital imaging of cell movement in tumours. *Nat Rev Cancer* **3**, 921–930.

18. Chishima, T., Miyagi, Y., Wang, X., Yamaoka, H., Shimada, H., Moossa, A.R., Hoffman, R.M. (1997) Cancer invasion and micrometastasis visualized in live tissue by green fluorescent protein expression. *Cancer Res.* **57**, 2042–2047.
19. Farina, K.L., Wyckoff, J.B., Rivera, J., Lee, H., Segall, J.E., Condeelis, J.S., Jones, J.G. (1998) Cell motility of tumor cells visualized in living intact primary tumors using green fluorescent protein. *Cancer Res.* **58**, 2528–2532.
20. Huang, M.S., Wang, T.J., Liang, C.L., Huang, H.M., Yang, I.C., Yi-Jan, H., Hsiao, M. (2002) Establishment of fluorescent lung carcinoma metastasis model and its real-time microscopic detection in SCID mice. *Clin Exp Metastasis* **19**, 359–368.
21. Wyckoff, J.B., Jones, J.G., Condeelis, J.S., Segall, J.E. (2000) A critical step in metastasis: in vivo analysis of intravasation at the primary tumor. *Cancer Res.* **60**, 2504–2511.
22. Yamauchi, K., Yang, M., Jiang, P., Yamamoto, N., Xu, M., Amoh, Y., et al. (2005) Real-time *in vivo* dual-color imaging of intracapillary cancer cell and nucleus deformation and migration. *Cancer Res.* **65**, 4246–4252.
23. Wang, W., Wyckoff, J.B., Frohlich, V.C., Oleynikov, Y., Hüttelmaier, S., Zavadil, J., et al. (2002) Single cell behavior in metastatic primary mammary tumors correlated with gene expression patterns revealed by molecular profiling. *Cancer Res.* **62**, 6278–6288.
24. Goswami, S., Sahai, E., Wyckoff, J.B., Cammer, M., Cox, D., Pixley, F.J., et al. (2005) Macrophages promote the invasion of breast carcinoma cells via a colony-stimulating factor-1/epidermal growth factor paracrine loop. *Cancer Res.* **65**, 5278–5283.
25. Naumov, G.N., Wilson, S.M., MacDonald, I.C., Schmidt, E.E., Morris, V.L., Groom, A.C., et al. (1999) Cellular expression of green fluorescent protein, coupled with high-resolution in vivo videomicroscopy, to monitor steps in tumor metastasis. *J Cell. Sci.* **112**, 1835–1842.
26. Mook, O.R.F, Van Marle, J.V., Vreeling-Sindelarova, H. et al. (2003) Visualization of early events in tumor formation of eGFP transfected rat colon cancer cells in liver. *Hepatology* **38**, 295–304.
27. Al-Mehdi, A.B., Tozawa, K., Fisher, A.B., Shientag, L., Lee, A., Muschel, R.J. (2000) Intravascular origin of metastasis from the proliferation of endothelium-attached tumor cells: a new model for metastasis. *Nat. Med.* **6**, 100–102.
28. Yamauchi, K, Yang, M., Jiang, P., Xu, M., Yamamoto, N., Tsuchiya, H., et al. (2006) Development of real-time subcellular dynamic multicolor imaging of cancer-cell trafficking in live mice with a variable-magnification whole-mouse imaging system. *Cancer Res.* **66**, 4208–4214.
29. Miller, M.J., Wei, S.H., Cahalan, M.D., Parker, I. (2003) Autonomous T cell trafficking examined in vivo with intravital two photon microscopy. *Proc. Natl. Acad. Sci. USA* **100**, 2604–2609.
30. von Andrian, U.H., Mempel, T.R. (2003) Homing and cellular traffic in lymph nodes. *Nat. Rev. Immunol.* **3**, 867–878.
31. Halin, C., Mora, J.R., Sumen, C., von Andrian, U.H. (2005) In vivo imaging of lymphocyte trafficking. *Annu Rev Cell Dev Biol* **21**, 581–603.
32. Stoll, S., Delon, J., Brotz, T.M., Germain, R.N. (2002) Dynamic imaging of T cell-dendritic cell interactions in lymph nodes. *Science* **296**, 1873–1876.
33. Dadiani, M., Kalchenko, V., Yosepovich, A., Margalit, R., Hassid, Y., Degani, H., Seger, D. (2006) Real-time imaging of lymphogenic metastasis in orthotopic human breast cancer. *Cancer Res* **66**, 8037–8041.
34. Hayashi, K., Jiang, P., Yamauchi, K., Yamamoto, N., Tsuchiya, H., Tomita, K., et al. (2007) Real-time imaging of tumor-cell shedding and trafficking in lymphatic channels. *Cancer Res.* **67**, 8223–8228.
35. Yang, M., Li, L., Jiang, P., Moossa, A.R., Penman, S., Hoffman, R.M. (2003) Dual-color fluorescence imaging distinguishes tumor cells from induced host angiogenic vessels and stromal cells. *Proc. Natl. Acad. Sci. USA* **100**, 14259–14262.
36. Yang, M., Reynoso, J., Jiang, P., Li, L., Moossa, A.R., Hoffman, R.M. (2004) Transgenic nude mouse with ubiquitous green fluorescent protein expression as a host for human tumors. *Cancer Res.* **64**, 8651–8656.
37. Amoh, Y., Li, L., Yang, M., Jiang, P., Moossa, A.R., Katsuoka, K., Hoffman, R.M. (2005) Hair follicle-derived blood vessels vascularize tumors in skin and are inhibited by doxorubicin. *Cancer Res.* **65**, 2337–2343.
38. Yamamoto, N., Yang, M., Jiang, P., Xu, M., Tsuchiya, H., Tomita, K., et al. (2003) Determination of clonality of metastasis by cell-specific color-coded fluorescent-protein imaging. *Cancer Res.* **63**, 7785–7790.
39. Yang, M., Jiang, P., Hoffman, R.M. (2007) Whole-body subcellular multicolor imaging of tumor-host interaction and drug response in real time. *Cancer Res.* **67**, 5195–5200.
40. Bouvet, M., Tsuji, K., Yang, M., Jiang, P., Moossa, A.R., Hoffman, R.M. (2006) *In vivo* color-coded imaging of the interaction of colon cancer cells and splenocytes in the

formation of liver metastases. *Cancer Res.* **66**, 11293–11297.

41. Tsuji, K., Yamauchi, K., Yang, M., Jiang, P., Bouvet, M., Endo, H., et al. (2006) Dual-color imaging of nuclear-cytoplasmic dynamics, viability, and proliferation of cancer cells in the portal vein area. *Cancer Res.* **66**, 303–306.
42. Yamauchi, K., Yang, M., Hayashi, K., Jiang, P., Yamamoto, N., Tsuchiya, H., et al. (2008) Induction of cancer metastasis by cyclophosphamide pretreatment of host mice: an opposite effect of chemotherapy. *Cancer Res.* **68**, 516–520.
43. Kimura, H., Hayashi, K., Yamauchi, K., Yamamoto, N., Tsuchiya, H., Tomita, K., et al. (2010) Real-time imaging of single cancer-cell dynamics of lung metastasis. *J. Cell. Biochem.* **109**, 58–64.
44. Kimura, H., Lee, C., Hayashi, K., Yamauchi, K., Yamamoto, N., Tsuchiya, H., et al. (2010) UV light killing efficacy of fluorescent protein-expressing cancer cells *in vitro* and *in vivo*. *J. Cell. Biochem.* **110**, 1439–1446.
45. Sakaue-Sawano, A., Kurokawa, H., Morimura, T., Hanyu, A., Hama, H., Osawa, H., et al. (2008) Visualizing spatiotemporal dynamics of multicellular cell-cycle progression. *Cell* **132**, 487–498.
46. Kimura, H., Momiyama, M., Tomita, K., Tsuchiya, H., Hoffman, R.M. (2010) Long-working-distance fluorescence microscope with high-numerical-aperture objectives for variable-magnification imaging in live mice from macro- to subcellular. *J. Biomed. Optics* **15(6)**, 066029.
47. Hoffman, R.M. (2008) *In vivo* real-time imaging of nuclear-cytoplasmic dynamics of dormancy, proliferation and death of cancer cells. *Acta Pathol.Microbiol. Immunol. Scand.* **116**, 716–729.
48. Hoffman, R.M. (2005) The multiple uses of fluorescent proteins to visualize cancer in vivo. *Nat. Rev. Cancer* **5**, 796–806.
49. Hoffman, R.M., Yang, M. (2006) Subcellular imaging in the live mouse. *Nat. Protoc.* **1**, 775–782.
50. Hoffman, R.M., Yang, M. (2006) Color-coded fluorescence imaging of tumor-host interactions. *Nat. Protoc.* **1**, 928–935.
51. Hoffman, R.M., Yang, M. (2006) Whole-body imaging with fluorescent proteins. *Nat. Protoc.* **1**, 1429–1438.
52. Yang, M., Reynoso, J., Bouvet, M., Hoffman, R.M. (2009) A transgenic red fluorescent protein-expressing nude mouse for color-coded imaging of the tumor microenvironment. *J. Cell. Biochem.* **106**, 279–284.
53. Tran Cao, H.S., Reynoso, J., Yang, M., Kimura, H., Kaushal, S., Snyder, C.S., et al. (2009) Development of the transgenic cyan fluorescent protein (CFP)-expressing nude mouse for "Technicolor" cancer imaging. *J. Cell. Biochem.* **107**, 328–334.

Index

A

B

C

D

Robert M. Hoffman (ed.), *In Vivo Cellular Imaging Using Fluorescent Proteins: Methods and Protocols*,
Methods in Molecular Biology, vol. 872, DOI 10.1007/978-1-61779-797-2, © Springer Science+Business Media, LLC 2012

E

F

G

H

I

L

M

N

O

P

R

S

T

V

W

MIX
Papier aus verantwortungsvollen Quellen
Paper from responsible sources
FSC® C105338

If you have any concerns about our products,
you can contact us on
ProductSafety@springernature.com

In case Publisher is established outside the EU,
the EU authorized representative is:
Springer Nature Customer Service Center GmbH
Europaplatz 3, 69115 Heidelberg, Germany

Printed by Libri Plureos GmbH
in Hamburg, Germany